OCEAN TIDES
Mathematical Models and Numerical Experiments

Related Pergamon Titles of Interest

Books

GORSHKOV
World Ocean Atlas
   Volume 1: Pacific Ocean
   Volume 2: Atlantic and Indian Oceans
   Volume 3: Arctic Ocean
MARCHUK
Differential Equations and Numerical Mathematics
MARCHUK & NISEVICH
Mathematical Methods in Clinical Practice
MELCHIOR
The Tides of the Planet Earth, 2nd edition

Journals

Computers & Geosciences
Continental Shelf Research
Deep-Sea Research & Oceanographic Abstracts
Progress in Oceanography

Full details of all Pergamon publications/free specimen copy of any Pergamon journal available on request from your nearest Pergamon office.

OCEAN TIDES

Mathematical Models and Numerical Experiments

by

G. I. MARCHUK

and

B. A. KAGAN

Translated by
E. V. Blinova and L. Ya. Yusina

Translation Editor
D. E. CARTWRIGHT
Institute of Oceanographic Sciences,
Bidston Observatory, Birkenhead, Merseyside, UK

PERGAMON PRESS
OXFORD · NEW YORK · TORONTO · SYDNEY · PARIS · FRANKFURT

U.K.	Pergamon Press Ltd., Headington Hill Hall, Oxford OX3 0BW, England
U.S.A.	Pergamon Press Inc., Maxwell House, Fairview Park, Elmsford, New York 10523, U.S.A.
CANADA	Pergamon Press Canada Ltd., Suite 104, 150 Consumers Road, Willowdale, Ontario M2J 1P9, Canada
AUSTRALIA	Pergamon Press (Aust.) Pty. Ltd., P.O. Box 544, Potts Point, NSW 2011, Australia
FRANCE	Pergamon Press SARL, 24 rue des Ecoles, 75240 Paris, Cedex 05, France
FEDERAL REPUBLIC OF GERMANY	Pergamon Press GmbH, Hammerweg 6, D-6242 Kronberg-Taunus, Federal Republic of Germany

Copyright © 1984 Pergamon Press Ltd.

All Rights Reserved. No part of this publication may be reproduced, stored in a retrieval system or transmitted in any form or by any means; electronic, electrostatic, magnetic tape, mechanical, photocopying, recording or otherwise, without permission in writing from the publishers.

First edition 1984

Library of Congress Cataloging in Publication Data
Marchuk, G. I. (Gurii Ivanovich), 1925—
Ocean tides.
Translation of: Okeanskie prilivy.
Bibliography: p.
1. Tides—Mathematical models. I. Kagan, B. A. (Boris Abramovich) II. Title.
GC305.5.M3M3713 1983 551.47'08'0724 82-18898

British Library Cataloguing in Publication Data
Marchuk, G. I.
Ocean tides.
1. Tides Mathematical models
I. Title II. Kayan, B. A.
III. Okeanshire prilivy. English
525': 6'0151 GC305.5.M5

ISBN 0-08-026236-8

Translated from
OKEANSKIE PRILIVY
Matematicheskie Modeli i Chislennye Eksperimenty
Published by
Gidrometsoizdat Leningrad

Printed in Hungary by Franklin Printing House

Foreword

AFTER more than a century of modest progress by unaided mathematicians in the theory of oceanic tides founded by Laplace, the 1960s and 1970s saw a new wave of impetus in the subject, largely based on the new computing technology. Important papers appeared from several quarters. To quote only the leading authors from well-known schools of colleagues, Pekeris in Israel, Zahel in Germany, Longuet-Higgins in Great Britain, Lambeck in France, Hendershott, Munk and Platzman in the USA, and Bogdanov, Kagan and Marchuk in the USSR, all attacked different aspects of the then outstanding problems from new directions. The resulting growth in knowledge and understanding of the behaviour of the tides in the ocean over these two decades has been extraordinary. This, together with new advances in oceanographic measuring techniques, has restored the ocean tides, previously thought of as a rather old-fashioned subject of interest only to hydrographic surveyors and a few mathematicians, to a place of interest in the wider field of geophysical science.

The Western literature documenting this progress is well known to English-speaking scientists, and comprehensive reviews have been published, the most recent being the chapter by Hendershott in *Evolution of Physical Oceanography—Scientific Surveys in honor of Henry Stommel*, pp. 292–341, MIT Press, 1981, and "Oceanic Tides" by Cartwright in *Reports on Progress in Physics* **40**, 665–708, 1977. The Russian literature, however, has been largely restricted to Russian periodicals, only some of which have been regularly translated into English. It has long been suspected that the part of this literature known in the West is only the tip of an iceberg. In particular, the extensive work by Marchuk on numerical modelling has been unavailable in English. The present book repairs this deficiency, bringing together the distilled work and views on tides by two of the leading Russian authorities in the subject.

I feel sure that Western workers in this field will find this work a valuable commentary on the subject with which to compare and contrast the more familiar trends in American and European researches. Some of the latter are, of course, referred to in the present text, but the bulk of the material consists of original work by the authors, their colleagues and other compatriots, of which most of us have been unaware.

D.E. CARTWRIGHT

Institute of Oceanographic Sciences, UK

Preface

STUDIES of tides have reached a critical period of development. Within a comparatively limited period of time impressive results have been obtained in the field of experiment: the basic net of coastal and island sea-level stations has been expanded; new deep-sea recorders equipped with sensitive pressure meters have been designed and tested, and soon we shall get data on tides in the open ocean from the artificial satellites of the Earth. All these facts lead one to the conclusion that in the not too distant future reliable and detailed information on tides on a global scale will be available.

Substantial results have also been achieved in the field of theory. The dynamical theory of tides by Laplace has been enriched by taking into account the essential factor—turbulence, which has made it possible to turn from the solution of tidal equations under idealized conditions to the development of a tidal theory for real water basins.

Successful tidal calculations in the marginal seas which would have been impossible without employing computers and numerical methods of the solution of differential equations, made many investigators believe that the basic difficulties hampering the quantitative description of tides in the World Ocean would soon be overcome. More advanced computers, it seemed, would be sufficient to solve the problem.

But do we not appear to be too hypnotized by the power of computers? Comparing the computed data on tides in the World Ocean with the empirical data reveals noticeable discrepancies. These can be attributed to the use of approximate meshes imposed by limited computer technique. Nevertheless, we are prepared to predict that the appearance of more advanced computers will not result in better agreement between computational and observational data. Incidentally, calculations of tides in the World Ocean by Pekeris and Accad with the use of a one-degree mesh prove that even in this case we are not rid of systematic errors. This is because we are unable to take into account the interaction of the earth- and ocean tides, the energy transfer to the internal waves and shelf effects, the latter including every aspect of the phenomena of trapping, refraction and dissipation of tidal waves in shallow water. These are not the only sources of error but in our opinion they are the main ones.

Thus, the discrepancy between the data calculated by the available models and the empirical data can be accounted for by transient inadequacy of our knowledge rather than by imperfection of the means of computation employed. The predicament calls for the investigators to halt for a while and to pay more attention to the physical aspects of the problem. Only after due consideration is given to these aspects shall we be able to reproduce the true picture of tides observed in the World Ocean, which is, in fact, the final goal of the theory.

However, on condition that we do not demand from the models more than they can give (they can now reliably guarantee to present only qualitatively valid structure), they can even now be useful to reach the second, and probably more important goal of the theory—the interpretation of physical peculiarities of tidal generation in the World Ocean, which is the main idea of the book.

We set the task of drawing certain conclusions at the present stage of development and of maintaining the reader's interest in this traditional subject of ocean dynamics. If we are sometimes too sceptical of the reality of progress, our scepticism is conditioned by our desire to show the vastness of the field still open to the investigator.

Contents

Introduction   *xi*

1. Indispensable information on the theory of tides   1

1.1 Forces inducing ocean tides   1
1.2 Tidal potential   11
1.3 Equations of tidal dynamics   23
1.4 Additional potentials of deformation   26
1.5 Boundary conditions   31
1.6 References   34

2. Studies on the equations of tidal dynamics   36

2.1 Formulation of the problem   36
2.2 Basic ideas and definitions   40
2.3 Uniqueness theorem   43
2.4 *A priori* estimates   45
2.5 Existence theorem   52
2.6 On the existence of a periodic solution of the equations of tidal dynamics   59
2.7 Conjugate equations of tidal dynamics   63
2.8 The perturbation theory   67
2.9 The spectral problem   71
2.10 References   75

3. Numerical methods for the solution of the equations of tidal dynamics   76

3.1 Method of boundary values   76
3.2 HN-method   84
3.3 Modified variant of the HN-method   89
3.4 The method of fractional steps   95
3.5 A modified variant of the method of fractional steps   107
3.6 References   108

4. Tides in the World Ocean   110

4.1 Empirical cotidal charts   110
4.2 Basic features of the spatial distribution of tides in the World Ocean   121
4.3 An example of numerical modelling of tides in the World Ocean   128
4.4 Some other calculations of tides in the World Ocean   143
4.5 Numerical experiments on tidal dynamics in the World Ocean   155
4.6 Estimation of the rate of tidal energy dissipation in the open ocean   171
4.7 References   175

5. The bottom boundary layer in tidal flows — 177

- 5.1 Some definitions — 177
- 5.2 Experimental data — 184
- 5.3 Theoretical models of the bottom boundary layer in tidal flows — 204
- 5.4 On the law of drag in tidal flow — 233
- 5.5 References — 246

6. Vertical structure of internal tidal waves — 248

- 6.1 Generation of internal tidal waves — 248
- 6.2 Qualitative analysis of the equations for internal waves — 256
- 6.3 Vertical structure of the internal tidal waves in a realistically stratified ocean — 262
- 6.4 References — 270

Bibliography — 273

Appendix — 285

Index — 291

Introduction

OF RECENT years the application of direct measurements of ocean tides has spread significantly. Apart from use in traditional oceanographic problems, such data are beginning to be applied in related areas of geophysics such as the Earth tides, the elastic properties of the Earth's crust, tidal variations of gravity, and in calculating the orbits of artificial satellites used for space exploration. New uses are also being found for tidal current data. All these applications call for higher standards in the completeness and accuracy of tidal data.

Such standards have been partially met in the deep-sea tidal measurement programmes initiated and developed by Working Group no. 27 of the Scientific Committee on Oceanic Research (SCOR). This Working Group envisaged recording tidal surface elevations at a set of stations covering an open area of the World Ocean including a continental slope. Fifteen years have passed since such plans were first formulated but the end of the research is not yet in sight. It therefore appears that we shall have to rely on future satellite measurements for much of the required tidal data.

Already, observations of the perturbation of orbital elements from some satellites have made it possible to determine the parameters of the second spherical harmonic of the ocean tides. These enable one to estimate the rate of global dissipation of tidal energy but the data are insufficient to reproduce any details of the global tidal map.

For tidal detail, satellite altimetry may prove rather useful, but unfortunately there are complications. Firstly, tidal amplitudes over much of the ocean are smaller than the accuracy available to determine the mean surface levels. Secondly, since satellite orbits vary in space, sequential measurements cannot be made at a given point but only within a certain area surrounding that point. Thirdly, the altimetry is subject to large errors in computing the altitude of the orbit. As a result, satellite altimetry is still inferior to deep sea and standard observations. Yet from the point of view of mass measurements in the open ocean, the altimetry surpasses all other methods and has a promising future.

Yet so far all data on ocean tides are supplied by calculations based on more or less reliable models of the phenomenon concerned. It goes without saying that the quality of such calculations largely depends on the completeness and correctness of description of multiple factors involved in tide formation.

INTRODUCTION

Since the book was written evolution of ocean tide theory has witnessed considerable progress. In this connection mention should be made of the way in which the problem of global interaction of ocean and earth tides is being solved. Important headway is made with regard to parameterization of shelf effects—an unfinalized problem which has become a stumbling-block for any attempt at numerical modelling of ocean tides. It appears that the spectral problem for the World ocean with real configuration will be solved in the near future. Its solution applicable to a wide frequency range including, *inter alia*, semi-diurnal and daily bands, will make it possible to give at last an exhaustive answer about the nature of ocean tides.

Not all these problems are reflected in our book but we hope to take them up later.

Revised from first English translation,
by D. E. CARTWRIGHT August 1982

CHAPTER 1

Indispensable Information on the Theory of Tides

1.1 Forces Inducing Ocean Tides

Tide-generating force

Consider a balance of forces affecting a unit mass at a point of the Earth's surface. This mass moves in the gravitational field conditioned by the attraction force of the Earth $\mathbf{G}(\mathcal{A})$, the Moon, the Sun and other perturbing bodies of the solar system. Denote the attractive forces of these bodies by $\sum_i \mathcal{T}_i(\mathcal{A})$. The displacement of the mass at the point $\mathcal{A}$ will be controlled also by the forces of pressure $\mathbf{P}(\mathcal{A})$ and friction $\mathbf{F}(\mathcal{A})$.

Introduce an inertial system of coordinates with the origin at a point O. Let $\mathcal{C}$ be the centre of the Earth, and $\mathcal{B}_i$ ($i = 1, 2, 3, \ldots$) be the centres of the perturbing bodies (Fig. 1.1). Then Newton's second law for the absolute motion of a unit mass at the point $\mathcal{A}$ can be represented by

$$\frac{d_a^2}{dt^2} \underline{O\mathcal{A}} = \mathbf{P}(\mathcal{A}) + \mathbf{G}(\mathcal{A}) + \sum_i \underline{\mathcal{T}_i(\mathcal{A})} + \mathbf{F}(\mathcal{A}). \qquad (1.1.1)$$

Here the operator d_a/dt refers to the inertial system of coordinates.

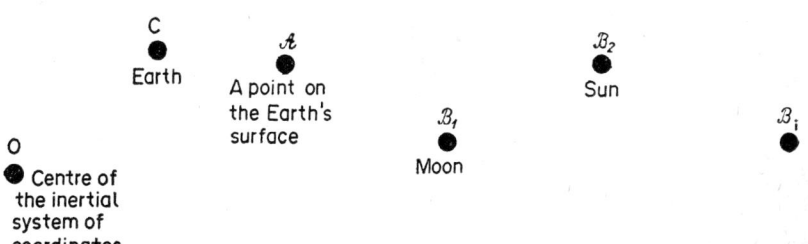

F I G. 1.1. Position of the considered point $\mathcal{A}$, the centre of the Earth $\mathcal{C}$ and perturbing bodies $\mathcal{B}_i$ with respect to the origin of the inertial system of coordinates.

Compare the orders of magnitude of $\mathcal{T}(\mathcal{A})$ and $\mathbf{G}(\mathcal{A})$. For the Moon and the Sun $r \ll R$ (Fig. 1.2). Hence $|\mathcal{T}(\mathcal{A})| \approx |\mathcal{T}(\mathcal{C})| \approx \gamma \mathcal{M}/R^2$ where γ is the gravitational constant, $\mathcal{M}$ is the mass of the perturbing body $\mathcal{B}$, R is its distance

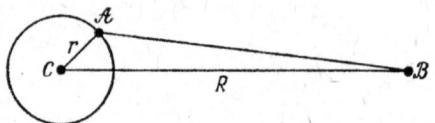

F I G. 1.2. On the definition of the force of a unit mass attracted by a body $\mathcal{B}$ at the point $\mathcal{A}$.

from the centre of the Earth. Since $|\mathbf{G}(\mathcal{A})| \approx \gamma (\mathcal{M}_\oplus/r^2)$ (here $\mathcal{M}_\oplus$ is the Earth's mass, r is the distance between the centre of the Earth and the point $\mathcal{A}$) then

$$|\mathcal{T}(\mathcal{A})|/|\mathbf{G}(\mathcal{A})| \approx (\mathcal{M}/\mathcal{M}_\oplus)(r/R)^2.$$

The mass ratios between the Moon and the Earth and the Sun and the Earth are known to be

$$\mathcal{M}_\mathbb{C}/\mathcal{M}_\oplus = 1/81.30; \qquad \mathcal{M}_\odot/\mathcal{M}_\oplus = 332{,}958, \qquad (1.1.2)$$

respectively.

If r is set equal to the equatorial radius of the Earth r_1, then r/R is the equatorial parallax, its mean value for the Moon and the Sun being

$$\frac{r_1}{c_\mathbb{C}} = 0.016\,593; \qquad \frac{r_1}{c_\odot} = 4.2615 \times 10^{-5}, \qquad (1.1.3)$$

respectively. Here r/c is the mean value of r/R with respect to the orbit. Thus

$$\frac{|\mathcal{T}|(\mathcal{A})|}{|\mathbf{G}(\mathcal{A})|} \approx \begin{cases} 3.4 \times 10^{-6} & \text{for the Moon,} \\ 6.0 \times 10^{-4} & \text{for the Sun.} \end{cases}$$

The Sun is seen to exert a greater attraction than the Moon.

Remember that when moving along its orbit, the Earth is subjected to acceleration. Use Newton's second law again and obtain

$$\frac{d_a^2}{dt^2} \mathcal{O}\mathcal{C} = \sum_i \mathcal{T}_i(\mathcal{C}).$$

Subtract this relation from equation (1.1.I) and have

$$\frac{d_a^2}{dt^2} \mathcal{O}\mathcal{A} = \mathbf{P}(\mathcal{A}) + \mathbf{G}(\mathcal{A}) + \sum_i \mathbf{T}_i(\mathcal{A}) + \mathbf{F}(\mathcal{A}), \qquad (1.1.4)$$

where $\mathbf{T}_i(\mathcal{A}) \equiv \overline{\mathcal{U}}_i(\mathcal{A}) - \overline{\mathcal{U}}_i(\mathcal{C})$ is the tide-generating force. Thus, the tide-generating force is determined as the resultant of the forces of attraction of a unit mass at the points $\mathcal{A}$ (arbitrary points on the surface) and (the Earth's centre) exerted by the perturbing body. The direction of $\mathbf{T}(\mathcal{A})$ coincides with that of the vector sum of $\overline{\mathcal{U}}_i(\mathcal{A})$ and $-\overline{\mathcal{U}}_i(\mathcal{A})$ forces. Figure 1.3 presents the direction and the magnitude of the tide-generating force at eight points of the meridional section of the sphere centred at the point $\mathcal{C}$.

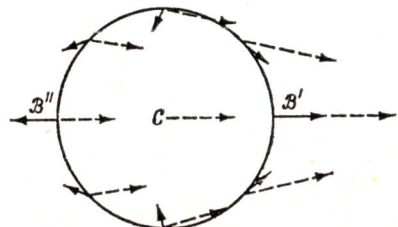

FIG. 1.3. Tide-generating force (full arrows) at eight points of the sphere meridional section. Dashed arrows denote the force of attraction at the point $\mathcal{A}$. In the centre of the sphere $\overline{\mathcal{U}}(\mathcal{A}) = \overline{\mathcal{U}}(\mathcal{C})$.

In the case discussed, when the distance from the perturbing body $\mathcal{B}$ (the Moon or the Sun) to the centre of the Earth greatly exceeds its radius (i.e. for small values of the parallax), the distributions $\mathbf{T}(\mathcal{A})$ in the sublunar and antilunar (analogously in subsolar and antisolar) Earth's hemispheres will be almost mirror images of each other.

The order of magnitude of $\mathbf{T}(\mathcal{A})$ can be evaluated by the difference between the vectors $\overline{\mathcal{U}}(\mathcal{A})$ and $\overline{\mathcal{U}}(\mathcal{C})$ at the point $\mathcal{B}'$ (Fig. 1.3) where their directions coincide. If $r(\equiv \mathcal{CB}') \ll R(\equiv \mathcal{CB})$, then

$$|\mathbf{T}(\mathcal{B}')| = \frac{\gamma \mathcal{M}}{(R-r)^2} - \frac{\gamma \mathcal{M}}{R^2} \approx \frac{2\gamma \mathcal{M} r}{R^3}.$$

Compare $|\mathbf{T}(\mathcal{B}')|$ with $|\mathbf{G}(\mathcal{B}')| \approx \gamma \mathcal{M}_\oplus/r^2$ and obtain

$$\frac{|\mathbf{T}(\mathcal{B}')|}{|\mathbf{G}(\mathcal{B}')|} \approx 2\left(\frac{\mathcal{M}}{\mathcal{M}_\oplus}\right)\left(\frac{r}{R}\right)^3.$$

Hence, unlike the Moon's and the Sun's attraction forces which when referred to those of the Earth are proportional to the squared parallax, the tide-generating force normalized in the same way is proportional to r_1/c cubed. Equalities

(1.1.2) and (1.1.3) afford

$$\frac{\mathcal{M}_{\mathbb{C}}}{\mathcal{M}_{\oplus}} \left(\frac{r_1}{c_{\mathbb{C}}}\right)^3 = 0.5603 \times 10^{-7},$$

$$\frac{\mathcal{M}_{\odot}}{\mathcal{M}_{\oplus}} \left(\frac{r_1}{c_{\odot}}\right)^3 = 0.2580 \times 10^{-7}. \qquad (1.1.5)$$

The tide-generating forces of the Moon and the Sun appear to be of the same order of magnitude and about 10^7 times less than the gravitational force of the Earth. However, despite this fact, they cannot be neglected. Indeed, in the ocean (as well as in both the atmosphere and lithosphere) the force **G** is counteracted by the force of the hydrostatic pressure which nearly equals **P**, so that the sum **P**+**G** can be negligible as compared to **G**.

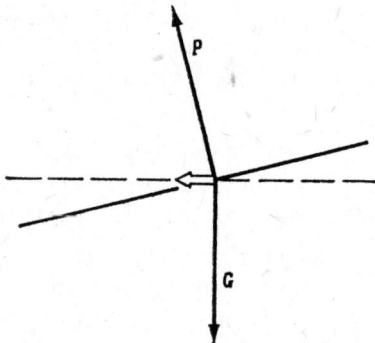

FIG. 1.4. Balance of forces of the Earth's attraction and hydrostatic pressure.

This conclusion is illustrated by Fig. 1.4 in which the surfaces normal to **G** and the hydrostatic component of the force **P** are presented in dashed lines, while the resultant **P**+**G**, equal in the order of magnitude to the tide-generating force and orientated almost perpendicular to **G**, is denoted by an arrow.

This discrepancy in the directions of the **G** and **T** forces results in the fact that the latter leads to the horizontal shift of the particles along the Earth's surface and hence to the accumulation of the mass in the vicinity of the points $\mathscr{B}'$ and $\mathscr{B}''$ (Fig. 1.3). The accompanying vertical (radial) shifts are of a kinematic nature. They are induced by the law of conservation of mass rather than by the radial component of the tide-generating force.

This is in outline the tide-generating mechanism in the atmosphere, hydrosphere and lithosphere of the Earth.

Rotational and gravitational forces

In equation (1.1.4) the acceleration of the vector $\mathcal{A}$ was presented in the inertial system of coordinates. The use of the relative system of coordinates rotating together with the Earth at the constant angular velocity ω seems, however, more natural. In case when a particle at the point $\mathcal{A}$ remains fixed with respect to the Earth, the absolute velocity of the shift of the vector $\mathbf{r} \equiv \mathcal{OA}$ must equal $\omega \times \mathbf{r}$. Hence, when the particle moves with respect to the Earth, the relative velocity of the shift of the vector $\mathbf{r}$ will be $(d_a\mathbf{r}/dt) - \omega \times \mathbf{r}$. Denote the relative velocity by $\mathbf{u}$ and the relative acceleration $((d_a\mathbf{u}/dt) - \omega \times \mathbf{u}))$ by $d\mathbf{u}/dt$. The definition of $d\mathbf{u}/dt$ affords

$$\frac{d\mathbf{u}}{dt} = \frac{d_a^2\mathbf{r}}{dt^2} - 2\omega \times \mathbf{u} - \omega \times (\omega \times \mathbf{r}).$$

Substituting the expression for $d_a^2\mathbf{r}/dt^2$ from equation (1.1.4) gives

$$\frac{d\mathbf{u}}{dt} = \mathbf{P} + \mathbf{G} + \mathbf{T} + \mathbf{F} - 2\omega \times \mathbf{u} - \omega \times (\omega \times \mathbf{r}). \tag{1.1.6}$$

In this equation the last two terms containing ω can be considered acceleration-induced "forces" in the rotating system of coordinates.

The Coriolis force $-2\omega \times \mathbf{u}$ is by its definition normal to both vectors ω and $\mathbf{u}$. The order of its magnitude equals $2\omega|\mathbf{u}|$, where $\omega = 7.29 \times 10^{-5}$ rad s^{-1}. At $|\mathbf{u}| = 1$ cm/s it amounts approximately to 1.5×10^{-4} cm/s^2, which is 10^7 times less than the G value. As we have already mentioned, the force $\mathbf{G}$ is, however, nearly compensated by that of the hydrostatic pressure, and thus the Coriolis force, as well as the tide-generating force $\mathbf{T}$, must be taken into account in the equation of motion.

The centrifugal force $-\omega \times (\omega \times \mathbf{r})$ is directed from the axis of rotation along the radius-vector connecting it with the point $\mathcal{A}$, and like $\mathbf{G}$ depends only upon the position of the considered particle. It is expedient thus to unite $\mathbf{G}$ and $-\omega \times (\omega \times \mathbf{r})$ into a single vector

$$\mathbf{g} = \mathbf{G} - \omega \times (\omega \times \mathbf{r})$$

which is called the gravitational force.

All the forces were earlier compared with the gravitational force $\mathbf{G}$. We have already mentioned, in particular, that the components of the tide-generating force, important from the standpoint of tide formation (the same is true for the Coriolis force as well), are normal to the force $\mathbf{G}$. Let us compare the said forces with $\mathbf{g}$. Though small, the difference between the vectors $\mathbf{G}$ and $\mathbf{g}$ causes the shift of particles towards the equator, which results in the fact that the Earth's shape is somewhat flattened at the poles and bulges at the equator.

The difference between the equatorial and polar radii of the Earth being only 21.4 km, to solve tidal problems with the accuracy necessary for practical purposes the Earth can be considered an ideal sphere with the radius $r = 6371$ km, which assumption allows one to neglect the difference between the two vectors and to substitute **G** by **g** everywhere.

The force **G**, however, cannot be completely abandoned. Indeed, horizontal and vertical particle shifts, in both the ocean and the Earth's crust, induced by the tide-generating forces, result in re-distribution of masses and thus in changes of the Earth's gravitational field. These changes are negligible as compared with the undisturbed value of **G**, but comparable with **T** as shown below. Let us present the gravitational force as a sum of the undisturbed value **G** and its deviations $\mathbf{G}^{(1)}$, $\mathbf{G}^{(2)}$ conditioned by the corresponding changes of the Earth's gravitational field due to oceanic and terrestrial tides. Introducing the expression for **g** into equation (1.1.6) gives

$$\frac{d\mathbf{u}}{dt} = \mathbf{P} + \mathbf{g} + \mathbf{G}^{(1)} + \mathbf{G}^{(2)} + \mathbf{T} + \mathbf{F} - 2\boldsymbol{\omega} \times \mathbf{u}. \qquad (1.1.7)$$

$\mathbf{G}^{(1)}$ and $\mathbf{G}^{(2)}$ will be considered later. Now let us obtain the expression for the frictional force **F**.

Frictional force

Motions in a real ocean are practically always of turbulent character. In a turbulent flow the impulse exchange is by two mechanisms: the mechanism of molecular viscosity stipulating the impulse exchange between liquid particles, and that of turbulent mixing causing the impulse transfer between liquid bodies and associated with the velocity pulsations existing in a liquid medium.

In a viscous liquid the expression for the density tensor of the average impulse flux per a mass unit can be presented as

$$\overline{\Pi}_{ij} = \bar{u}_i \bar{u}_j - (\bar{\sigma}_{ij} - \overline{u'_i u'_j}) \qquad (1.1.8)$$

where u_i, u_j are the components of the velocity vector **u**; the subscripts i, j can assume values 1, 2, 3 with the account taken of the fact that, according to the rules of tensor analysis, the terms containing a repeated index must be summed over all the three possible index values; the velocity pulsations are denoted by primes; the upper bar denotes averaging. We may consider the tensor of viscous stresses $\bar{\sigma}_{ij}$ as a sum of the isotropic part $-(\bar{p}/\varrho_0)\delta_{ij}$ which has the same form as the tensor of viscous stresses in an undisturbed liquid (here $\bar{p}$ and ϱ_0 are the pressure and the mean density of the liquid, respectively, δ_{ij} is the Kronecker's delta-tensor equal to unity at $i = j$ and to zero at $i \neq j$)

and the anisotropic part d_{ij} which is called the deviatoric stress tensor. The latter includes tangential stresses and some diagonal terms amounting to zero.

The tensor of the Reynolds stresses $\overline{u_i' u_j'}$, which serves as the tensor of additional stresses with respect to $\bar{\sigma}_{ij}$, can be expressed in a similar fashion as

$$-\overline{u_i' u_j'} = -\tfrac{1}{3}\overline{u_i' u_j'} + \bar{\tau}_{ij}. \qquad (1.1.8')$$

In this expression the term $\tfrac{1}{3}\overline{u_i' u_j'} = \tfrac{2}{3} b \delta_{ij}$ (where b is the kinetic energy of velocity pulsations) describes the istropic part of the tensor $\overline{u_j' u_i'}$, while $\bar{\tau}_{ij}$ pertains to the anisotropic part associated solely with the shift of neighbouring liquid bodies with respect to one another (the deviator of Reynolds stresses).

Introduce the local Cartesian system of coordinates with the axes $x_1 = x$, $x_2 = y$, $x_3 = z$. Locate the axes x, y in the plane tangential to the Earth's surface and orientate them to the east and north, respectively, with the axis z orientated to the zenith. Denote the velocity components along the given axes through $u_1 = u, u_2 = v, u_3 = w$, respectively.

In the above system of coordinates the tensor $\bar{\tau}_{ij}$ (the upper bar henceforth omitted) has nine components

$$\begin{bmatrix} \tau_{xx} & \tau_{xy} & \tau_{xz} \\ \tau_{yx} & \tau_{yy} & \tau_{yz} \\ \tau_{zx} & \tau_{zy} & \tau_{zz} \end{bmatrix},$$

not all of them being, however, independent. According to the stress duality law $\tau_{ij} = \tau_{ji}$, the number of independent components of the tensor τ_{ij} is reduced to six. Of these, three diagonal components $\tau_{xx}, \tau_{yy}, \tau_{zz}$ are normal stresses, while the three off-diagonal components $\tau_{yz} = \tau_{zy}, \tau_{xz} = \tau_{zx}, \tau_{xy} = \tau_{yx}$, are tangential or shear-stresses.

As we have already pointed out, the deviator of the Reynolds stresses τ_{ij} is non-zero only when the field of mean velocity is non-homogeneous in space (x, y, z). Hence, τ_{ij} must depend upon the derivatives of the mean velocity with respect to coordinates. The local velocity gradients being small, one can assume the dependence of τ_{ij} upon the velocity derivatives $\partial u_i/\partial x_j$ to be linear.

At the same time, however, τ_{ij} must become zero also in the case of "solid" rotation of the liquid, i.e. with all its elements rotating uniformly. In the case of "solid" rotation, τ_{ij} goes to zero under the condition

$$e_{k_l} = \frac{1}{2}\left(\frac{\partial u_k}{\partial x_l} + \frac{\partial u_l}{\partial x_k}\right), \qquad (1.1.9)$$

which is called the tensor of deformation velocities, with the indices k and l passing through the values 1, 2, 3. Therefore, the deviatoric, Reynolds stresses τ_{ij} must be linear functions of e_{kl}.

By analogy with the theory of elasticity the six components of the deviatoric stresses τ_{ij} at any point of the liquid can be associated with the six components of the tensor of the deformation velocity e_{kl} at the same point by the expressions

$$\tau_{xx} = k_{11}e_{xx} + k_{12}e_{yy} + k_{13}e_{zz} + k_{14}e_{yz} + k_{15}e_{xz} + k_{16}e_{xy};$$

$$\cdots \cdots \cdots \cdots \cdots \cdots \cdots \cdots \cdots \cdots \cdots$$

$$\tau_{yz} = \tau_{zy} = k_{11}e_{xx} + k_{12}e_{yy} + k_{13}e_{zz} + k_{14}e_{yz} + k_{15}e_{xz} + k_{16}e_{xy}:$$

or the same expressed in matrix form

$$\begin{bmatrix} \tau_{xx} \\ \tau_{yy} \\ \tau_{zz} \\ \tau_{yz} = \tau_{zy} \\ \tau_{xz} = \tau_{zx} \\ \tau_{xy} = \tau_{yz} \end{bmatrix} = \begin{bmatrix} k_{11} & k_{12} & k_{13} & k_{14} & k_{15} & k_{16} \\ k_{21} & k_{22} & k_{23} & k_{24} & k_{25} & k_{26} \\ k_{31} & k_{32} & k_{33} & k_{34} & k_{35} & k_{36} \\ k_{41} & k_{42} & k_{43} & k_{44} & k_{45} & k_{46} \\ k_{51} & k_{52} & k_{53} & k_{54} & k_{55} & k_{56} \\ k_{61} & k_{62} & k_{63} & k_{64} & k_{65} & k_{66} \end{bmatrix} \begin{bmatrix} e_{xx} \\ e_{yy} \\ e_{zz} \\ e_{yz} \\ e_{xz} \\ e_{xy} \end{bmatrix}, \quad (1.1.10)$$

where k_{rs} ($r, s = 1, 2, \ldots, 6$) are the coefficients of the "turbulent viscosity".

To ensure the existence of the quadratic form (which denotes here the dissipation of the mean motion energy, while in the theory of elasticity it is the elasticity potential), the elements of the matrix of the coefficients k_{rs} must obey the condition

$$k_{rs} = k_{sr}. \quad (1.1.11)$$

This condition reduces the number of turbulent viscosity coefficients (whose analogues in the theory of elasticity are k_{rs} elasticity constants) from 36 to 21, this number being typical of any anisotropic medium. While studying motions of planetary scale, however, we can assume them to have the property of axial symmetry with respect to the vertical direction. In this case the matrix of the coefficients k_{rs} relation (1.1.10) can be written as

$$\begin{bmatrix} k_{11} & k_{12} & k_{13} & 0 & 0 & 0 \\ k_{12} & k_{11} & k_{13} & 0 & 0 & 0 \\ k_{13} & k_{13} & k_{33} & 0 & 0 & 0 \\ 0 & 0 & 0 & k_{44} & 0 & 0 \\ 0 & 0 & 0 & 0 & k_{44} & 0 \\ 0 & 0 & 0 & 0 & 0 & \frac{1}{2}(k_{11} - k_{12}) \end{bmatrix}. \quad (1.1.12)$$

Thus, with the condition of axial symmetry fulfilled, there remain only five independent coefficients of turbulent viscosity.

Developing then relation (1.1.10), and substituting the expressions for the elements of the deformation velocity tensor from relation (1.1.9) in the relations obtained, we have

$$\tau_{xx} = k_{11}u_x + k_{12}v_y + k_{13}w;$$

$$\tau_{yy} = k_{12}u_x + k_{11}v_y + k_{13}w;$$

$$\tau_{zz} = k_{13}u_x + k_{13}v_y + k_{33}w;$$

$$\tau_{yz} = \tau_{zy} = k_{44}(v_z + w_y);$$

$$\tau_{xz} = \tau_{zx} = k_{44}(u_z + w_x);$$

$$\tau_{xy} = \tau_{yx} = \tfrac{1}{2}(k_{11} - k_{12})(u_y + v_x), \qquad (1.1.13)$$

where the corresponding derivatives are marked by subscripts of the velocity components.

It will be recalled that according to the definition of the deviatoric Reynolds stresses, $\tau_{xx} + \tau_{yy} + \tau_{zz} = 0$; hence, with the use made of the first three relations of relation (1.1.13) for the case of an incompressible liquid, we obtain

$$k_{11} + k_{12} = k_{13} + k_{33}.$$

Introduce further notations for the following combinations of coefficients:

$$\tfrac{1}{2}(k_{11} - k_{12}) = \kappa_h;$$

$$k_{44} = \kappa_v;$$

$$k_{13} - k_{12} = \kappa.$$

Then relations (1.1.13) finally take the form

$$\tau_{xx} = 2\kappa_h u_x + \kappa w;$$

$$\tau_{yy} = 2\kappa_h u_x + \kappa w;$$

$$\tau_{zz} = 2(\kappa_h - \kappa)w;$$

$$\tau_{yz} = \tau_{zy} = \kappa(v_z + w_y);$$

$$\tau_{xz} = \tau_{zx} = \kappa_v(u_z + w_x);$$

$$\tau_{xy} = \tau_{yx} = \kappa_h(u_y + v_x). \qquad (1.1.14)$$

Thus, the expressions for the tensor τ_{ij} components include three coefficients of turbulent viscosity, κ_h, κ_v and κ. The values of the first two, which are coefficients of horizontal and vertical turbulent viscosity, respectively, are known to an accuracy of at least two orders, e.g. according to direct measurements and indirect evaluation in the ocean $\kappa_h = 10^7$ to 10^9 cm/s, $\kappa_v = 1$ to 10^2 cm/s. As to the third coefficient κ, there are no data on it at all. One can only suggest that κ is within the range $(0, \kappa_h)$ and dependent on the intensities of the vertical velocity pulsations $\overline{w'^2}$, (cf. the third relation in (1.1.14) with expression (1.1.8')). Under the condition of free convection, when vertical velocity fluctuations predominate over the horizontal ones, it is apparent that $\kappa \approx \kappa_h$. In case of a stably stratified ocean, on the contrary, one may expect $\kappa \approx 0$.

Let us assume equals zero. Then, with the isotropic part of the tensor $\overline{u_i u}$ included in the pressure and the viscosity stresses neglected as compared to the Reynolds stress, one can obtain the following expressions for the components of the frictional force $\mathbf{F}$ ($= \partial \tau_{ij} / \partial x_j$):

$$F_x = \kappa_h \Delta u + \frac{\partial}{\partial z} \kappa_v \frac{\partial u}{\partial z} - \kappa_h \frac{\partial^2 w}{\partial x \partial z} + \frac{\partial}{\partial z} \kappa_v \frac{\partial w}{\partial x};$$

$$F_y = \kappa_h \Delta v + \frac{\partial}{\partial z} \kappa_v \frac{\partial v}{\partial z} - \kappa_h \frac{\partial^2 w}{\partial y \partial z} + \frac{\partial}{\partial z} \kappa_v \frac{\partial w}{\partial y}; \qquad (1.1.15)$$

$$F_z = \frac{\partial}{\partial x} \kappa_v \frac{\partial w}{\partial x} + \frac{\partial}{\partial y} \kappa_v \frac{\partial w}{\partial y} + 2\kappa_h \frac{\partial^2 w}{\partial z^2} + \frac{\partial}{\partial x} \kappa_v \frac{\partial u}{\partial z} + \frac{\partial}{\partial y} \kappa_v \frac{\partial v}{\partial z}$$

Here $\kappa_h = $ const.

The above relations obtained for F_x, F_y differ from the corresponding expressions for the frictional force components

$$F_x = \kappa_h \Delta u + \frac{\partial}{\partial z} \kappa_v \frac{\partial u}{\partial z};$$

$$F_y = \kappa_h \Delta v + \frac{\partial}{\partial z} \kappa_v \frac{\partial v}{\partial z}, \qquad (1.1.16)$$

written in a traditional form with two additional terms.

The last term in each of the first two relations in (1.1.15) can be in fact neglected since in the case of large-scale oceanic motions

$$\left| \frac{\partial w}{\partial x} \right| \ll \left| \frac{\partial u}{\partial z} \right| \quad \text{and} \quad \left| \frac{\partial w}{\partial y} \right| \ll \left| \frac{\partial v}{\partial z} \right|.$$

However, since

$$|\Delta u| \approx \left| \frac{\partial^2 w}{\partial x \partial z} \right|, \quad |\Delta v| \approx \left| \frac{\partial^2 w}{\partial y \partial z} \right|,$$

then there are generally speaking no grounds to neglect the terms

$$\kappa_h \frac{\partial^2 w}{\partial x \, \partial z}, \quad \kappa_h \frac{\partial^2 w}{\partial y \, \partial z}$$

in relations (1.1.15). Nevertheless, this is a common practice under the pretext that the values of the turbulent viscosity coefficients are known to an accuracy of one or two orders only and, therefore, even with the above terms excluded, the error in the determination of frictional forces does not increase at all.

By way of conclusion, let us evaluate the order of magnitude of the frictional force. At $|\mathbf{u}| = 1$ cm/s, $\kappa_h = 10^8$ cm²/s, $\kappa_v = 10^2$ cm²/s for the planetary-scale motions (horizontal and vertical scales being equal to 10^8 and 10^5 cm, respectively) we obtain $|\mathbf{F}| \approx 10^{-8}$ cm/s². In this way, both the tide-generating and the Coriolis forces appear to be approximately 10^3 times greater than the frictional force. We do not, however, intend to neglect it in equation (1.1.7) for the effects of tidal energy dissipation to be considered later.

1.2. Tidal Potential

Definition

Denote the potential of the attractive force effect of a body $\mathcal{B}$ upon a unit mass at a point $\mathcal{A}$ by $\mathcal{V}$, and the potential of the tide-generating force $\mathbf{T}$ (tidal potential) at the same point by $\Omega(\mathcal{A})$. Let the point $\mathcal{A}$ shift along $\mathbf{r}$ by $\delta\mathbf{r}$ due to the tidal potential variations $\delta\Omega$ (Fig. 1.5). In this case it is obvious that

$$\delta\Omega(\mathcal{A}) = \mathbf{T}(\mathcal{A}) \cdot \delta\mathbf{r} \equiv \underline{\mathcal{V}}(\mathcal{A}) \cdot \delta\mathbf{r} - \underline{\mathcal{V}}(\mathcal{C}) \cdot \delta\mathbf{r}.$$

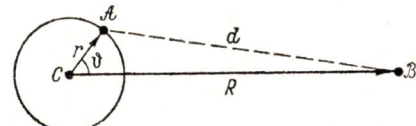

FIG. 1.5. On the definition of the tidal potential.

However $\underline{\mathcal{V}}(\mathcal{A}) \cdot \delta\mathbf{r} = \delta\mathcal{V}(\mathcal{A})$, $\underline{\mathcal{V}}(\mathcal{C}) = 0$ and hence

$$\Omega(\mathcal{A}) = \mathcal{V}(\mathcal{A}) - \underline{\mathcal{V}}(\mathcal{C}) \cdot \mathbf{r} + \text{const}, \tag{1.2.1}$$

where the constant is independent of the position of the point $\mathcal{A}$.

Assume the body $\mathcal{B}$ to be a homogeneous sphere, in which case the potential $\mathcal{V}(\mathcal{A})$ is known to equal $\gamma \mathcal{M}/d$ (see Fig. 1.5 for the definition of d). Then, bearing in mind that $\bar{\mathcal{C}}(\mathcal{C}) = \gamma \mathcal{M} \mathbf{R}/R^3$, equation (1.2.1) can be rewritten as

$$\Omega(\mathcal{A}) = \frac{\gamma \mathcal{M}}{R} \left(\frac{R}{d} - 1 - \frac{\mathbf{R} \cdot \mathbf{r}}{R^2} \right),$$

where the additive constant has been set equal to $-\gamma \mathcal{M}/R$.
Setting $r/R < 1$ (in fact $r/R \ll 1$) gives

$$\frac{R}{d} = \left[1 + \left(\frac{r}{R} \right)^2 - 2\frac{r}{R} \cos \vartheta \right]^{-1/2}.$$

Expand the function on the right-hand side of this relation in a power series of r/R:

$$\left[1 + \left(\frac{r}{R} \right)^2 - 2\frac{r}{R} \cos \vartheta \right]^{-1/2}$$
$$= P_0(\cos \vartheta) + P_1(\cos \vartheta) \frac{r}{R} + P^2(\cos \vartheta) \frac{r^2}{R^2} + \ldots,$$

where $P_n(\cos \vartheta)$ is the Legendre polynomial.

Since the first two terms of the series, $P_0(\cos \vartheta)$ and $P_1(\cos \vartheta) r/R$, equal 1 and $(r/R) \cos \vartheta$, respectively, and $\mathbf{R} \cdot \mathbf{r}/R^2 = (r/R) \cos \vartheta$, the expression for $\Omega(\mathcal{A})$ can be written as

$$\Omega(\mathcal{A}) = \Omega_2(\mathcal{A}) + \Omega_3(\mathcal{A}) + \ldots + \Omega_n(\mathcal{A}) + \ldots, \qquad (1.2.2)$$

where

$$\Omega_n(\mathcal{A}) \equiv \left(\frac{\gamma \mathcal{M}}{R} \right) \left(\frac{r}{R} \right)^n P_n(\cos \vartheta).$$

At $n = 2, 3, \ldots$ the expression takes the form

$$P_2(\cos \vartheta) = \tfrac{1}{2}(3 \cos^2 \vartheta - 1);$$
$$P_3(\cos \vartheta) = \tfrac{1}{2}(5 \cos^3 \vartheta - 3 \cos \vartheta);$$
$$\cdot \ \cdot \ \cdot \ \cdot \ \cdot \ \cdot \ \cdot \ \cdot \ \cdot \ \cdot$$

Thus, the tidal potential can be presented as an infinite series which begins with the term Ω_2 and converges with the rate r/R. Since r/R is approximately 4×10^{-5} for the Sun and from 0.0157 to 0.0180 for the Moon, all the terms in the right-hand side of (1.2.2) except for the first one $-\Omega_2(\mathcal{A})$ can be omitted to a high degree of accuracy.

Equipotential surfaces of the tidal potential are hyperboloids of rotation with respect to the axis $\mathcal{CB}$ (Fig. 1.6), these being the hyperboloids of two sheets when $\Omega_2 > 0$ and of one sheet when $\Omega_2 < 0$.

Thus the potential Ω can be presented in the following way:

$$\Omega_2 = \left(\frac{\gamma\mathcal{M}}{R}\right)\left(\frac{r}{R}\right)^2 P_2(\cos\vartheta)$$

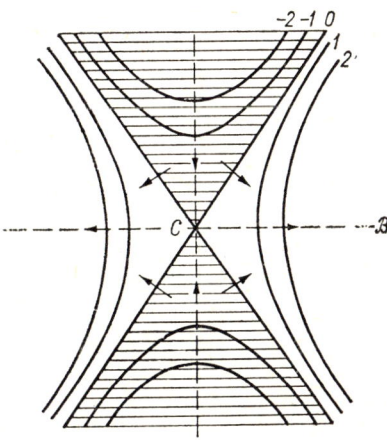

FIG. 1.6. Equipotential surfaces of the tidal potential Ω_2 of the second order (by Platzman, 1971). Arrows show direction of the tide-generating force. The region of negative values of Ω_2 is shaded.

and appears to be symmetrical with respect to the plane perpendicular to $\mathcal{BC}$ at the point $\mathcal{C}$.

The above expression for the potential Ω_2 includes the zenith angle ϑ. To get rid of the necessity to use the local coordinate ϑ of the perturbating body considered, introduce ordinary equatorial coordinates (hour-angle T and declination δ) and astronomical coordinates (latitude φ and longitude λ) of the given point $\mathcal{A}$. Then, as applied to the spherical triangle $\mathcal{PAB}$ (Fig. 1.7), we may write

$$\cos\vartheta = \cos\mathcal{PA}\cos\mathcal{PB} + \sin\mathcal{PA}\sin\mathcal{PB}\cos(T+\lambda)$$
$$= \sin\varphi\sin\delta + \cos\varphi\cos\delta\cos(T+\lambda). \quad (1.2.3)$$

The substitution of relation (1.2.3) into $P_2(\cos\vartheta)$ and multiplying it by $\gamma\mathcal{M}r^2/R^3$ yields

$$\Omega_2(\mathcal{A}) = \frac{3}{4}\frac{\gamma\mathcal{M}r^2}{R^3}\left[\left(\frac{1}{2} - \frac{3}{2}\sin^2\varphi\right)\left(\frac{2}{3} - 2\sin^2\delta\right)\right.$$
$$\left. + \sin 2\varphi\sin 2\delta\cos(T+\lambda) + \cos^2\varphi\cos^2\delta\cos 2(T+\lambda)\right]. \quad (1.2.4)$$

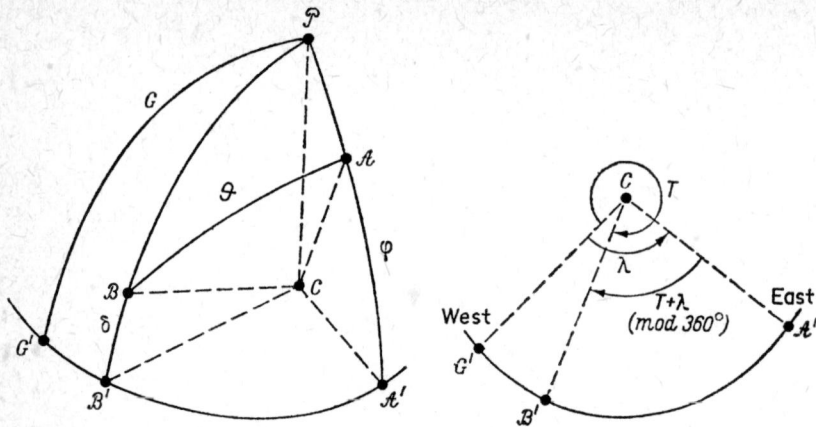

FIG. 1.7. Picture of the celestial sphere with respect to the Earth's centre (to the left) and the plane of the celestial equator (to the right).

Points: $\mathcal{A}$ is a considered point on the Earth's surface, $\mathcal{B}$ is a perturbing body, $\mathcal{P}$ is the North Pole.

Circumference segments: $\mathcal{PAA'}$ is the meridian of the point $\mathcal{A}$, $\mathcal{PBB'}$ is the meridian of the body $\mathcal{B}$, $\mathcal{PGG'}$ is the Greenwich meridian, $\mathcal{G'B'A'}$ is the celestial equator. The position of $\mathcal{A}$ and $\mathcal{B}$ is determined in the following way: point $\mathcal{A}$: $\mathcal{A'CA} \equiv \varphi$ is a latitude, $\mathcal{GCA'} \equiv \lambda$ is a longitude; point $\mathcal{B}$: $\mathcal{B'CB} \equiv \delta$ is a declination, $\mathcal{GCB'} \equiv \lambda$ is an hour-angle. The position of $\mathcal{B}$ with respect to $\mathcal{A}$ is determined via $\mathcal{ACB} \equiv \vartheta$ which is a zenith angle, $\mathcal{A'CB'} \equiv T+\lambda \pmod{360°}$ is an hour-angle. The longitude λ is read eastward and the hour-angle T westward from the Greenwich meridian.

The potential $\Omega_2(\mathcal{A})$ is seen to be the sum of three terms, each depending in its way upon the hour-angle, which increases by 360° daily due to the Earth's rotation. The first term is T-independent and, since the declination and the parallax of the perturbing body change slowly with time, the tides caused by them are called long-period tides (or the tides of the "first species"). The second term includes $\cos(T+\lambda)$ as a factor. Its period is a day, if one neglects small deviations arising in the course of the orbital motion of the body. The tides generated by this part of the potential can be referred to as diurnal (or the tides of the "second species"). Finally, the third term, including $\cos 2(T+\lambda)$ and having the period equal in fact to 12 hours, generates semidiurnal tides (or the tides of the "third species")[†].

The three terms in equation (1.2.4) correspond to three types of the second-order spherical harmonics (Fig. 1.8). The first of them—the zonal harmonic—has nodal lines coinciding with the circles of latitudes $\pm 35°16'$, where $\sin^2 \varphi = 1/3$. Beyond the zonal belt limited by these parallels, the tidal potential

---

[†] In Western literature, these "species" are denoted by 0, 1 and 2. (Ed.).

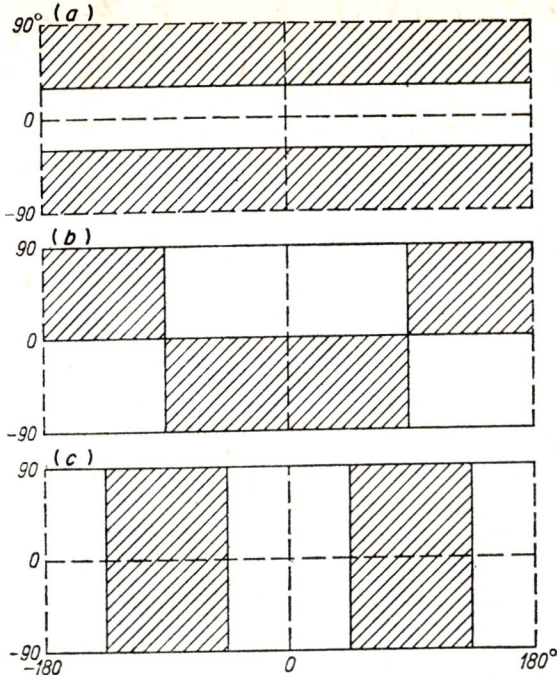

FIG. 1.8. Schematic picture of the three species of tides on the earth's surface. Nodes are presented by full lines, regions of negative values of the tidal potential are shaded. A perturbing body is located at the Greenwich meridian ($\lambda = 0°$) and has a positive declination (for a diurnal term). (a) A zonal harmonic, (b) a tesseral harmonic, (c) a sectorial harmonic.

corresponding to this function becomes negative because $\sin^2 \delta$ never exceeds 1/3.† The period of the harmonic is determined by the evolution in time of the declination and the parallax.

In the second function—a tesseral harmonic—the meridians, normal to the perturbing body meridian, and the equator serve as the nodal lines. Values of Ω_2 in different regions of the sphere change their signs depending upon the sign of declination. The period of the harmonic equals a day, its amplitude being maximum at the latitudes $\pm 45°$ at the maximum values of the declination of the perturbing body and zero at the equator and poles.

The nodal lines of the third function—a sectorial harmonic—coincide with

---

† Remember that the Moon's orbit is inclined to the ecliptic by about 5°, this angle changing during 173 days from 4°59′ to 5°18′. Thus the Moon's declination changes within the range (28°45′, −28°45′), the period being limited by the tropical month (27.327 days). The Sun's declination changes its sign twice a year in the range (23°27′, −23°27′).

the meridians 45° on either side of the meridian beneath the perturbing body. These lines divide the sphere into four sectors in which the function assumes by turn positive and negative values. The corresponding tides are of semidiurnal character, the amplitude being maximum at the equator (at zero declination of the perturbing body) and zero at the poles.

Thus $\Omega_2(A)$ can be presented as the sum of the three terms

$$\Omega_2(A) = \Omega_2^0(A) + \Omega_2^1(A) + \Omega_2^2(A);$$

$$\Omega_2^0(A) \equiv \mathcal{D}\left[\left(\frac{r}{a}\right)^2 \left(\frac{1}{2} - \frac{3}{2}\sin^2\varphi\right)\right]\left[\left(\frac{c}{R}\right)^3 \left(\frac{2}{3} - 2\sin^2\delta\right)\right];$$

$$\Omega_2^1(\mathcal{A}) \equiv \mathcal{D}\left[\left(\frac{r}{a}\right)^2 \sin 2\varphi\right]\left[\left(\frac{c}{R}\right)^3 \sin 2\delta \cos(T+\lambda)\right];$$

$$\Omega_2^2(\mathcal{A}) = \mathcal{D}\left[\left(\frac{r}{a}\right)^2 \cos^2\varphi\right]\left[\left(\frac{c}{R}\right)^3 \cos^2\delta \cos 2(T+\lambda)\right]. \tag{1.2.5}$$

Here a and c are the values of r and R chosen to obey the condition that r/a and c/R are of unit order. r/a and c/R will equal unity, provided, for instance, a is equal to the mean value of r, and c^{-1} is equal to the mean value of R^{-1}.

Relations (1.2.5) for the potentials of the zero, the first and the second types involve three factors. The first of them, $\mathcal{D} = 3\gamma \mathcal{M}a^2/4c^3$, common for Ω_2^0, Ω_2^1 and Ω_2^2, is a dimensional constant (the Doodson constant). It characterizes the order of magnitude of the tidal potential. The second (the first square brackets) is used in the sense of a "geodetic" factor. It is time-independent and associated solely with the position of the considered point $\mathcal{A}$ on the sphere. The third factor (the second brackets) can be identified as an "astronomical" factor conditioned by the motion of the perturbating body along the orbit and by the Earth's rotation.

For numerical evaluation, it is convenient to express the Doodson constant in the form

$$\mathcal{D} = \frac{3}{4} \frac{\gamma \mathcal{M}_\oplus}{a} \left(\frac{a}{r_1}\right)^3 \cdot \frac{\mathcal{M}}{\mathcal{M}_\oplus} \left(\frac{r_1}{c}\right)^3,$$

where r_1 is the Earth's equatorial radius.

As known from the theory of the Earth's shape (see Jeffreys, 1970),

$$\frac{\gamma \mathcal{M}_\oplus}{a} = ga\bigg/\left(1 - \frac{2}{3}m\right);$$

$$\left(\frac{a}{r_1}\right)^3 = 1 - e,$$

where g is the acceleration of the gravitational force at $r = a$; m and e are the Clairaut constant, and the longitudinal contraction of the Earth, equal to 0.345×10^{-2} and 0.3353×10^{-2}, respectively. Hence

$$\frac{\mathcal{D}}{ga} = \frac{3}{4} \cdot \frac{1-e}{1-\frac{2}{3}m} \cdot \frac{\mathcal{M}}{\mathcal{M}_\oplus} \left(\frac{r_1}{c}\right)^3.$$

Substituting in this relation the value of the second factor from equation (1.1.5) and the above values of m and e, we obtain

$$\frac{\mathcal{D}_\mathbb{C}}{ga} = 0.4198 \times 10^{-7};$$

$$\frac{\mathcal{D}_\odot}{\mathcal{D}_\mathbb{C}} = 0.4605, \tag{1.2.6}$$

which gives

$$\mathcal{D}_\mathbb{C} = 2.621 \times 10^4 \text{ cm}^2/\text{s}^2 \tag{1.2.7}$$

at $g = 979.76$ cm/s² and $a = 6371.27$ km.

The value of the Doodson constant referring to solar tides can be obtained from the ratio given in (1.2.6).

It is sometimes convenient to replace the tidal potential by the height of the "static tide" ζ^+, which can be determined as follows. As common practice in the static theory of tides, let the ocean surface respond instantaneously to the tide-generating force and assume the "equilibrium" shape of the equipotential surface. The latter is known to possess the following property: at all its points the magnitude of the potential of the gravitational forces $-g\zeta^+$ and tide-generating forces Ω_2 will remain constant, i. e. $\Omega_2 - g\zeta^+ = $ const, the constant being only time-dependent. In this case

$$\zeta^+ \equiv \frac{\Omega_2}{g} + c(t). \tag{1.2.8}$$

Here $c(t)$ is a new constant sometimes referred to as the "Darwin correction". Its value is obtained under the condition that the total volume of tidal deformations in the World Ocean is zero, i.e.

$$\int_S \zeta^+ \, dS = 0,$$

with the integration carried out over the entire surface of the World Ocean.

Thus, we have

$$c = -\frac{1}{gS} \int_S \Omega_2 \, dS.$$

In the case of an ocean covering the Earth completely, $c = 0$ and, hence, the height of the static tide ζ^+ differs from the tidal potential Ω only by a constant factor. If continents, assumed are $c \neq 0$, resulting in a general change of the location of the ocean-free surface and in the difference between the moments of the occurrence of high water and of the maxima of the perturbing force. The value of the correction $c(t)$ is only a few percent of the Ω_2/g value, and hence it can be neglected when evaluating the height of the static tide ζ^+.

Harmonic analysis of the tidal potential

As noted above, the declination and the distance to the perturbating body involved in relations for "astronomic" factors, as well as $\cos s(T+\lambda)$ (where s is the constant equal to 0, 1, 2 depending on the tidal type[†]) are periodic time functions. Since the "astronomic" factors depend upon the above parameters, in a non-linear way, the components of the tidal potential cannot be presented as a single harmonic or a strictly periodic time function. The frequency spectrum of Ω_2^0, Ω_2^1 and Ω_2^2 can be expected to have, besides the main peak at the frequency $S\omega$ (ω is the angular velocity of the Earth's rotation), some additional peaks at modulation frequencies. It can be illustrated by the following example.

Let the declination δ and the hour angle T of the perturbating body be

$$\delta \approx \varepsilon \sin \omega_1 t$$

$$T \approx \omega t$$

where ω_1 is the frequency of the declination changes; $\omega(=\omega_0-\omega_1)$ is the angular velocity of the Earth's rotation; ω_0 is the sidereal rotation velocity; ε is a numerical parameter.

Consider the three factors $\sin^2 \delta$, $\sin 2\delta \cos(T+\lambda)$ and $\cos^2 \delta \cos 2(T+\lambda)$ in equation (1.2.5) and try to evaluate the perturbation frequencies induced by them. If the parameter ε is small, then

$$\sin^2 \delta \approx \sin^2(\varepsilon \sin \omega_1 t) \approx \varepsilon^2 \sin^2 \omega t \approx \tfrac{1}{2}\varepsilon^2(1-\cos 2\omega_1 t).$$

[†] The tides of the first type correspond to $s = 0$, those of the second type to $s = 1$, and those of the third type to $s = 2$.

Thus, the first of the above factors is associated with the appearance of lunar half-monthly (or solar half-yearly) perturbations. In an analogous way

$$\sin 2\delta \cos (T+\lambda) \approx \sin (2\varepsilon \sin \omega_1 t) \cos (T+\lambda)$$
$$\approx 2\varepsilon \sin \omega_1 t \cos (T+\lambda)$$
$$\approx \varepsilon \{\sin (\omega_0 t + \lambda) - \sin [(\omega_0 - 2\omega_1)t + \lambda]\}.$$

The component of the tidal potential Ω_2^1 is seen to induce two oscillations with new frequencies ω_0 and $\omega_0 - 2\omega_1$ equidistant from $\omega \equiv \omega_0 - \omega_1$.

Finally, the third term can be presented as

$$\cos^2 \delta \cos 2(T+\lambda) \approx \cos^2 (\varepsilon \sin \omega_1 t) \cos 2(T+\lambda)$$
$$\approx (1 - \varepsilon^2 \sin^2 \omega_1 t) \cos 2(T+\lambda)$$
$$\approx \left(1 - \frac{\varepsilon^2}{2}\right) \cos 2[(\omega_0 - \omega_1)t + \lambda]$$
$$+ \frac{\varepsilon^2}{4} \{\cos 2(\omega_0 t + \lambda) + \cos 2[(\omega_0 - 2\omega_1)t + \lambda]\}.$$

Therefore, the potential Ω_2^2 induces not only lunar (or solar) tides but also an additional pair of waves with frequencies $2\omega_0$ and $2(\omega_0 - 2\omega_1)$, which are symmetrically located along the frequency axis with respect to the frequency of the main oscillation. $2\omega = 2(\omega_0 - \omega_1)$.

So far only the time changes of the perturbing body's declination have been investigated. However, more generally, when not only the declination but also the distance to the perturbating body are time functions, any of the "astronomic" factors in (1.2.5) can be presented as a sum of harmonics

$$\sum_j C_j \cos (\sigma_j t + s\lambda + q_j).$$

Each of the above harmonics is characterized by its amplitude C and the argument $\sigma t + s\lambda + q$ linearly dependent on the Greenwich mean solar time t and the eastern longitude λ. The harmonic argument equals q at $t = 0$ at the Greenwich meridian. The harmonic frequency σ is a linear combination of the angular velocity of the Earth's rotation ω and the sum and the difference of angular velocities $\omega_\kappa (\kappa = 1, \ldots, 5)$, to which correspond the five fundamental astronomic periods given in Table 1.1. According to this definition

$$\sigma = s\omega + \sum_{\kappa=1}^{5} m_\kappa \omega_\kappa,$$

TABLE 1.1. *Fundamental periods of the Earth's and the Moon's orbital motion*
(from Bartels, 1957)

Period (of mean solar days or years)	Nomenclature
$360°/\omega_1 = 27.321\ 582$ days	Period of lunar declination
$360°/\omega_2 = 365.242\ 199$ days	Period of solar declination
$360°/\omega_3 = 8.847$ years	Period of lunar perigee rotation
$360°/\omega_4 = 8.613$ years	Period of lunar node rotation
$360°/\omega_5 = 20.940$ years	Period of perihelion rotation

where $m_\kappa = 0, \pm 1, \pm 2, \ldots$ while ω can be assumed to equal either $\omega \equiv \omega_0 - \omega_1$ or $\omega_0 \equiv \omega_0 - \omega_2$. The sense of ω_1 and ω_2 is seen from Table 1.1.

All practically significant harmonics of the tidal potential Ω_2, whose amplitudes exceed 0.001, are grouped into seven frequency bands, their nomenclature and half-width $|\Delta\sigma|$ listed in Table 1.2. If, however, one confines oneself to

TABLE 1.2. *Frequency bands in the tidal potential spectrum,* $C > 0.001$
(by Platzman, 1971)

$|\Delta\sigma|$ degree/hour	$360°/|\Delta\sigma|$		Nomenclature
$\omega_4 = 0.002\ 2064$	18.613	years	Nodal (lunar)
$\omega_2 - \omega_5 = 0.041\ 0667$	365.260	days	Solar elliptical
$2\omega_2 = 0.082\ 1372$	182.621	days	Solar declinational
$\omega_1 - 2\omega_2 + \omega_3 = 0.471\ 5211$	31.812	days	Evectional (lunar)
$\omega_1 - \omega_3 = 0.544\ 3747$	27.555	days	Lunar elliptical
$2(\omega_1 - \omega_2) = 1.015\ 8958$	14.765	days	Variational (lunar)
$2\omega_1 = 1.098\ 0330$	13.661	days	Lunar declinational

harmonics with $C > 0.05$ only, then the number of spectral bands will reduce to four: two declinational, the nodal and the lunar elliptical. In this case the total number of harmonic terms, reduces from 65 to 18, seven of them being long-period, seven diurnal and four semidiurnal. They are listed in Table 1.3 in order of increasing frequency (see also Fig. 1.9).

Table 1.3 includes also the harmonic K_2^S, the amplitude thereof being less than 0.05. Its frequency coincides with that of the harmonic K_2^M, hence they are completely indistinguishable. On this ground they are combined into a single lunar-solar semidiurnal wave denoted by K_2.

TABLE 1.3. *Tidal harmonics with amplitude coefficients C > 0.05*
(by Bartels, 1957)

Coefficient C	Frequency σ degree/hour	Period $360°/\sigma$	Denotation and nomenclature
Long period tides			
0.2341	0		S_0 (solar constant)
0.5046	0		M_0 (lunar constant)
0.0655	$\omega_4 = 0.002\,21$	18.613 years	— (nodal M_0)
0.0729	$2\omega_2 = 0.082\,14$	182.621 days	SSa (declinational S_0)
0.0825	$\omega_1-\omega_3 = 0.544\,37$	27.555	Mm (elliptical M_0)
0.1564	$2\omega_1 = 1.098\,03$	13.661	Mf (declinational M_0)
0.0648	$2\omega_1+\omega_4 = 1.100\,24$	13.663	— (nodal Mf)
Diurnal tides			
0.0722	$(\omega_{\mathbb{C}}\omega_1)-(\omega_1-\omega_3) = 13.398\,66$	26.868 h	Q_1 (elliptical O_1)
0.0710	$-(\omega_{\mathbb{C}}-\omega_1)-\omega_4 = 13.940\,83$	25.823	— (nodal O_1)
0.3769	$(\omega_{\mathbb{C}}-\omega_1) = 13.943\,04$	25.819	O_1 (basic linar)
0.1755	$(\omega_{\odot}-\omega_2) = 14.958\,93$	24.066	P_1 (basic solar)
0.1682	$(\omega_{\odot}-\omega_2)+2\omega_2 = 15.041\,07$	23.934	K_1^S (declinational P_1)
0.3623	$(\omega_{\mathbb{C}}-\omega_1)+2\omega_1 = 15.041\,07$	23.934	K_1^M (declinational O_1)
0.0718	$(\omega_{\mathbb{C}}-\omega_1)+\omega_4 = 15.043\,28$	23.931	— (nodal K_1^M)
Semidiurnal tides			
0.1739	$2\omega_{\mathbb{C}}-(\omega_1-\omega_3) = 28.439\,73$	12.658 h	N_2 (elliptical M_2)
0.9081	$2\omega_{\mathbb{C}} = 28.984\,10$	12.421	M_2 (basic lunar)
0.4229	$2\omega_{\odot} = 30$	12	S_2 (basic solar)
0.0365	$2\omega_{\odot}+2\omega_2 = 30.082\,14$	11.967	K_2^S (declinational S_2)
0.0786	$2\omega_{\mathbb{C}}+2\omega_1 = 30.082\,14$	11.967	K_2^M (declinational M_2)
Combined tides			
0.5305	$\omega_0 = 15.041\,07$	23.934 h	K_1 (lunar-solar declinational)
0.1151	$2\omega_0 = 30.082\,14$	11.967	K_2 (lunar-solar declinational)

It is expedient to note that tide generation in the ocean takes place against the background of other time-dependent processes of non-tidal nature. In spectral analysis of the tidal components under consideration, these act as noise, of a level comparable with the signal. This means that one must analyse at least 19 years of observation to resolve the harmonics belonging to the extremely narrow "nodal" modulations.

It is common practice, however, to conduct observations throughout several years only. In this case an annual series of observations (e.g. a sample length

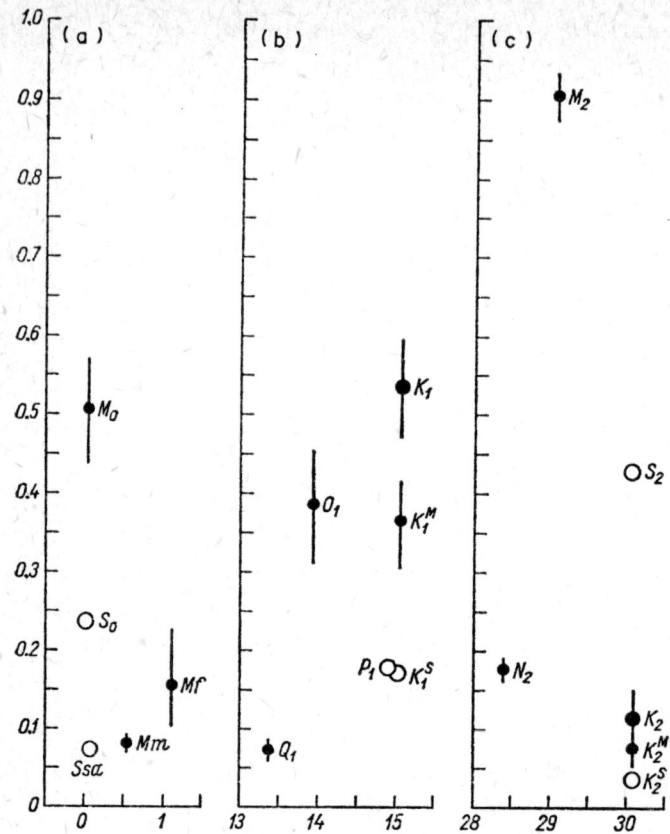

FIG. 1.9. Tidal harmonics with amplitude coefficients C exceeding 0.05 (from Platzman, 1971). The coefficient C is plotted along the ordinate, the frequency (in degrees/hour)—along the abscissa. Vertical bars denote the nodal modulation range. Tides: (a) are long-period, (b) are diurnal, (c) are semidiurnal.

comprising 355 or 369 days) is quite enough to distinguish the basic diurnal and semidiurnal harmonics. For such a short interval the harmonics of a nodal modulation can be considered as stationary. Their influence, however, cannot be neglected. Otherwise, as seen from Table 1.3, all the other harmonics, including the basic ones M_0, S_0, O_1, P_1, K_1, M_2 and S_2, δ with frequencies located in the centres of modulating frequency bands, will be noticeably distorted.

We do not intend to describe the conventional method of allowing for nodal modulations, which entails some corrections to the values of the amplitude and the phase of the basic harmonics. Note only that with the nodal terms excluded, the general number of harmonics in Table 1.3 reduces from 19 to 15. Overall, with the constant terms S_0 and M_0 excluded and K_1^S, K_1^M and K_2^S, K_2^M replaced by their combinations K_1 and K_2, the whole set of harmonics reduces to eleven basic ones,

long-period: S_{sa}, M_m, M_f;
diurnal: Q_1, O_1, P_1, K_1;
semidiurnal: N_2, M_2, S_2, K_2.

The relationships between these harmonics are summarized in Table 1.4.

TABLE 1.4. *Frequency band distribution of basic tidal harmonics*
(by Platzman, 1971)

Type of tides	Lunar harmonics			Solar harmonics	
	basic	elliptical	declinational	basic	declinational
Long-period	M_0	Mm	Mf	S_0	Ssa
Diurnal	O_1	Q_1	K_1^M	P_1	K_1^S
Semidiurnal	M_2	N_2	K_2^M	S_2	K_2^S

1.3. Equations of Tidal Dynamics

Basic harmonics of the tidal potential generate in the ocean forced oscillations with frequencies concentrated within the range 0.6×10^{-7} to 0.3×10^{-5} c/s. However, due to the presence of high-frequency wind and even acoustic waves, the spectrum of oscillations observed in the ocean comprises a much wider frequency range.

Since general hydrodynamic equations have solutions corresponding to all components of the spectrum, they are to be modified so that the new equations should not contain solutions pertaining to non-tidal waves and, at the same time, not distort the oscillations of the limited frequency range we are interested in. Thus, there is need to apply suitable filters. The assumption of sea-water incompressibility may serve as one such filter, to suppress acoustic waves. To distinguish long (including tidal) waves from a more general class of gravitational oscillations, one can use the condition of smallness of the ratio between vertical and horizontal motions. It follows from the equation of continuity that vertical velocities are small compared with horizontal velocities.

The above criteria allow one to neglect the quadratic inertia terms and the component of the Coriolis force which depends on the vertical velocity, in the horizontal equations of motion, and to reduce the z-component of the equation of motion to the equation of statics. Besides, in this context, ocean tidal motions can be shown to be practically independent of the density stratification of sea water.

In fact, in a stratified ocean, the pressure p at any level z can be presented as

$$p = (p_s + g\varrho_0\zeta_s) - g\varrho_0 z - p', \qquad (1.3.1)$$

where p_s is the atmospheric pressure at the free oceanic surface, ϱ_0 is the average sea-water density, ζ_s is the deviation of the free ocean surface from its undisturbed level.

In line with equtions (1.3.1) and (1.1.7) for the vertical component of velocity the pressure p', in stratified sea water, must obey the condition

$$\frac{1}{\varrho_0}\frac{\partial p'}{\partial z} = Q_z + 2\omega_y u - \frac{\partial w}{\partial t} + F_z, \qquad (1.3.2)$$

where Q_z is the projection of the vector $\mathbf{G}^{(1)} + \mathbf{G}^{(2)} + \mathbf{T}$ on the axis z, and $\omega_y = \omega \cos\varphi$.

Integrate equation (1.3.2) vertically from the free surface ($p' = 0$) to an arbitrary level z. Note that Q_z is of the same order of magnitude as projections of $\mathbf{Q}$ on the horizontal axes, the latter being comparable with the corresponding components of the Coriolis force. Hence, neglecting the value F_z, which is small compared with the other terms, gives

$$\frac{O(p')}{\varrho_0} = H \cdot O\left(2\omega_y u - \frac{\partial w}{\partial t}\right) < H \cdot \left[2|\omega_y| O(u) + O\left(\frac{\partial w}{\partial t}\right)\right], \qquad (1.3.3)$$

where the symbol $O(\ )$ denotes the order of magnitude of the function given in round brackets, and H is the vertical scale of motion, i.e. the depth scale.

As $O(\partial w/\partial t) = \sigma \cdot O(\omega)$ and $O(\omega) = O(\partial \zeta_s/\partial t) = \sigma \cdot O(\zeta_s)$ then $O(\partial w/\partial t) = \sigma^2 \cdot O(\zeta_s)$. Here σ is a typical oscillation frequency of the motion under discussion. To evaluate the order of the horizontal velocity component, use equation (1.1.7) projected on axis x. Setting $O(\partial u/\partial t) = \sigma \cdot O(u)$ and introducing the horizontal scale of motion (wave length) L, one obtains $O(u) = (g/\sigma L) O(\zeta_s)$. Substituting these expressions in inequality (1.3.4) gives

$$\frac{O(p')}{\varrho_0 g \cdot O(\zeta_s)} < \left(\frac{2|\omega_y|}{\sigma}\right)\frac{H}{L} + \frac{\sigma^2 H}{g}. \qquad (1.3.4)$$

With $\sigma/2|\omega_y| \approx 1$ and H equal to the average ocean depth, the second term in the right-hand side of inequality (1.3.4) is much less than the first term which is of the order $H/L \ll 1$. Thus, this may lead to the conclusion that the ocean behaves as a quasi-homogeneous system in relation to tidal motions induced by basic harmonics of the tidal potential. This conclusion can be easily verified to be valid within the entire frequency range of forced tidal oscillations.

Owing to a certain correlation between the terms of equation (1.1.7), the latter can be written as

$$\frac{\partial \mathbf{v}}{\partial t} = -g \cdot \nabla \left(\zeta_s + \frac{p_s}{\varrho_0 g} - \frac{\Omega_2}{g} \right) + \mathbf{G}^{(1)} + \mathbf{G}^{(2)} - A_1 \mathbf{v} + \mathcal{F}, \qquad (1.3.5)$$

where $\mathbf{v} = \{u, v\}$ and $\mathcal{F} = \{F_x, F_y\}$ are the velocity and the friction force per a mass unit, ∇ is the operator of the horizontal gradient, A_1 is the coefficient matrix equal to

$$A_1 = \begin{bmatrix} 0 & -2\omega_z \\ 2\omega_z & 0 \end{bmatrix},$$

and $2\omega_z = 2\omega \sin \varphi$ is the Coriolis parameter.

Combine equation (1.3.5) and the equation of continuity

$$\text{div } \mathbf{v} + \frac{\partial w}{\partial z} = 0, \qquad (1.3.6)$$

which represents the law of mass conservation in an elementary volume of incompressible liquid. In this relation we have neglected the small term which allows for the fact that neighbouring vertical planes are, strictly speaking, not quite parallel to one another. Compared to $\partial w/\partial z$ this term is of order H/a $\sim 10^{-3}$ (a is the Earth's mean radius).

Now integrate equation (1.3.6) vertically from the surface ($z = \zeta_s$) to the bottom ($z = -D + \zeta_b$, where ζ_b is the shift of the ocean bottom) and use the kinematic conditions on corresponding boundary surfaces. Then

$$\frac{\partial}{\partial t}(\zeta_s - \zeta_b) = -\text{div} \int_{-D+\zeta_b}^{\zeta_s} \mathbf{v} \, dz,$$

or, introducing the definition of the total transport vector

$$\mathbf{w} = \int_{-D+\zeta_b}^{\zeta_s} \mathbf{v} \, dz \qquad (1.3.7)$$

and the deviations of free surface with respect to the ocean bottom,

$$\zeta = \zeta_s - \zeta_b, \qquad (1.3.8)$$

we have

$$\frac{\partial \zeta}{\partial t} + \operatorname{div} \mathbf{w} = 0. \tag{1.3.9}$$

Subjecting equation of motion (1.3.5) to analogous transformations and defining $\mathcal{F}$ by equation (1.1.16) gives

$$\frac{\partial \mathbf{w}}{\partial t} = -gD \cdot \nabla \left(\zeta + \zeta_b + \frac{p_s}{\varrho_0 g} - \frac{\Omega_2}{g}\right) + D(\mathbf{G}^{(1)} + \mathbf{G}^{(2)})$$
$$- A_1 \mathbf{w} + \kappa_h \Delta \mathbf{w} - \boldsymbol{\tau}_b, \tag{1.3.10}$$

where $\boldsymbol{\tau}_b = \{\tau_b^x, \tau_b^y\}$ is the bottom friction stress.

Note that in equation (1.3.10) we have set the friction stress on the free ocean surface to equal zero and have omitted the terms $(\mathbf{v}_s \cdot \Delta \zeta_s + \nabla \zeta_s \cdot \nabla \mathbf{v}_s)$,[†] which are three or four orders less than $\Delta \mathbf{w}$. Also, we assumed $\mathbf{G}^{(1)}$, $\mathbf{G}^{(2)}$ not to vary with depth. In fact, they are functions of the vertical coordinate, which is expressed in the equation for $\mathbf{G}$ by the factor $(r/a)^2 \approx 1 + 3z/a$. The error due to replacing $(r/a)^2$ by unity is, however, of order H/a, which is equal to that of other terms omitted in equation (1.3.10).

1.4. Additional Potentials of Deformation

The above equations (1.3.9), (1.3.10) are the basic equations in tidal dynamics. Their terms

$$gD \cdot \nabla \left(\zeta_b + \frac{p_s}{\varrho_0 g}\right)$$

describe the effects of terrestrial and atmospheric tides on ocean tides, whose influence is, as it will be shown later, quite appreciable.

The latter statement refers more importantly to terrestrial tides. Indeed, the amplitude of tidal oscillations of atmospheric pressure at the sea level does not exceed 1 mb (10^3 din/cm^2), which corresponds to $p_s/\varrho_0 g \approx 1$ cm. Since the static tide amplitude for at least the basic (lunar and solar) components is approximately an order greater than the above value, the influence of atmospheric tides upon ocean ones can be neglected.

Terrestial tides (often called "Earth tides"), however, may not be neglected. The Earth's shape would not change if the Earth were an absolutely solid body. In reality, the Earth's crust is elastic and viscous and, hence, under the

---

[†] Here $\mathbf{v}_s$ is the velocity at the ocean's surface.

influence of the potential of lunar-solar attraction, the earth's surface rises and falls by $h_L(\Omega_2/g)$, where $h_L \approx 0.6$ is the Love number. Thus, the amplitude of terrestrial tides is only half that of static tides in the ocean. In studying the ocean tides it is obviously desirable to take into account the effect of vertical shift of the ocean bottom.

It could be readily done if the effect of terrestrial tides was not indirectly contained in $\mathbf{G}^{(1)} + \mathbf{G}^{(2)}$, where, as has already been mentioned, $\mathbf{G}^{(1)}$ and $\mathbf{G}^{(2)}$ are variations in the Earth's gravitational field conditioned by tides in the ocean and in the Earth's crust, respectively.

The period of free oscillations of a liquid sphere, having the dimensions and the mean density of the Earth, is known to approximately equal 1.5 hours while the period of the seismic impulse transfer along the Earth's diameter is about 20 minutes, i.e. they are both much less than the shortest (semidiurnal) period of the perturbating force. Consequently, in this case the tides in the solid body of the Earth can be considered static.

Vertical shifts of the Earth's surface and changes in the gravitational force induced by them will then be equal to

$$h_L \frac{\Omega_2}{g} \quad \text{and} \quad \mathbf{G}^{(2)} = \kappa_L \cdot \nabla \Omega_2, \tag{1.4.1}$$

where h_L and κ_L are the Love numbers characterizing the ratio of the height of the terrestrial tide to that of the static tide in the ocean $\zeta^+ = \Omega_2/g$, and the ratio of the additional potential induced by deformations of the Earth's crust to the tidal potential Ω_2, respectively. Quantities h_L and κ_L are determined by the elasticity and density of the matter inside the Earth. In particular, according to Longman (1963) who calculated the Love numbers for the Gutenberg model of a non-homogeneous Earth, $h_L = 0.612$, $\kappa_L = 0.302$.

Consider now the effect of gravitational forces of the Earth's liquid mantle deformed by tides. Tidal rises of the ocean surface level result in the increase of the mass of a unit water column and, hence, in additional loading on the ocean bottom and in its deformation. Besides, the surplus mass of water itself attracts the Earth, leading to additional deformation of the Earth's surface.

As a result of the redistribution of water masses, let the free ocean surface be lifted with respect to the bottom by ζ. Then the corresponding perturbating potential will be equal to $g\alpha_n \zeta_n$, and the displacement of the ocean bottom resulting from the combined effect of loading and attraction will equal $-h'_n \alpha_n \zeta_n$. These bottom deformations will, in their turn, result in the additional potential $\kappa'_n g \alpha_n \zeta_n$. Here κ'_n and h'_n are the Love numbers of the nth order (their values are given in Table 1.5), $\alpha_n = (3/2n+1) \times (\varrho_0/\varrho_\oplus)$, $\varrho_0/\varrho_\oplus \approx 0.18$ is the ratio between the mean density of sea water and that of the Earth. n is the number of the term

TABLE 1.5. *The Love-numbers of the nth order (from Longman, 1963)*

n	$-h'_n$	$-k'_n$	n	$-h'_n$	$-k'_n$	n	$-h'_n$	$-k'_n$
2	1.007	0.310	10	1.439	0.069	18	1.902	0.054
3	1.059	0.197	11	1.506	0.066	19	1.949	0.052
4	1.059	0.133	12	1.572	0.064	20	1.994	0.051
5	1.093	0.104	13	1.631	0.062	21	2.037	0.050
6	1.152	0.090	14	1.691	0.060	22	2.078	0.049
7	1.223	0.082	15	1.747	0.058	23	2.117	0.048
8	1.296	0.076	16	1.798	0.056	24	2.156	0.047
9	1.369	0.072	17	1.852	0.055	25	2.194	0.046

in the expansion of ζ in spherical harmonics:

$$\zeta = \sum_n \zeta_n \equiv \sum_n \sum_m P_n^m (\sin \varphi) \binom{a_{nm} \cos m\lambda}{b_{nm} \sin m\lambda}, \quad (1.4.2)$$

where P_n^m is the associated Legendre function. a_{nm} and b_{nm} are the series coefficients obtained under the condition of orthogonality

$$\int_0^{2\pi} \int_{-\pi/2}^{\pi/2} \left(P_n^m (\sin \varphi') \binom{\cos m\lambda'}{\sin m\lambda'} \right)^2 \cos \varphi' \, d\varphi' \, d\lambda'$$

$$= \begin{cases} \dfrac{4\pi}{2n+1} & \text{for} \quad m = 0, \\ \dfrac{2\pi(n-m)!\,(n+m)!}{(2n+1)\,(n!)^2} & \text{for} \quad m > 0 \end{cases} \quad (1.4.3)$$

from the formulae

$$a_n^0 = \frac{2n+1}{4\pi} \int_0^{2\pi} \int_{-\pi/2}^{\pi/2} \zeta(\lambda', \varphi') P_n^0 (\sin \varphi') \cos \varphi' \, d\varphi' \, d\lambda'$$

for $m = 0$ and

$$\binom{a_{nm}}{b_{nm}} = \frac{(2n+1)\,(n!)^2}{2\pi(n-m)!\,(n+m)!} \int_0^{2\pi} \int_{-\pi/2}^{\pi/2} \zeta(\lambda', \varphi') P_n^m (\sin \varphi') \times$$

$$\times \binom{\cos m\lambda'}{\sin m\lambda'} \cos \varphi' \, d\varphi' \, d\lambda' \quad (1.4.4)$$

for $m > 0$.

Let us introduce the following definitions:

$$\zeta_b = h_L \zeta^+ + \sum_n h'_n \alpha_n \zeta_n,$$
$$\mathbf{G}^{(1)} = g \cdot \nabla \sum_n (1 + \kappa'_n) \alpha_n \zeta_n \qquad (1.4.5)$$

and substitute equality (1.4.1) and definition (1.4.5) in equation of motion (1.3.10). Then, returning to series (1.4.2) and formula (1.4.4) and denoting the right-hand side of (1.4.3) by $\mathcal{H}_{nm}$, we obtain

$$\frac{\partial \mathbf{w}}{\partial t} = -gD \cdot \nabla(\zeta - \gamma_L \zeta^+ - \zeta^\oplus) - \mathbf{A}_1 \mathbf{w} + \kappa_h \Delta \mathbf{w} - \tau_b, \qquad (1.4.6)$$

where $\gamma_L (= 1 + \kappa_L - h_L)$ – is the Love reduction factor, equal approximately to 0.7,

$$\zeta^\oplus = \int_0^{2\pi} \int_{-\pi/2}^{\pi/2} \zeta(\lambda', \varphi') G(\lambda', \varphi'; \lambda, \varphi) \cos \varphi' \, d\varphi' \, d\lambda';$$

$$G(\lambda', \varphi'; \lambda, \varphi) = \sum_n (1 + \kappa'_n - h'_n) \alpha_n \sum_m \mathcal{H}_{nm}^{-1} P_n^m (\sin \varphi) \times$$

$$\times P_n^m (\sin \varphi') (\cos m\lambda \cos m\lambda' + \sin m\lambda \sin m\lambda'). \qquad (1.4.7)$$

In the resulting integro-differential equation (1.4.6) the terms $\gamma_L gD \cdot \nabla \zeta^+$ and $gD \cdot \nabla \zeta^\oplus$ describe the effect of tide-generating forces on tides in the ocean with an underlying elastic surface, as well as the self-attraction of these tides and the bottom deformations due to loading, respectively.

To understand the need to account for the direct influence of the terrestrial tides upon oceanic tides, it suffices to consider the magnitude of the Love reduction factor. Evaluation of the reaction to the effects of self-attraction and loading is a much more difficult problem, since to solve it one has to know the space distribution of the oceanic tide. Nevertheless, some estimations are available. For instance, according to preliminary data by Farrel (1973), the total potential of the tide-generating forces $g(1 + \kappa_L) \zeta^+$ and the resulting potential of the ocean tides $\sum_n g(1 + \kappa'_n) \alpha_n \zeta_n$ are of the same order (Fig. 1.10).

Thus, in order to have a completely reliable picture of ocean tides, the terms $\gamma_L \zeta^+$ and $\zeta^\oplus$ must be present in the equation of motion (1.4.6).

We now see that the conventional presentation of oceanic tides as an isolated phenomenon, induced only by the tide-generating forces of the Moon and the Sun, is quite inadequate. However, how to treat the coupled equations is one of the many problems still to be solved in modern investigations of tides in the World Ocean.

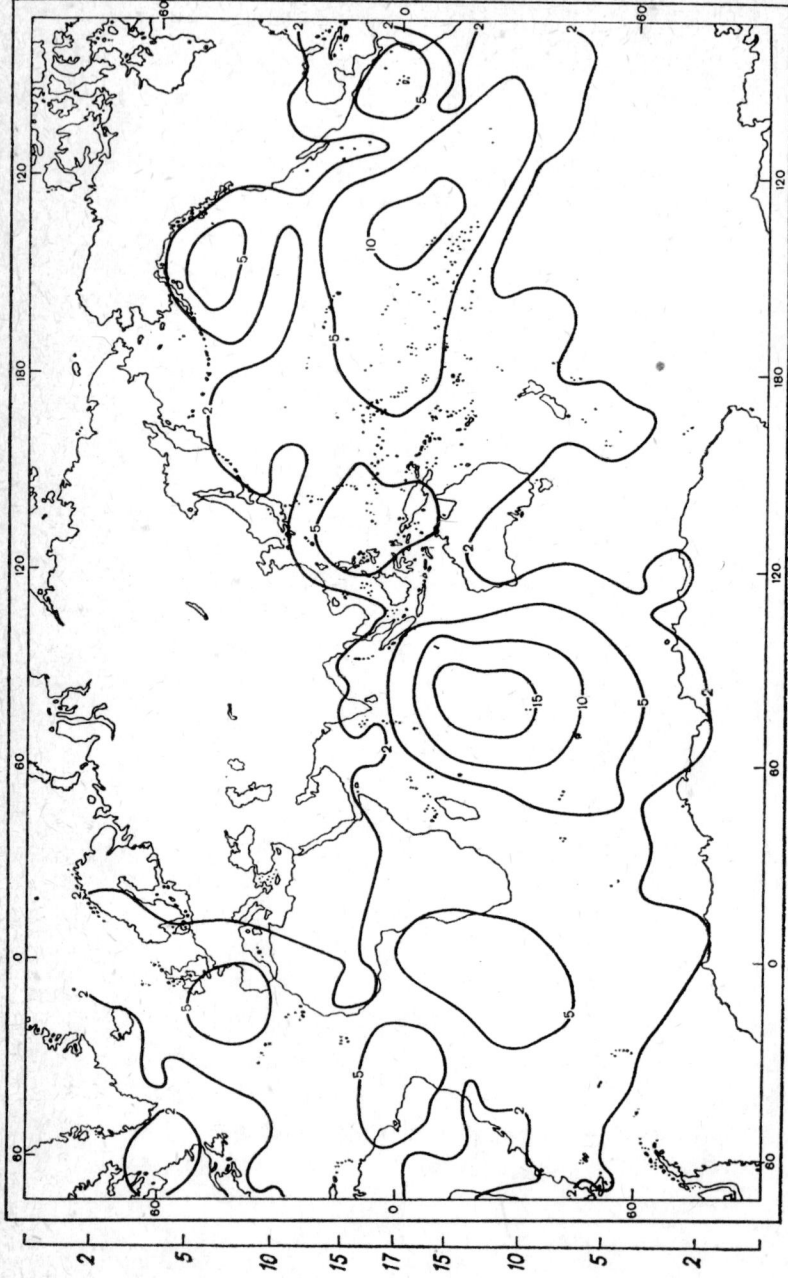

FIG. 1.10. Isoamplitudes of the total additional potential induced by oceanic tides and elastic deformations of the Earth's crust under load (from Farrel, 1973). Values of the potential amplitude are divided by the gravity. Numbers in the left-hand side of the picture denote the height of the static tide in the ocean with allowances made for the static effect of the terrestrial tides.

1.5. Boundary Conditions

It might seem at first sight that to formulate the boundary conditions for the equations of tidal dynamics (1.4.6), (1.3.9) is not at all difficult. This would be true if the boundary conditions were set at the ocean-land boundary, where the water depth equals zero. In this case the total transport is also zero and the conditions of no-slip and no-transport (the latter taken at $k_h = 0$, i.e. when the macroturbulence effect is neglected), are automatically fulfilled. However, for reasons which will be later clarified, the oceanic boundary is usually located along the shelf edge. It is this transfer of the actual ocean boundary to the shelf edge that is a source of serious problems, which are still unsolved.

For simplicity, transfer the above natural boundary conditions to the shelf edge. In other words, let, at the contour of the ocean basin Γ, the conditions

$$\mathbf{w}|_\Gamma = 0 \quad (\text{or} \quad \mathbf{w}_n|_\Gamma = 0), \tag{1.5.1}$$

hold, where w_n is normal to the boundary component of the total transport.

This condition is equivalent to the assumption that there is no energy transfer between the shelf zone and the open ocean. Its validity is, however, open to argument. Indeed, a considerable portion of the tidal energy is known to dissipate in the zone of the coastal shelf which is also the place of the tidal "wave-trapping", consisting of refraction and multiple reflection of waves from the coastline and the shelf edge. Under the latter condition the shelf may be viewed as a resonator. The whole scope of these effects can play quite an appreciable role in tidal generation in the open ocean and especially in the regions adjacent to well-developed shelf zones. Thus, we face the problem of parametric description of shelf effects.

At present all attempts to consider shelf effects are reduced to the use of boundary conditions conforming to two limiting cases, namely to the total energy dissipation of the shelf, and to the shelf "trapping" of tidal waves without dissipation.

The first variant of the boundary conditions can be obtained in the following way. Let the depth within the shelf zone be constant and equal to $\mathcal{H}$, the corresponding value of the surface oscillations and the component of the vertically averaged velocity normal to the shelf edge being equal to ζ and u, respectively. The relation

$$\frac{u}{c} = \frac{\zeta}{\mathcal{H}},$$

where $c \left(= \sqrt{g\mathcal{H}} \right)$ is the phase wave velocity, is valid for the case of progressively long waves, transferring the energy in the direction of their propagation.

Rewrite this relation taking into account continuity of the surface and integral transport at the shelf edge

$$\zeta|_\Gamma = \bar{\mathfrak{z}}; \qquad w_n|_\Gamma = u\mathcal{H}. \tag{1.5.2}$$

As a result, we have

$$w_n|_\Gamma = \sqrt{g\mathcal{H}} \cdot \zeta|_\Gamma, \tag{1.5.3}$$

which demonstrates the fact that the entire tidal energy induced by the ocean is transferred throughout the shelf towards the coastline where it completely dissipates either due to turbulent friction or due to wave-breaking.

Consider the second variant of the boundary conditions. For simplicity, confine the case to that of a linear coastline with a bordering shelf of constant width $\mathcal{L}$ and constant depth $\mathcal{H}$. Introduce a local Cartesian system of coordinates with origin at the coastline and axes x, y, orientated to the east (towards the ocean) and to the north (along the coastline), respectively.

Assume the shelf waves to have the character of free oscillations unaffected by the mechanism of energy dissipation. Let these waves belong to the class of Kelvin (or Poincaré) waves moving along the coastline towards the south. Then the surface elevation on the shelf can be described by the real part of the relation

$$\mathfrak{z} = \bar{\mathfrak{z}}(x) \exp[i(l_1 y + \sigma t)],$$

where l_1 is a coastal wave number, with $2\pi/l_1$ approximately equal to a or, to be more exact, to the perimeter of the ocean, comparable with the Earth's radius a.

We have the following equations to determine oscillation amplitudes of the surface $\bar{\mathfrak{z}}(x)$ and of the flow velocity $\bar{u}(x)$

$$\frac{d^2\bar{\mathfrak{z}}}{dx^2} + l^2\bar{\mathfrak{z}} = 0;$$

$$\bar{u} = \frac{ig}{\sigma^2 - 4\omega_z^2}\left(2\omega_z l_1\bar{\mathfrak{z}} + \sigma\frac{d\bar{\mathfrak{z}}}{dx}\right). \tag{1.5.4}$$

where $l^2 = \dfrac{\sigma^2 - 4\omega_z^2}{g\mathcal{H}} - l_1^2.$

Integrating the first relation and applying the condition of continuity of elevation and of transport at the shelf edge, as well as the condition of no transport of liquid through the ocean-earth boundary, we obtain

$$\bar{\mathfrak{z}} = \zeta_\Gamma \frac{\left(1 - \frac{2\omega_z l_1}{\sigma l}\tan lx\right)}{\left(1 - \frac{2\omega_z l_1}{\sigma l}\tan l\mathcal{L}\right)} \frac{\cos lx}{\cos l\mathcal{L}};$$

(1.5.5)

$$\bar{w}_n|_\Gamma = -i\frac{g\mathcal{H}l}{\sigma^2 - 4\omega_z^2}\sigma\zeta_\Gamma \frac{\left(1 + \frac{4\omega_z^2 l_1^2}{\sigma^2 l^2}\right)\tan l\mathcal{L}}{1 - \frac{2\omega_z l_1}{\sigma l}\tan l\mathcal{L}}.$$

Note that for narrow shelves ($O(\mathcal{L}) = 100$ km), $l\mathcal{L} \ll 1$. Hence, the expression for the amplitude of the transport $\bar{w}_n|_\Gamma$, normal to the contour Γ, can be rewritten as

$$|\bar{w}_n|_\Gamma \approx \sigma\zeta_\Gamma\mathcal{L}\left(1 - \frac{g\mathcal{H}l_1^2}{\sigma^2 - 4\omega_z^2}\right)\frac{1 + \frac{4\omega_z^2 l_1^2}{\sigma^2 l^2}}{1 - \frac{2\omega_z l_1}{\sigma l}l\mathcal{L}}[1 + O(l\mathcal{L})^2].$$

Setting $l^2 > 0$ and $|l_1/l| < 1$ (these conditions are fulfilled, for instance, for waves of semidiurnal period within a shallow shelf) one finally obtains

$$|\bar{w}_n|_\Gamma \approx \sigma\zeta_\Gamma\mathcal{L}.$$

(1.5.6)

The amplitude of the integral transport at the boundary with the open ocean, as is seen from the above equations, is proportional to that of the surface elevation at the shelf edge ζ_Γ and to the shelf-width $\mathcal{L}$.

There are other variants of boundary conditions allowing for shelf effects. For instance, in the case of partial reflection of progressive waves from the shelf edge, when the ratio between the amplitudes of a reflected and an approaching waves is

$$\mathfrak{r} = \frac{\sqrt{D} - \sqrt{\mathcal{H}}}{\sqrt{D} + \sqrt{\mathcal{H}}},$$

the condition (1.5.3) can be replaced by

$$w_n|_\Gamma = \frac{1-\mathfrak{r}}{1+\mathfrak{r}}\sqrt{g\mathcal{H}}\cdot\zeta_\Gamma.$$

(1.5.7)

At $\mathfrak{r} = 0$ ($\mathcal{H} = D$) and $\mathfrak{r} = 1$ ($\mathcal{H} = 0$) the latter condition can be reduced to the boundary conditions (1.5.3) and (1.5.1), respectively.

In cases when the energy of the wave, passing across the shelf, is only partially dissipated, the reflection from the slope will be supplemented by the "shelf radiation" which is a combined effect of multiple reflection within the shelf. The initial single reflection from the coastal slope and the "shelf radiation" result in the total reflection from the shelf zone, its corresponding wave having a phase shift with respect to the incoming wave. In this case boundary condition (1.5.7) can be presented in the form

$$w_n|_\Gamma = \frac{1-\Re}{1+\Re} \sqrt{g\mathcal{H}} \cdot \zeta_\Gamma. \qquad (1.5.8)$$

The modulus and the argument of the total reflection coefficient $\Re = \Re_0 \exp(i \arg \Re)$ depend on the morphology of the shelf zone and the coastal slope in a fairly complex way. Therefore, unless this dependence is determined, the boundary condition (1.5.8) is unlikely to be more meaningful in parameterizing shelf effects than the boundary condition (1.5.3) and relation (1.5.6).

1.6. References

Quite an accessible treatment of the problems concerning the analysis of tide-generating forces as well as the forces of rotation and gravity can be found in monographs by Defant (1961), Zubov (1933) and Proudman (1953). In this chapter we have mainly followed the paper by Platzman (1971).

It was Richardson (1922) who suggested the idea of analogy with the theory of elasticity while deducing equations for the components of the Reynolds stress deviator in motions of planetary scale. This idea was later realized by Williams (1972). The equation for the matrix of "turbulent viscosity" coefficients (1.1.12) can be found in the monograph by Love (1927).

A somewhat different approach to the determination of "turbulent viscosity" coefficients for large-scale motions of the atmosphere and ocean was suggested by Kamenkovich (1967) (see also Kirwan (1969)). By analogy with molecular viscosity he set the tensor of turbulent stresses to be a linear non-homogeneous function of the tensor of deformation velocities, in which case the coefficients of turbulent viscosity yield the tensor of the fourth order. To determine its components, the hypothesis of axial symmetry of the tensor with respect to the vertical was used.

Harmonic analysis of the tidal potential is described in detail in the monographs by Berezkin (1947) and Duvanin (1960) as well as in the review by Bartels (1957). In section 1.2 of this chapter we made use of the work of Platzman (1971).

The correction to the equation of the height of the static tide was given by

Thomson and Tait (1883). Its evaluation obtained by Turner can be found in the reference to the paper by Darwin (1886).

Derivation of the equations of tidal dynamics from general hydrodynamic equations and substantiation of relevant simplifications is available in the monographs by Voltsinger and Pjaskovsky (1968), Kagan (1968), Stretensky (1936) and Stoker (1957). The argument that one can neglect the effect of density variations on tidal motions in the ocean was borrowed from the work by Platzman (1971).

Interdependence of terrestrial and ocean tides as well as the determination of the Love numbers of the second and nth orders are discussed in the book by Munk and McDonald (1960). The effects of the additional gravitational potential induced by ocean tides are discussed in Lamb (1945). The idea of the need to allow for this potential as well as for the effects associated with elastic deformations of the Earth's crust, in the oceanic tidal equations was first suggested by Poincaré (1895) and later fully formulated by Hendershott (1972).

Some preliminary considerations on calculating shelf effects are given in the work by Heaps (1969b). Condition (1.5.3) was obtained by Proudman (1941). A condition at the shelf edge similar in sense to equation (15.6) is discussed by Baines (1973). Conditions (1.5.7) and (1.5.8) were proposed by Nekrasov (1973).

CHAPTER 2

Studies on the Equations of Tidal Dynamics

2.1. Formulation of the Problem

It is expedient, if perhaps rather unexpected, to begin the chapter on mathematical principles of the equations with some simplification of the general problem of tidal dynamics discussed in detail in the previous chapter.

As has been noted earlier, initial relations (1.4.6) are quasi-linear integrodifferential equations in which the integral term describing the effect of the gravitational potential of ocean tides and loading contains an unknown value of the surface elevation. Difficulties in solving equations of a similar type have been thoroughly studied. Of course, one can try to solve the problem by the method of iteration setting in the first approximation this term equal to zero. Then, after the determination of the elevation they can be substituted into equation (1.4.7), which yields new equations (1.4.6), (1.3.9) for the next approximation, etc. Generally speaking, this method makes it possible to calculate the functions in question, i.e. the components of the total transport and the elevation, under the supposition that the method of iterations converges. Its convergence, however, in the case when the integral term and the potential of the perturbing force have the same order of magnitude has not yet been proved. This is not the only difficulty impeding the solution of the problem. It should be noted that we cannot yet confidently describe the shelf effects, and usually have to confine ourselves to the conditions of type (1.5.3) or (1.5.6) which do not at all describe the whole variety of phenomena taking place in the shelf zone and appreciably affecting tide generation in mid-ocean.

The above naturally leads to the necessity for serious simplifications of the problem formulated in Chapter 1. This is the way we have treated the problem, having confined ourselves to the analysis of the "classical" problem of tidal dynamics which results from the general problem under the following suppositions: (i) the Earth is a perfectly solid body, (ii) tides in the ocean do not change the Earth's gravitational field and (iii) there is no energy exchange between the mid-ocean and the shelf zone.

EQUATIONS OF TIDAL DYNAMICS

As a result we obtain a model combining in fact all the up-to-date formulations of the problem of oceanic tides. The equations of this problem are written as

$$\frac{\partial \mathbf{w}}{\partial t} + \mathbf{A}_1 \mathbf{w} - \kappa_h \Delta \mathbf{w} + \frac{\iota}{D^2} |\mathbf{w}| \mathbf{w} = -gD \cdot \nabla \zeta + \mathbf{f}; \qquad (2.1.1)$$

$$\frac{\partial \zeta}{\partial t} + \text{div } \mathbf{w} = 0, \qquad (2.1.2)$$

where $\mathbf{f} \equiv gD \cdot \nabla \zeta^+$ is a generating function; ι is the bottom friction coefficient equal to a numerical constant; the rest of the notations are the same.

On the domain contour, which in the general case consists of two parts, a solid part Γ_1 coinciding with the shelf edge and the open boundary Γ_2, the vector of full flow is considered a known function of horizontal coordinates and time

$$\mathbf{w}|_\Gamma = \mathbf{w}^0(x, y, t), \qquad (2.1.3)$$

in which case

$$\mathbf{w}^0|_{\Gamma_1} = 0 \qquad (2.1.4)$$

and

$$\oint dt \int_{\Gamma_2} \mathbf{w}^0 \cdot \mathbf{n} \, d\Gamma = 0, \qquad (2.1.5)$$

where $\mathbf{n}$ is a normal to the contour Γ_2.

The first of these relations demonstrates the fact that the no-slip condition is fulfilled on the solid part of the contour Γ_1, the second relation means that the transfer via the open boundary of the ocean Γ_2, integrated during the tidal cycle, turns to zero. Note that condition (2.1.5) must by all means be fulfilled, otherwise, in the case when the normal component of the full flow across the contour Γ_2 does not obey condition (2.1.5), the law of mass conservation in the basin under investigation will undoubtedly be violated.

Let us set the initial conditions in order to complete the formulation of the problem. With the behaviour of the tidal characteristics $\mathbf{w}$ and ζ in time not fixed beforehand, specify that initially (at $t = 0$) the fields of tidal flow and elevation are

$$\begin{aligned} \mathbf{w} &= \mathbf{w}_0(x, y); \\ \zeta &= \zeta_0(x, y) \quad \text{at} \quad t = 0. \end{aligned} \qquad (2.1.6)$$

In particular, w_0 and ζ_0 can be set equal to zero, which indicates the fact that initially the ocean is at rest.

Note in conclusion that the quadratic law of resistance used in equation (2.1.1) does not take into account real phase shear between the tangential stress at the bottom and the tidal flow velocity. One more method of describing the bottom friction stress, free from the above limitation, will be considered in Chapter 5. This chapter will also deal with the fact that the bottom friction coefficient r is not at all a universal constant but depends in a complex fashion on the external parameters determining the vertical structure of the bottom boundary layer.

Remark. The possibility of an arbitrary choice of initial conditions is associated with the fact that the solution of equations (2.1.1), (2.1.2) becomes in the long run independent of the initial conditions. It can be illustrated, for example, by the following simple model.

Assume that the depth D and the Coriolis parameter $2\omega_z$ are constant within the whole investigated domain, macroturbulence effects are not considered ($\kappa_h = 0$), the bottom friction is described by the linear law of resistance ($\varkappa|w|/D^2 = r$) and the generating function $\mathbf{f}$ is equal to zero. In this case the system of equations (2.1.1), (2.1.2) can be presented as a single vector equation

$$\frac{\partial \mathfrak{w}}{\partial t} + \mathfrak{B}\mathfrak{w} = 0, \qquad (2.1.7)$$

where $\mathfrak{w}$ is the vector-function with the components (u, v, ζ); $\mathfrak{B}$ is a differential matrix operator equal to

$$\begin{bmatrix} r & -2\omega_z & g\dfrac{\partial}{\partial x} \\ 2\omega_z & r & g\dfrac{\partial}{\partial y} \\ D\dfrac{\partial}{\partial x} & D\dfrac{\partial}{\partial y} & 0 \end{bmatrix}.$$

The solution of equation (2.1.7) must obey the initial condition

$$\mathfrak{w} = \mathfrak{w}_0(x, y) \quad \text{at} \quad t = 0 \qquad (2.1.8)$$

and certain boundary conditions on the domain contour. It is not obligatory to set the latter condition if the vector components $\mathfrak{w}$ are assumed to be harmonic functions of horizontal coordinates, for instance

$$\mathfrak{w} = \overline{\mathfrak{w}}(t) \cdot \exp i\kappa \cdot x,$$

where $\overline{\mathfrak{w}}(t)$ is a vector whose components are solely time functions; $\kappa = (\kappa_x, \kappa_y)$, κ_x, κ_y are real wave numbers along x and y axes.

Make the same assumption for the vector $\mathfrak{w}_0$. Then substituting the relation for $\mathfrak{w}$ and $\mathfrak{w}_0$ into equations (2.1.7), (2.1.8), we obtain the following equation for $\overline{\mathfrak{w}}(t)$:

$$\frac{d\overline{\mathfrak{w}}}{dt} + \mathfrak{B}\overline{\mathfrak{w}} = 0; \qquad (2.1.9)$$

$$\overline{\mathfrak{w}} = \overline{\mathfrak{w}}_0 \quad \text{at} \quad t = 0 \qquad (2.1.10)$$

where

$$\mathfrak{B} = \begin{bmatrix} r & -2\omega_z & i\kappa_x g \\ 2\omega_z & r & i\kappa_y g \\ i\kappa_x D & i\kappa_y D & 0 \end{bmatrix}.$$

The solution of the uniform equations (2.1.9), (2.1.10) at $r = \text{const}$ is

$$\overline{\mathfrak{w}} = \overline{\mathfrak{w}}_0 e^{-\mathfrak{B}t}, \qquad (2.1.11)$$

which means that if the real part of the matrix $\mathfrak{B}$ is positive, the solution of equation (2.1.11) will decrease in amplitude with increasing time.

To determine the eigenvalues λ of the matrix $\mathfrak{B}$ consider the spectral problem

$$\mathfrak{B}\psi = \lambda\psi. \qquad (2.1.12)$$

Set the determinant equal to zero and develop it. As a result obtain a characteristic equation for λ

$$(r-\lambda)[\lambda(r-\lambda) - |\kappa|^2 gD] + 4\omega_z^2 \lambda = 0, \qquad (2.1.13)$$

where $|\kappa| = (\kappa_x^2 + \kappa_y^2)^{1/2}$ is a horizontal wave number.

If in equation (2.1.13) $4\omega_z^2 r$ is neglected as compared other terms, the roots of this equation are

$$\lambda_1 = r; \quad \lambda_{2,3} = \frac{r}{2} \pm i \sqrt{4\omega_z^2 + |\kappa|^2 gD - \frac{r^2}{4}}. \qquad (2.1.14)$$

Since the bottom friction coefficient r is always positive, it follows from equation (2.1.14) at $(4\omega_z^2 + |\kappa|^2 gD) > r^2/4$ that the real parts of all eigenvalues of the matrix $\mathfrak{B}$ are also positive. It allows one to calculate the velocity components of the tidal flow and the elevation using the system of equations (2.1.1)—(2.1.6), even in the case when there is no detailed information about the initial distribution of tidal characteristics in the open part of the domain under consideration. The latter fact substantially simplifies calculations in real physico-geographical conditions, as it allows one to use arbitrary initial values of the functions w and ζ instead of scarce data from direct measurements.

2.2. Basic Notions and Definitions

Before investigating the univalent solvability of the problem (2.1.1)—(2.1.3) within the finite time interval $[0, T]$, we introduce some definitions. Let S be a domain in the Cartesian space $\mathbf{R}_2$, Γ the boundary of this domain, $\bar{S}$ its closure, $\mathbf{x} = (x, y)$ a point on the plane with coordinates x, y.

Further we shall need various functional spaces. Let us consider them in brief.

The Sobolev spaces $W_2^\kappa(S)$ consist of all measurable functions $u(\mathbf{x})$ within S, having all possible derivatives $D_\mathbf{x}^\kappa$ with respect to $\mathbf{x}$ of the order $m < \kappa$. The scalar product and norm are determined here in the following way:

$$(u, v)^{(\kappa)} = \int_S \sum_{m=0}^{\kappa} \sum_{(m)} D_\mathbf{x}^m u \cdot D_\mathbf{x}^m v \, dx;$$

$$\|u\|_{2,S}^{(\kappa)} = \sqrt{(u, u)^{(\kappa)}},$$

where $\sum_{(m)}$ has the sense of summation over all possible derivatives $D_\mathbf{x}^m u$ of the order of m.

At $\kappa = 0$, spaces $L_2(S)$ are $W_2^\kappa(S)$ elements. To denote the scalar product and norm within $L_2(S)$ we use the symbols $(\,,\,)_{2,S}$ and $\|\cdot\|_{2,S}$, where the indices are sometimes omitted.

Then, consider $W_2^{\kappa,m}(Q_T)$ (here $Q_T = S \times [0, T]$, $\kappa, m > 0$) to be the space of the functions $\mathbf{u}(\mathbf{x}, t)$ quadratically summed together with their derivatives with respect to $\mathbf{x}$ up to the order κ and with respect to t up to the order of m inclusive. In this case the norm can be defined by the equality

$$|u|_{W_2^{\kappa,m}(Q_T)} = \left[\int_{Q_T} \left(\sum_{i=0}^{\kappa} \sum_{(i)} |D_\mathbf{x}^i u|^2 + \sum_{i=1}^{m} |D_t^i u|^2 \right) dx \, dt \right]^{1/2}.$$

Spaces consisting of vector-functions $\mathbf{w} = (u, v)$, whose components belong to the spaces described, will be denoted by the same symbols but in bold print. The same notations will be preserved to denote the norms of these domains while the symbol $|\mathbf{w}|$ will characterize the vector length

$$|\mathbf{w}| = (u^2 + v^2)^{1/2}.$$

EQUATIONS OF TIDAL DYNAMICS

The following notations will be also used

$$\left|\frac{\partial \mathbf{w}}{\partial \mathbf{x}}\right| = \left[\left(\frac{\partial u}{\partial x}\right)^2 + \left(\frac{\partial u}{\partial y}\right)^2 + \left(\frac{\partial v}{\partial x}\right)^2 + \left(\frac{\partial v}{\partial y}\right)^2\right]^{1/2};$$

$$\left\|\frac{\partial \mathbf{w}}{\partial \mathbf{x}}\right\|_{2,S} = \left\|\left|\frac{\partial \mathbf{w}}{\partial \mathbf{x}}\right|\right\|_{2,S};$$

$$\left|\frac{\partial^2 \mathbf{w}}{\partial \mathbf{x}^2}\right| = \left[\sum_{(2)} (D_\mathbf{x}^2 u)^2 + \sum_{(2)} (D_\mathbf{x}^2 v^2)\right]^{1/2};$$

$$\left\|\frac{\partial^2 \mathbf{w}}{\partial \mathbf{x}^2}\right\|_{2,S} = \left\|\left|\frac{\partial^2 \mathbf{w}}{\partial \mathbf{x}^2}\right|\right\|_{2,S}.$$

The symbol C with various indices will denote all the constants occurring in this chapter and depending only on T, as well as the initial parameters and the constants from the enclosure theorems.

As in many other problems of mathematical physics, let us operate with a weak solution. To determine it, let us first reduce the boundary-value problem (2.1.1)—(2.1.3) to that with uniform boundary conditions. For this purpose introduce some new unknown functions by formulae

$$\mathbf{w}'(\mathbf{x}, t) = \mathbf{w}(\mathbf{x}, t) - \mathbf{w}^0(\mathbf{x}, t);$$

$$\zeta'(\mathbf{x}, t) = \zeta(\mathbf{x}, t) + \int_0^t \operatorname{div} \mathbf{w}^0 \, dt,$$

where $\mathbf{w}^0(\mathbf{x}, t)$ is a smooth continuation of boundary conditions inside the considered domain S.

In this case $\mathbf{w}'$, ζ' will be determined by the following system of equations and boundary conditions:

$$\mathfrak{L}\mathbf{w}' \equiv \frac{\partial \mathbf{w}'}{\partial t} + \mathbf{A}_1 \mathbf{w}' - \kappa_h \, \varDelta \mathbf{w}' + \frac{t}{D^2} |\mathbf{w}' + \mathbf{w}^0| \, (\mathbf{w}' + \mathbf{w}^0)$$

$$= -gD \cdot \nabla \zeta' + \mathbf{f}'(\mathbf{x}, t); \tag{2.2.1}$$

$$\frac{\partial \zeta'}{\partial t} + \operatorname{div} \mathbf{w}' = 0, \tag{2.2.2}$$

$$\mathbf{w}'|_\Gamma = 0;$$

$$\mathbf{w}'|_{t=0} = \mathbf{w}'_0(\mathbf{x});$$
$$\zeta'|_{t=0} = \zeta'_0(\mathbf{x}), \tag{2.2.3}$$

where

$$f'(x, t) = gD \cdot \nabla \zeta + \frac{\partial w^0}{\partial t} + gD \cdot \nabla \int_0^t \operatorname{div} w^0 \, dt + \kappa_h \Delta w^0 - A_1 w^0;$$

$$w_0'(x) = w_0(x) - w^0(x, 0),$$

$$\zeta_0'(x) \equiv \zeta_0(x).$$

For simplicity, primes will be henceforth omitted.

Let the weak solution of the problem (2.2.1)—(2.2.3) be a pair of functions w, ζ, belonging to classes† $\overset{0}{W}{}_2^{1,1}(Q_T)$, and $W_2^{0,1}(Q_T)$, respectively, and obeying the relations

$$\int_{Q_T} \left[\frac{\partial w}{\partial t} \mathfrak{M} + \kappa_h \frac{\partial w}{\partial x} \frac{\partial \mathfrak{M}}{\partial x} + \frac{\iota}{D^2} |w + w^0| (w + w^0) \mathfrak{M} \right.$$

$$\left. - g\zeta \cdot \operatorname{div}(D\mathfrak{M}) \right] dx \, dt = \int_{Q_T} f\mathfrak{M} \, dx \, dt; \qquad (2.2.4)$$

$$\int_{Q_T} \mathfrak{z} \left(\frac{\partial \zeta}{\partial t} + \operatorname{div} w \right) dx \, dt = 0 \qquad (2.2.5)$$

for all $\mathfrak{M} \in \overset{0}{W}{}_2^{1,1}(Q_T)$ and $\mathfrak{z} \in W_2^{0,1}(Q_T)$.

To verify the validity of this definition it is necessary to make sure that all the integrals of equations (2.2.4), (2.2.5) are finite for any $w, \zeta, \mathfrak{M}, \mathfrak{z}$ of the given classes.

Let $f \in L_2(Q_T)$, and $w_0(x)$ and $\zeta_0(x)$ be quadratically summed in the domain S. Besides, assume the depth $D(x)$ to be a continuously differentiated function x, nowhere becoming zero, so that

$$\min_{x \in S} D \equiv \mu > 0, \qquad \max_{x \in S} |\nabla D| \leqslant M$$

(here M is a some positive constant which equals zero at a constant ocean depth).

In this case all the terms in equations (2.2.4), (2.2.5) except for

$$J = \iota \int_{Q_T} \frac{\mathfrak{M}}{D^2} |w + w^0| (w + w^0) \, dx \, dt,$$

---

† The class $\overset{0}{W}{}_2^{1,1}(Q_T)$ is obtained by closure of a set of smooth functions, equal to zero on $\varGamma$ in the norm $W_2^{1,1}(Q_T)$.

will be finite owing to the Cauchy inequality. As to the integral J, its finiteness follows from

$$\int_{Q_T} |\mathbf{w}|^2 |\mathfrak{M}| \, d\mathbf{x} \, dt \leq \|\mathbf{w}\|^2_{L_4(Q_T)} \|\mathfrak{M}\|_{2,Q_T}$$

$$\leq \sqrt{2} \max_{0 \leq t \leq T} \|\mathbf{w}\| \cdot \left\|\frac{\partial \mathbf{w}}{\partial \mathbf{x}}\right\|_{2,Q_T} \cdot \|\mathfrak{M}\|_{2,Q_T} < \infty,$$

resulting from the Helder inequality, and

$$\|\mathbf{w}\|^2_{4,Q_T} \leq 2 \max_{0 \leq t \leq T} \|\mathbf{w}\|^2 \cdot \left\|\frac{\partial \mathbf{w}}{\partial \mathbf{x}}\right\|^2_{2,Q_T}.$$

2.3. Uniqueness Theorem

Let us now prove the following uniqueness theorem.

THEOREM. *Problem (2.2.1)—(2.2.3) can have but the only weak solution* $\mathbf{w} \in \overset{0}{\mathbf{W}}{}^{1,1}_2(Q_T), \zeta \in W^{0,1}_2(Q_T)$.

Assume that there exist two solutions $(\mathbf{w}_1, \zeta_1)$, $(\mathbf{w}_2, \zeta_2)$ of problem (2.2.1)—(2.2.3). Their difference $\mathbf{w} = \mathbf{w}_1 - \mathbf{w}_2$, $\zeta = \zeta_1 - \zeta_2$ will, in all likelihood, obey the relations

$$\int_{Q_T} \left[\frac{\partial \mathbf{w}}{\partial t} \mathfrak{M} + \kappa_h \frac{\partial \mathbf{w}}{\partial \mathbf{x}} \frac{\partial \mathfrak{M}}{\partial \mathbf{x}} + \frac{\iota}{D^2} \mathfrak{M}(|\mathbf{w}_1 + \mathbf{w}^0| (\mathbf{w}_1 + \mathbf{w}^0)\right.$$

$$\left. - |\mathbf{w}_2 + \mathbf{w}^0| (\mathbf{w}_2 + \mathbf{w}^0)) - g\zeta \cdot \operatorname{div}(D\mathfrak{M})\right] d\mathbf{x} \, dt = 0; \qquad (2.3.1)$$

$$\int_{Q_T} \mathfrak{z} \left(\frac{\partial \zeta}{\partial t} + \operatorname{div} \mathbf{w}\right) d\mathbf{x} \, dt = 0. \qquad (2.3.2)$$

Set $\mathfrak{M} = \mathbf{w}$ and $\mathfrak{z} = \zeta$. Then, with equation (2.3.2) taken into account, equality (2.3.1) can be rewritten as follows:

$$\frac{1}{2} \|\mathbf{w}\|^2 + \frac{1}{2} g \int_S D\zeta^2 \, d\mathbf{x} + \kappa_h \int_0^t \left\|\frac{\partial \mathbf{w}}{\partial \mathbf{x}}\right\|^2 dt$$

$$+ \iota \int_{Q_T} \frac{\mathbf{w}}{D^2} (|\mathbf{w}_1 + \mathbf{w}^0| (\mathbf{w}_1 + \mathbf{w}^0) - |\mathbf{w}_2 + \mathbf{w}^0| (\mathbf{w}_2 + \mathbf{w}^0)) \, d\mathbf{x} \, dt$$

$$= g \int_{Q_\tau} \zeta \mathbf{w} \cdot \nabla D \, d\mathbf{x} \, dt. \qquad (2.3.3)$$

Evaluate the fourth term in the left-hand part of this relation using a generalized Cauchy inequality

$$2ab \leq \varepsilon a^2 + \varepsilon^{-1} b^2,$$

which is valid for any positive values of a, b and ε. As a result we obtain

$$t \left| \int_{Q_T} \frac{\mathbf{w}}{D^2} \left(|\mathbf{w}_1 + \mathbf{w}^0| (\mathbf{w}_1 + \mathbf{w}^0) - |\mathbf{w}_2 + \mathbf{w}^0| (\mathbf{w}_2 + \mathbf{w}^0) \right) d\mathbf{x} \, dt \right|$$

$$\leq t\mu^{-2} \int_{Q_T} |\mathbf{w}|^2 \cdot |\mathbf{w}_1 + \mathbf{w}_2 + 2\mathbf{w}^0| \, d\mathbf{x} \, dt$$

$$\leq t\mu^{-2} ||\,|\mathbf{w}_1| + |\mathbf{w}_2| + 2|\mathbf{w}^0|\,||_{2, Q_T} ||\mathbf{w}||_{4, Q_T}^2$$

$$\leq C \left(\varepsilon \left\| \frac{\partial \mathbf{w}}{\partial \mathbf{x}} \right\|_{2, Q_T}^2 + \varepsilon^{-1} ||\mathbf{w}||_{2, Q_T}^2 \right). \qquad (2.3.4)$$

Estimate the right-hand part of equality (2.3.3) in the following way

$$g \left| \int_{Q_T} \zeta \mathbf{w} \cdot \nabla D \, d\mathbf{x} \, dt \right| \leq \frac{gM}{\mu^{1/2}} ||\mathbf{w}||_{2, Q_T} \left(\int_{Q_T} D\zeta^2 \, d\mathbf{x} \, dt \right)^{1/2}$$

$$\leq \frac{M}{2} \left(\frac{g}{\mu} \right)^{1/2} \left(||\mathbf{w}||_{2, Q_T}^2 + g \int_{Q_T} D\zeta^2 \, d\mathbf{x} \, dt \right). \qquad (2.3.5)$$

Substitute equations (2.3.4) and (2.3.5) into equation (2.3.3), and setting $\varepsilon = \kappa_h/2C$ introduce the expression for the total energy

$$\Im(t) = \frac{1}{2} ||\mathbf{w}(x, t)||^2 + \frac{g}{2} \int_S D(\mathbf{x}) \zeta^2(\mathbf{x}, t) \, d\mathbf{x}. \qquad (2.3.6)$$

Then, omitting the positive terms in the left-hand part of the resulting expression, obtain the inequality

$$\Im(t) \leq C_1 \int_0^t \Im(t) \, dt. \qquad (2.3.7)$$

Inequality (2.3.7) guarantees $\Im(t)$ and hence w, ζ becoming zero throughout the whole space-time domain Q_T. Thus the theorem is proved.

2.4. A priori Estimates

Let us deduce some *a priori* estimations for exact solutions of problem (2.2.1)—(2.2.3) which will be further necessary while proving the existence theorem.

Relations (2.2.1), (2.2.2) give

$$\frac{d}{dt}\frac{1}{2}\left(\|\mathbf{w}\|^2 + g\int_S D\zeta^2\,dx\right) + \kappa_h\left\|\frac{\partial\mathbf{w}}{\partial x}\right\|^2$$

$$+ \iota\int_S \frac{1}{D^2}|\mathbf{w}+\mathbf{w}^0|(|\mathbf{w}|^2+\mathbf{w}\mathbf{w}^0)\,dx = \int_S \mathbf{f}\cdot\mathbf{w}\,dx + g\int_S \zeta\mathbf{w}\cdot\nabla D\,dx; \quad (2.4.1)$$

$$\frac{d}{dt}\frac{1}{2}\left(\left\|\frac{\partial\mathbf{w}}{\partial t}\right\|^2 + g\int_S \left(\frac{\partial\zeta}{\partial t}\right)^2 D\,dx\right) + \kappa_h\left\|\frac{\partial^2\mathbf{w}}{\partial x\,\partial t}\right\|$$

$$+ \iota\int_S \frac{1}{D^2}\frac{\partial}{\partial t}(|\mathbf{w}+\mathbf{w}^0|(\mathbf{w}+\mathbf{w}^0))\frac{\partial\mathbf{w}}{\partial t}\,dx$$

$$= \int_S \frac{\partial\mathbf{f}}{\partial t}\frac{\partial\mathbf{w}}{\partial x}\,dx + g\int_S \frac{\partial\zeta}{\partial t}\frac{\partial\mathbf{w}}{\partial t}\cdot\nabla D\,dx. \quad (2.4.2)$$

They are obtained from identities $\int_S \mathfrak{L}\mathbf{w}\cdot\mathbf{w}\,dx = \int_S (-gD\cdot\nabla\zeta+\mathbf{f})\mathbf{w}\,dx;$

$$\int_S \mathfrak{L}\frac{\partial\mathbf{w}}{\partial t}\cdot\frac{\partial\mathbf{w}}{\partial t}\,dx = \int_S \left(-gD\cdot\nabla\frac{\partial\zeta}{\partial t} + \frac{\partial\mathbf{f}}{\partial t}\right)\frac{\partial\mathbf{w}}{\partial t}\,dx$$

integrated by parts and conditions (2.2.2), (2.2.3).

Take advantage of the fact that $\mathbf{w}$ and ζ obey conditions (2.4.1), (2.4.2) and assume that these identities are fulfilled for all $t \in [0, T]$;

$$\text{values}\quad \|\mathbf{w}\|^2,\quad \left\|\frac{\partial\mathbf{w}}{\partial t}\right\|^2,\quad \|\zeta\|^2,\quad \left\|\frac{\partial\zeta}{\partial t}\right\|^2$$

are absolutely continuous with respect to t and all the terms in (2.4.1), (2.4.2) are $L_1(0, T)$ elements. One may then formulate the following lemma.

Lemma 1. *If* w, ζ *for* $t \in [0, T]$ *obey relations* (2.4.1), (2.4.2) *and* $w(x, 0) \in L_2(S)$, $\zeta(x, 0) \in L_2(S)$, *then at any* $t \in [0, T]$ *the estimates take the form*

$$\|w\|^2 + g \int_S D\zeta^2 \, dx + \kappa_h \int_0^T \left\|\frac{\partial w}{\partial x}\right\|^2 dt \leq C_2;$$

$$\left\|\frac{\partial w}{\partial t}\right\|^2 + g \int_S D\left(\frac{\partial \zeta}{\partial t}\right)^2 dx + \kappa_h \int_0^T \left\|\frac{\partial^2 w}{\partial x \, \partial t}\right\|^2 dt \leq C_3;$$

$$\left\|\frac{\partial w}{\partial x}\right\| \leq C_4. \tag{2.4.3}$$

To prove the lemma, consider equation (2.4.1) and evaluate the integral

$$\iota \int_S |w + w^0| \frac{w \cdot w^0}{D^2} \, dx,$$

in the left-hand part of this relation through the use of the Holder's inequality

$$\|w\|_{4, S}^2 \leq \varepsilon^2 \left\|\frac{\partial w}{\partial x}\right\|^2 + \frac{\varepsilon^{-2}}{2} \|w\|^2$$

with arbitrary $w \in \overset{0}{W}{}_2^1(S)$ and $\varepsilon > 0$. Use a generalized Cauchy inequality and obtain

$$\iota \left| \int_S |w + w^0| \frac{w \cdot w^0}{D^2} \, dx \right| \leq \frac{\kappa_h}{2} \left\|\frac{\partial w}{\partial x}\right\|^2 + C_5 \|w\|^2 + C_6 \|w\|. \tag{2.4.4}$$

Evaluate the second term in the right-hand part of equation (2.4.1) using $\max_S |\nabla D|$ boundedness

$$g \left| \int_S \zeta w \cdot \nabla D \, dx \right| \leq \frac{M}{2} \left(\frac{g}{\mu}\right)^{1/2} \left(\|w\|^2 + g \int_S D\zeta^2 \, dx\right). \tag{2.4.5}$$

Substitute relations (2.4.4), (2.4.5) into (2.4.1) and obtain the inequality

$$\frac{d\vartheta}{dt} + \kappa_h \left\|\frac{\partial w}{\partial x}\right\|^2 \leq C_7 \vartheta + C_8, \tag{2.4.6}$$

where ϑ was determined in equation (2.3.6).

EQUATIONS OF TIDAL DYNAMICS

Relation (2.4.6) affords the first estimate of the lemma.

Evaluate now the third term in the left-hand part of relation (2.4.2). For this purpose use Hölder's and Young's inequalities and the expression found above for $\|\mathbf{w}\|$ [see the first inequality in equation (2.4.3)]. As a result

$$\varepsilon \left| \int_S \frac{1}{D^2} \frac{\partial}{\partial t} (|\mathbf{w}+\mathbf{w}^0|(\mathbf{w}+\mathbf{w}^0)) \cdot \frac{\partial \mathbf{w}}{\partial t} d\mathbf{x} \right|$$

$$\leq \frac{\kappa_h}{2} \left\| \frac{\partial^2 \mathbf{w}}{\partial \mathbf{x} \partial t} \right\| + C_9 \left\| \frac{\partial \mathbf{w}}{\partial t} \right\|^2 + C_{10} \|\mathbf{w}\|^2 + C_{11}. \qquad (2.4.7)$$

Taking into consideration relation (2.4.7) and

$$g \left| \int_S \frac{\partial \zeta}{\partial t} \frac{\partial \mathbf{w}}{\partial t} \cdot \nabla D \, d\mathbf{x} \right|$$

$$\leq \frac{M}{2} \left(\frac{g}{\mu} \right)^{1/2} \left(\left\| \frac{\partial \mathbf{w}}{\partial t} \right\|^2 + g \int_S D \left(\frac{\partial \zeta}{\partial t} \right)^2 d\mathbf{x} \right), \qquad (2.4.8)$$

which follows from relation (2.4.2), we obtain the inequality

$$\frac{d\Im_t}{\partial t} + \kappa_h \left\| \frac{\partial^2 \mathbf{w}}{\partial \mathbf{x} \, dt} \right\|^2 \leq C_{12} \Im_t + C_{13}, \qquad (2.4.9)$$

where

$$\Im_t = \frac{1}{2} \left(\left\| \frac{\partial \mathbf{w}}{\partial t} \right\|^2 + g \int_S D \left(\frac{\partial \zeta}{\partial t} \right)^2 d\mathbf{x} \right).$$

Relation (2.4.9) yields the second estimate of the lemma.

Finally, rewrite relation (2.4.1) as

$$\kappa_h \left\| \frac{\partial \mathbf{w}}{\partial \mathbf{x}} \right\|^2 + \varepsilon \int_S \frac{1}{D^2} |\mathbf{w}+\mathbf{w}^0| \cdot |\mathbf{w}|^2 \, d\mathbf{x}$$

$$= -\int_S \left[\mathbf{w} \left(\frac{\partial \mathbf{w}}{\partial t} - \mathbf{f} - g\zeta \cdot \nabla D + \frac{\varepsilon}{D^2} |\mathbf{w}+\mathbf{w}^0| \mathbf{w}^0 \right) + \zeta \frac{\partial \zeta}{\partial t} \right] d\mathbf{x}$$

and evaluate the right-hand part of this relation

$$\int_S \left[\mathbf{w} \left(\frac{\partial \mathbf{w}}{\partial t} - \mathbf{f} - g\zeta \cdot \nabla D + \frac{\iota}{\mu^2} |\mathbf{w} + \mathbf{w}^0| \mathbf{w}^0 \right) + \zeta \frac{\partial \zeta}{\partial t} \right] d\mathbf{x}$$

$$\leq \|\mathbf{w}\| \left[\left\| \frac{\partial \mathbf{w}}{\partial t} \right\| + \|\mathbf{f}\| + Mg \|\zeta\| + \frac{\iota}{\mu^2} \left(\max_S |\mathbf{w}^0| \cdot \|\mathbf{w}\| + \|\mathbf{w}^0\|_{4,S}^2 \right) \right]$$

$$+ g \|\zeta\| \cdot \left\| \frac{\partial \zeta}{\partial t} \right\| \max_S D \leq C_{14}.$$

Thus

$$\kappa_h \left\| \frac{\partial \mathbf{w}}{\partial \mathbf{x}} \right\|^2 + \iota \int_S \frac{1}{D^2} |\mathbf{w} + \mathbf{w}^0| \cdot |\mathbf{w}|^2 \, d\mathbf{x} \leq C_{15}, \qquad (2.4.10)$$

which proves the last statement of the lemma.

Let us obtain one more *a priori* estimate, necessary to prove the functions $\mathbf{w}$ and $\nabla \zeta$ belonging to spaces $\mathbf{W}_2^2(S)$ and $\mathbf{L}_2(S)$, respectively, for any $t \in [0, T]$. Assume that the domain S has a twice-boundedly differentiated boundary Γ.

Apply the gradient operation to equation (2.2.2) and substitute the resulting relation into equation (2.2.1) differentiated with respect to time. Multiply the resulting expression in a scalar way by $\Delta \mathbf{w}$ and integrate it over the domain S and in time from 0 to t. Then

$$\int_0^t \int_S \frac{\partial^2 \mathbf{w}}{\partial t^2} \Delta \mathbf{w} \, d\mathbf{x} \, dt + \iota \int_0^t \int_S \frac{1}{D^2} \frac{\partial}{\partial t} (|\mathbf{w} + \mathbf{w}^0| (\mathbf{w} + \mathbf{w}^0)) \Delta \mathbf{w} \, d\mathbf{x} \, dt$$

$$+ \int_0^t \int_S \left(A_1 \frac{\partial \mathbf{w}}{\partial t} \right) \Delta \mathbf{w} \, d\mathbf{x} \, dt = g \int_0^t \int_S D \Delta \mathbf{w} \cdot \nabla \operatorname{div} \mathbf{w} \, d\mathbf{x} \, dt$$

$$+ \frac{1}{2} \|\Delta \mathbf{w}\|^2 - \frac{1}{2} \|\Delta \mathbf{w}(\mathbf{x}, 0)\|^2. \qquad (2.4.11)$$

Denote the first integral on the left-hand side of this equality by J_1 and rewrite it as

$$J_1 = -\int_0^t \int_S \frac{\partial w}{\partial t} \cdot \Delta \frac{\partial w}{\partial t} \, dx \, dt + \int_S \frac{\partial w}{\partial t} \Delta w \, dx$$

$$- \int_S \frac{\partial}{\partial t} w(x,0) \cdot \Delta w(x,0) \, dx = \int_0^t \left\| \frac{\partial^2 w}{\partial x \, \partial t} \right\|^2 dt\,]$$

$$+ \int_S \left(\frac{\partial w}{\partial t} \Delta w - \frac{\partial}{\partial t} w(x,0) \cdot \Delta w(x,0) \right) dx.$$

Taking into account that $\partial^2 w / \partial x \, \partial t$ belongs to $L_2(Q_T)$ and $\partial w / \partial t$ to $L_2(S)$ at $t \in [0, T]$, estimate J_1 as

$$|J_1| \leq C_{16} + C_{17} \|\Delta w\| + C_{18} \|\Delta w\|^2 \qquad (2.4.12)$$

Estimate the second integral J_2 in the left-hand part of equality (2.4.11) using Hölder's inequality and w belonging to the space $L_4(Q_T)$. As a result

$$|J_2| \leq C_{19} + C_{20} \int_0^t \|\Delta w\| \, dt. \qquad (2.4.13)$$

Transform now the third integral

$$J_3 = \int_0^t \int_S \left(A_1 \frac{\partial w}{\partial t} \right) \Delta w \, dx \, dt,$$

taking it by parts only once. In this case

$$|J_3| = \left| \int_0^t \int_S \left(A_1 \frac{\partial^2 w}{\partial x \, \partial t} \right) \frac{\partial w}{\partial x} \, dx \, dt \right| \leq C_{21}. \qquad (2.4.14)$$

Transform the integral J_4 on the right-hand side of equality (2.4.11) in a similar fashion. Then

$$J_4 = -\int_0^t \int_S \operatorname{div} w \cdot \operatorname{div}(D \Delta w) \, dx \, dt + \int_0^t \int_\Gamma D \Delta w \cdot \operatorname{div} w \cdot \mathbf{n} \, d\Gamma \, dt$$

or

$$J_4 = \int_0^t \int_S D |\nabla \operatorname{div} \mathbf{w}|^2 \, d\mathbf{x} \, dt - \int_0^t \int_S \operatorname{div} \mathbf{w} \cdot \nabla D \cdot (\Delta \mathbf{w} - \nabla \operatorname{div} \mathbf{w}) \, d\mathbf{x} \, dt$$

$$+ \int_0^t \int_\Gamma D \operatorname{div} \mathbf{w} \cdot (\Delta \mathbf{w} - \nabla \operatorname{div} \mathbf{w}) \cdot \mathbf{n} \, d\Gamma \, dt. \quad (2.4.15)$$

To evaluate the term containing ∇D use $\max |\nabla D|$ boundedness within the domain S, the first estimate in equation (2.4.3) and the inequality

$$\| \mathbf{w} \|_{2,S}^{(2)} \leqslant C_{22} \| \Delta \mathbf{w} \|_{2,S}, \quad (2.4.16)$$

which is valid for any $\mathbf{w} \in \mathbf{W}_2^2(S) \cap \overset{0}{\mathbf{W}}{}_2^1(S)$ functions within the domain S, having twice-boundedly differentiated boundary Γ. The symbol $\| \ \|_{2,S}^{(2)}$ denotes here the norm in the Sobolev's space $\mathbf{W}_2^2(S)$. As a result

$$\left| \int_0^t \int_S \operatorname{div} \mathbf{w} \cdot \nabla D \cdot (\Delta \mathbf{w} - \nabla \operatorname{div} \mathbf{w}) \, d\mathbf{x} \, dt \right| \leqslant M C_{23} \int_0^t \left\| \frac{\partial \mathbf{w}}{\partial \mathbf{x}} \right\| \cdot \| \mathbf{w} \|_{2,S}^{(2)} \, dt$$

$$\leqslant C_{24} + C_{25} \int_0^t \| \Delta \mathbf{w} \|^2 \, dt. \quad (2.4.17)$$

Consider the last term in equation (2.4.15) containing the integral with respect to the boundary. Take a point $0 \in \Gamma$ and bring it into coincidence with the centre of the local Cartesian system of coordinates $(x_1; y_1)$, the axis x_1 being tangential and the axis y_1 being normal to Γ at the point 0.

In the accepted coordinate system, derivatives with respect to x_1 of the vector-function components $\mathbf{w} = (u, v)$ at the point 0 are equal to zero by virtue of zero boundary conditions. Hence the integrand

$$D \operatorname{div} \mathbf{w} \cdot (\Delta \mathbf{w} - \nabla \operatorname{div} \mathbf{w})$$

can be presented as

$$I = D \frac{\partial v}{\partial y_1} \left(\frac{\partial^2 v}{\partial x_1^2} - \frac{\partial^2 u}{\partial x_1 \, \partial y_1} \right).$$

Let us now verify that the mixed derivative $\partial^2 u/(\partial x_1 \, \partial y_1)$ equals zero at the point $0(x_1; y_1 = 0)$, and the derivative $\partial^2 v/\partial x_1$ is related to $\partial v/\partial n$. Indeed, let $y_1 = \mathfrak{S}(x_1)$ be the equation of the segment of the curve Γ in the vicinity of the point 0. Differentiate equality $\mathbf{w}(x_1, \mathfrak{S}(x_1)) = 0$ twice with respect to x_1 and take the mixed derivative $\partial^2 w/(\partial x_1 \partial y_1)$. In this case

$$\frac{\partial w}{\partial x_1} + \frac{\partial w}{\partial y_1} \frac{\partial \mathfrak{S}}{\partial x_1} = 0;$$

$$\frac{\partial^2 w}{\partial x_1^2} + 2 \frac{\partial \mathfrak{S}}{\partial x_1} \frac{\partial^2 w}{\partial x_1 \partial y_1} + \left(\frac{\partial \mathfrak{S}}{\partial x_1}\right)^2 \frac{\partial^2 w}{\partial y_1^2} + \frac{\partial^2 \mathfrak{S}}{\partial x_1^2} \frac{\partial w}{\partial y_1} = 0;$$

$$\frac{\partial^2 w}{\partial x_1 \partial y_1} + \frac{\partial \mathfrak{S}}{\partial x_1} \frac{\partial^2 w}{\partial y_1^2} = 0.$$

At the point 0 (here $\partial \mathfrak{S}/\partial x_1 = 0$) these equations yield

$$\frac{\partial w}{\partial x_1} = 0;$$

$$\frac{\partial^2 w}{\partial x_1^2} = -\frac{\partial^2 \mathfrak{S}}{\partial x_1^2} \frac{\partial w}{\partial y_1};$$

$$\frac{\partial^2 w}{\partial x_1 \partial y_1} = 0.$$

Hence,

$$I = -D \left(\frac{\partial v}{\partial n}\right)^2 \frac{\partial^2 \mathfrak{S}}{\partial x_1^2}.$$

If the curve Γ is convex, then, as is known, $\partial^2 \mathfrak{S}(\Gamma)/\partial x^2 = 0$, and in this case $I \geq 0$.

In a general case when

$$\left| \frac{\partial^2 \mathfrak{S}(\Gamma)}{\partial x_1^2} = C(\Gamma) \right| \leq C_0,$$

the integral over Γ can be estimated in the following way:

$$\left| \int_0^t \int_\Gamma DC(\Gamma) \left(\frac{\partial v}{\partial n}\right)^2 d\Gamma \right| \leq C_0 \int_0^t \int_\Gamma D \left(\frac{\partial w}{\partial n}\right)^2 d\Gamma$$

$$\leq C \int_0^t \int_S \left(\left|\frac{\partial w}{\partial x}\right|^2 + 2 \left|\frac{\partial w}{\partial x}\right| \cdot \left|\frac{\partial^2 w}{\partial x^2}\right| \right) dx\, dt$$

$$\leq C_{23} + C_{24} \int_0^t \|\Delta w\|^2\, dt. \qquad (2.4.18)$$

Note, that while deducing equation (2.4.18) we used the known inequality

$$\int_\Gamma \left|\frac{\partial \mathbf{w}}{\partial n}\right|^2 d\Gamma \leq C \int_S \left(\left|\frac{\partial \mathbf{w}}{\partial \mathbf{x}}\right|^2 + 2\left|\frac{\partial \mathbf{w}}{\partial \mathbf{x}}\right|\cdot\left|\frac{\partial^2 \mathbf{w}}{\partial \mathbf{x}^2}\right|\right) d\mathbf{x}$$

and estimate (2.4.16).

Taking account of expressions (2.4.12)—(2.4.14) as well as (2.4.17), (2.4.18) and with the positive terms on the right-hand side of equation (2.4.11) omitted, we obtain the inequality

$$||\Delta \mathbf{w}||^2 \leq C_{25} + C_{26} \int_0^t ||\Delta \mathbf{w}||^2 \, dt,$$

which together with inequality (2.4.16) affords a required evaluation for $\mathbf{w}$

$$||\mathbf{w}||_{2,S}^{(2)} \leq \text{const.} \qquad (2.4.19)$$

Now estimation of $\nabla \zeta$ presents no difficulty, for

$$||\nabla \zeta||^2 \leq \text{const.} \qquad (2.4.20)$$

with use made of the equality (2.2.1).

2.5. Existence Theorem

To prove the solvability of problem (2.2.1)—(2.2.3) let us use Galerkin's method. Let a fundamental system of vectors $\{\mathbf{a}^\kappa(\mathbf{x})\}$ be in the Hilbert space $\mathbf{W}_2^2(S) \cap \overset{0}{\mathbf{W}}_2^1(S)$, and a fundamental system of functions $\{b^\kappa(\mathbf{x})\}$ in the space $W_2^2(S)$. For simplicity consider the systems $\{\mathbf{a}^\kappa(\mathbf{x})\}$ and $\{b^\kappa(\mathbf{x})\}$ to be orthonormalized in spaces $\mathbf{L}_2(S)$ and $L_2(S)$, respectively. Assume that the initial values of $\mathbf{a}(\mathbf{x})$ and $b(\mathbf{x})$ are the elements of corresponding spaces. In this case there existis a sequence of functions of the type

$$\mathbf{a}_n = \sum_{\kappa=1}^n c_{\kappa n}^0 \mathbf{a}^\kappa(\mathbf{x});$$

$$b_n = \sum_{\kappa=1}^n d_{\kappa n}^0 b^\kappa(\mathbf{x}),$$

which at $n \to \infty$ converges to $\mathbf{a}(\mathbf{x})$ by the norm $\mathbf{W}_2^2(S)$ and to $b(\mathbf{x})$ by the norm $W_2^2(S)$, respectively. And, finally, assume that the free term $\mathbf{f}$ belongs to the space $\mathbf{L}_2(S)$.

EQUATIONS OF TIDAL DYNAMICS

Present an approximate solution $\mathbf{w}^n$, ζ^n of the problem (2.2.1)—(2.2.3) as

$$\mathbf{w}^n(\mathbf{x}, t) = \sum_{\kappa=1}^{n} c_{\kappa n}(t)\, \mathbf{a}^\kappa(\mathbf{x});$$

$$\zeta^n(\mathbf{x}, t) = \sum_{\kappa=1}^{n} d_{\kappa n}(t)\, b^\kappa(\mathbf{x}).$$

Functions $c_{\kappa n}(t)$ and $d_{\kappa n}(t)$ are determined here from equations [†]

$$\left(\frac{\partial \mathbf{w}^n}{\partial t}, \mathbf{a}^\kappa\right) + \kappa_h \left(\frac{\partial \mathbf{w}^n}{\partial \mathbf{x}}, \frac{\partial \mathbf{a}^\kappa}{\partial \mathbf{x}}\right) + (A_1 \mathbf{w}^n, \mathbf{a}^\kappa)$$

$$+ \iota\left(\frac{1}{D^2}|\mathbf{w}^n + \mathbf{w}^0|(\mathbf{w}^n + \mathbf{w}^0), \mathbf{a}^\kappa\right) = g(\zeta^n, \operatorname{div}(D\mathbf{a}^\kappa)) + (\mathbf{f}, \mathbf{a}^\kappa); \quad (2.5.1)$$

$$\left(\frac{\partial \zeta^n}{\partial t}, b^\kappa\right) + (\operatorname{div} \mathbf{w}^n, b^\kappa) = 0, \quad (2.5.2)$$

obeying the initial conditions

$$c_{\kappa n}\big|_{t=0} = c^0_{\kappa n},$$
$$d_{\kappa n}\big|_{t=0} = d^0_{\kappa n}, \quad (2.5.3)$$

where, as well as in equations (2.5.1), (2.5.2), $\kappa = 1, 2, \ldots, n$.

Equalities (2.5.1), (2.5.2) can be viewed as a system of ordinary differential equations with respect to $c_{\kappa n}(t)$ and $d_{\kappa n}(t)$

$$\frac{d}{dt} c_{\kappa n}(t) + \kappa_h \sum_{i=1}^{n} a_{\kappa i} c_{in}(t) + \sum_{i=1}^{n} a'_{\kappa i} c_{in}(t)$$

$$+ \iota\left(\frac{1}{D^2}|\mathbf{w}^n + \mathbf{w}^0|(\mathbf{w}^n + \mathbf{w}^0), \mathbf{a}^\kappa\right) + \sum_{i=1}^{n} b_{\kappa i} s d_{in}(t) = f_\kappa(t);$$

$$\frac{d}{dt} d_{\kappa n}(t) + \sum_{i=1}^{n} b'_{\kappa i} c_{in}(t) = 0,$$

where $\kappa = 1, 2, \ldots, n$; $a_{\kappa i}$, $a'_{\kappa i}$, $b_{\kappa i}$, $b'_{\kappa i}$ are constants depending on the choice of $\{\mathbf{a}^\kappa\}$, $\{b^\kappa\}$, their derivatives and the problem parameters; $f_\kappa = (\mathbf{f}, \mathbf{a}^\kappa)$.

[†] These equations are formally obtained from equations (2.2.1)—(2.2.3), setting $\mathbf{w} = \mathbf{w}^n$, $\zeta = \zeta^n$ whereupon the first and the second equations of the system are scalarly multiplied by $\mathbf{a}^\kappa$, b^κ, respectively, with the following integration of the resulting expressions over S, bearing in mind that $\mathbf{a}^\kappa \in \mathbf{W}_2^2(S) \cap \mathbf{W}_2^1(S)$.

We can obtain the same result by another method. For instance, one can use equalities (2.2.4), (2.2.5) setting $\mathbf{w} = \mathbf{w}^n$, $\zeta = \zeta^n$, $\mathfrak{W}(\mathbf{x}, t) = \mathbf{a}^\kappa(\mathbf{x})\, \Psi(t)$ and $\mathfrak{z}(\mathbf{x}, t) = b^\kappa(\mathbf{x})\, \Psi(t)$, where $\Psi(t)$ is an arbitrary continuous time function. In this case, without integrating over t [this is possible due to sufficient arbitrariness in the choice of $\Psi(t)$], we obtain equations (2.5.1), (2.5.2) again.

A priori estimates of Lemma 1 lead one to conclude that the system of equations (2.5.1), (2.5.2) is unambiguously solvable, provided the conditions (2.5.3) are fulfilled for any $t \in [0, T]$. Indeed, since all the terms in equations (2.5.1), (2.5.2) are smoothly dependent on $c_{\kappa n}(t)$, $d_{\kappa n}(t)$ then, to prove this statement, it is sufficient to make sure of $\{c_{\kappa n}(t)\}$ and $\{d_{\kappa n}(t)\}$ *a priori* boundedness.

As a consequence of $\{\mathbf{a}^\kappa\}$ and $\{b^\kappa\}$ orthonormalization we have

$$\|\mathbf{w}^n\|^2 = \sum_{i=1}^{n} c_{in}^2(t);$$

$$\|\zeta^n\|^2 = \sum_{i=1}^{n} d_{in}^2(t).$$

Multiply each of the equalities of Equation (2.5.1) by corresponding $d_{\kappa n}(t)$ and sum the resulting expressions over κ from 1 to n. As a result

$$\frac{d}{dt} \frac{1}{2} \left(\|\mathbf{w}^n\|^2 + g \int_S D(\zeta^n)^2 \, dx \right) + \kappa_h \left\| \frac{\partial \mathbf{w}^n}{\partial x} \right\|^2$$

$$+ \varepsilon \int_S \frac{1}{D^2} |\mathbf{w}^n + \mathbf{w}^0| (\mathbf{w}^n + \mathbf{w}^0) \mathbf{w}^n \, dx + \int_S \mathbf{A}_1 \mathbf{w}^n \cdot \mathbf{w}^n \, dx$$

$$= \int_S \mathbf{f} \cdot \mathbf{w}^n \, dx + \int_S \zeta^n \mathbf{w}^n \cdot \nabla D \, dx; \qquad (2.5.4)$$

$$\frac{d}{dt} \|\zeta^n\|^2 + (\zeta^n, \operatorname{div} \mathbf{w}^n) = 0. \qquad (2.5.5)$$

From this it is inferred, as has been proved in Lemma 1, that

$$\max_{0 \leq t \leq T} \|\mathbf{w}^n(\mathbf{x}, t)\|, \quad \max_{0 \leq t \leq T} \|\zeta^n(\mathbf{x}, t)\|,$$

and, consequently,

$$\max_{0 \leq t \leq T} |c_{\kappa n}(t)|, \quad \max_{0 \leq t \leq T} |d_{\kappa n}(t)|$$

are bounded. Thus, the approximate solutions $\mathbf{w}^n$, ζ^n are unambiguously determined by differential equations (2.5.1), (2.5.2) and initial conditions (2.5.3) for any $t \in [0, T]$.

Prove now that functions $\mathbf{w}^n$, ζ^n in the limit (at $n \to \infty$) provide a required solution of the problem (2.2.1)—(2.2.3).

EQUATIONS OF TIDAL DYNAMICS

THEOREM. *Let S be a limited domain of (x, y) variations,*

$$\mathbf{w}|_{t=0} \equiv \mathbf{a}(x) \in \mathbf{W}_2^2(S) \cap \overset{0}{\mathbf{W}}{}_2^1(S), \quad \zeta|_{t=0} \equiv b(x) \in W_2^2(S),$$

$$\mathbf{f} \in \mathbf{L}_2(Q_T), \quad \int_0^T \left\| \frac{\partial \mathbf{f}}{\partial t} \right\| dt < \infty, \quad Q_T = S \times [0, T].$$

Then the problem (2.2.1)–(2.2.3) *in Q_T has the only weak solution* $\mathbf{w} \in \mathbf{L}_2(Q_T)$, $\zeta \in L_2(Q_T)$, *this solution possessing the following properties:*

1). $\mathbf{w}^2, \dfrac{\partial \mathbf{w}}{\partial t}, \dfrac{\partial \mathbf{w}}{\partial x}, \dfrac{\partial^2 \mathbf{w}}{\partial x \, \partial t} \in \mathbf{L}_2(Q_T), \dfrac{\partial \zeta}{\partial t} \in L_2(Q_T),$

2). $\dfrac{\partial^2 \mathbf{w}}{\partial x^2}, \nabla \zeta \in \mathbf{L}_2(Q_T),$

if the domain S has a limitedly differentiated boundary[†], and

3). *Lemma 1 estimates are valid for it.*

The proof of the third statement reduces to verifying the validity of equalities (2.4.1), (2.4.2) for $\mathbf{w} = \mathbf{w}^n$, $\zeta = \zeta^n$ and uniform boundedness of

$$\|\mathbf{w}^n(x, 0)\|, \quad \|\partial \mathbf{w}^n(x, 0)/\partial t\|, \quad \|\zeta^n(x, 0)\|, \quad \|\partial \zeta^n(x, 0)/\partial t\|.$$

The latter is a consequence of the above assumptions in reference to $\mathbf{a}$ and b as well as the choice of functions of $\mathbf{a}^\kappa$, b^κ and approximations $\mathbf{a}_n$, b_n.

Equality (2.4.1) coincides with equality (2.5.4). As to equality (2.4.2), it can be obtained from equations (2.5.1), (2.5.2) by differentiating them by t followed by multiplication of the resulting expressions by $(d/dt) c_{\kappa n}$, $(d/dt) d_{\kappa n}$, respectively, and summation over κ from 1 to n. Hence, equalities (2.4.1), (2.4.2) hold at $\mathbf{w} = \mathbf{w}^n$, $\zeta = \zeta^n$, $n = 1, 2, \ldots$ and thus, in keeping with Lemma 1, the following estimates are valid for $\mathbf{w}^n$, ζ^n:

$$\max_{0 \leq t \leq T} \|\mathbf{w}^n(x, t)\|_{2, S}^2 + \max_{0 \leq t \leq T} \|\zeta^n(x, t)\|_{2, S}^2$$
$$+ \max_{0 \leq t \leq T} \left\| \frac{\partial \mathbf{w}^n(x, t)}{\partial x} \right\|_{2, S}^2 \leq C_{27}; \quad (2.5.6)$$

$$\max_{0 \leq t \leq T} \left\| \frac{\partial \mathbf{w}^n(x, t)}{\partial t} \right\|_{2, S}^2 + \max_{0 \leq t \leq T} \left\| \frac{\partial \zeta^n(x, t)}{\partial t} \right\|_{2, S}^2$$
$$+ \left\| \frac{\partial^2 \mathbf{w}^n}{\partial x \, \partial t} \right\|_{2, Q_T}^2 \leq C_{28}. \quad (2.5.7)$$

[†] It is in fact sufficient that the domain S has a piecewise smooth boundary, since all the considerations used while deducing equation. (2.4.19) estimate were of local character.

Using the estimates of equations (2.5.6), (2.5.7) and the theorem of weak compactness of sets bounded in the Hilbert space as the basis, one can single out from $\{w^n\}$ and $\{\zeta^n\}$ subsequences $\{w^{n\kappa}\}$ and $\{\zeta^{n\kappa}\}$ for which $w^{n\kappa}$, $\partial w^{n\kappa}/\partial t$, $\partial w^{n\kappa}/\partial x$, $\partial^2 w^{n\kappa}/\partial x\,\partial t$ and $\zeta^{n\kappa}$, $\partial \zeta^{n\kappa}/\partial t$ weakly converge in $\mathbf{L}_2(Q_T)$ and $L_2(Q_T)$ to $\mathbf{w}$, $\partial \mathbf{w}/\partial t$, $\partial \mathbf{w}/\partial x$, $\partial^2 \mathbf{w}/\partial x\,\partial t$ and ζ, $\partial\zeta/\partial t$, respectively. From these facts it transpires that $\mathbf{w}^{n\kappa}$ and $\partial \mathbf{w}^{n\kappa}/\partial x$ will weakly converge in $\mathbf{L}_2(S)$, and $\zeta^{n\kappa}$ in $L_2(S)$ to $\mathbf{w}$, $\partial \mathbf{w}/\partial x$ and ζ for any $t \in [0, T]$.

Besides, one can show that for $\{\mathbf{w}^{n\kappa}\}$, $\{\zeta^{n\kappa}\}$ (or perhaps for some of their subconsequences) the derivatives $\partial/\partial t\,\{\mathbf{w}^{n\kappa}\}$, $\partial/\partial t\,\{\zeta^{n\kappa}\}$ weakly and uniformly converge by t to $\partial \mathbf{w}/\partial t$, $\partial\zeta/\partial t$ in spaces $\mathbf{L}_2(S)$ and $L_2(S)$, respectively. Indeed, denoting

$$\mathfrak{X}_{n,m} = \left(\frac{\partial \mathbf{w}^n}{\partial t},\, \mathbf{a}^m\right)$$

and integrating the previously differentiated by time relations (2.5.1) with respect to t from t to $t+\Delta t$, obtain

$$|\mathfrak{X}_{n,m}(t+\Delta t) - \mathfrak{X}_{n,m}(t)| \leq \kappa_h \int_t^{t+\Delta t} \left\|\frac{\partial^2 \mathbf{w}^n}{\partial \mathbf{x}\,\partial t}\right\| \cdot \|\mathbf{a}^m\|\,dt$$

$$+\,\varepsilon \int_t^{t+\Delta t}\int_S \frac{\partial}{\partial t}\left(|\mathbf{w}^n + \mathbf{w}^0|\,(\mathbf{w}^n + \mathbf{w}^0)\right) \frac{\mathbf{a}^m}{D^2}\,d\mathbf{x}\,dt$$

$$+\,|2\omega_z| \int_t^{t+\Delta t} \left\|\frac{\partial \mathbf{w}^n}{\partial t}\right\| \cdot \|\mathbf{a}^m\|\,dt$$

$$+\int_t^{t\Delta+t} \left\|\frac{\partial \zeta^n}{\partial t}\right\| \cdot \|\sqrt{D}\,\mathrm{div}\,\mathbf{a}^m\|\,dt$$

$$+\int_t^{t+\Delta t} \left\|\frac{\partial \mathbf{f}}{\partial t}\right\| \cdot \|\mathbf{a}^m\|\,dt \leq \kappa_h\sqrt{\Delta t}\,\|\mathbf{a}^m\| \cdot \left\|\frac{\partial^2 \mathbf{w}}{\partial \mathbf{x}\,\partial t}\right\|_{2,Q_T}$$

$$+\,\frac{\varepsilon\,\Delta t}{\mu^2}\|\mathbf{a}^m\| \cdot \max_{0 \leq t \leq T}\left\|\frac{\partial}{\partial t}\left(|\mathbf{w}^n + \mathbf{w}^0|\,(\mathbf{w}^n + \mathbf{w}^0)\right)\right\|$$

$$+\,|2\omega_z|\,\Delta t\,\|\mathbf{a}^m\| \cdot \max_{0 \leq t \leq T}\left\|\frac{\partial \mathbf{w}^n}{\partial t}\right\|$$

$$+\,\Delta t\,\|\sqrt{D}\,\mathrm{div}\,\mathbf{a}^m\| \cdot \max_{0 \leq t \leq T}\left\|\frac{\partial \zeta^n}{\partial t}\right\|$$

$$+ \int_t^{t+\Delta t} \left\| \frac{\partial \mathbf{f}}{\partial t} \right\| \cdot \|\mathbf{a}^m\| \, dt \leq C(m) \left(C_{29} \sqrt{\Delta t} + C_{30} \Delta t \right.$$

$$\left. + \int_t^{t+\Delta t} \left\| \frac{\partial \mathbf{f}}{\partial t} \right\| dt \right). \tag{2.5.8}$$

The above relations indicate that at fixed m and $n > m$ the right-hand part of the inequality tends to zero with respect to n uniformly, when $\Delta t \to 0$.

Choose a subsequence $\mathfrak{X}_{n_\kappa, m}$ uniformly converging at fixed m and $k \to \infty$ to a certain continuous function $\mathfrak{X}_m(t)$. The latter function known, determine the limiting function $(\partial/\partial t)\,\mathbf{w}(\mathbf{x}, t)$ in the following way:

$$\frac{\partial}{\partial t} \mathbf{w}(\mathbf{x}, t) = \sum_{m=1}^{\infty} \mathfrak{X}_m(t)\, \mathbf{a}^m(\mathbf{x}).$$

At $\kappa \to \infty$ the scalar product

$$\left(\frac{\partial \mathbf{w}^{n_\kappa}}{\partial t} - \frac{\partial \mathbf{w}}{\partial t},\, \mathfrak{X}(\mathbf{x}) \right)$$

is seen to tend to zero with respect to $t \in [0, T]$ uniformly at any element from $\overset{0}{W_2^1}(S)$, and $\partial \mathbf{w}/\partial t$ is continuous with respect to t in weak topology of $\overset{0}{W_2^1}(S)$. Hence $\partial \mathbf{w}^{n_\kappa}/\partial t$ weakly and uniformly converges by t to $\partial \mathbf{w}/\partial t$ in the space $L_2(S)$.[†]

Weak convergence of $\partial \zeta^{n_\kappa}/\partial t$ to $\partial \zeta/\partial t$ in the space $L_2(S)$ can be proved in an analogous way.

Prove now that $\mathbf{w}, \zeta$ are the required weak solutions. It is obvious that they obey conditions (2.2.3). That is, we shall see of equalities (2.4.1), (2.4.2) hold. For this purpose it is sufficient to verify the functions of the type[‡]

$$\mathfrak{W}^m(\mathbf{x}, t) = \sum_{\kappa=1}^m c_\kappa(t)\, \mathbf{a}^\kappa(\mathbf{x}); \tag{2.5.9}$$

$$\mathfrak{z}^m(\mathbf{x}, t) = \sum_{\kappa=1}^m d_\kappa(t)\, b^\kappa(\mathbf{x}), \tag{2.5.10}$$

where $c_\kappa(t)$ and $d_\kappa(t)$ are arbitrary continuous time functions.

[†] Note, that the evaluations of equations (2.5.6), (2.5.7) demonstrate strong compactness of $\{\mathbf{w}^{n_\kappa}\}$ in $\mathbf{L}_4(S)$. Hence, the subsequence $\{\mathbf{w}^{n_\kappa}\}$ may be considered to converge strongly to $\mathbf{w}$ in $\mathbf{L}_4(S)$.

[‡] The given sets of the functions $\mathfrak{W}^m(\mathbf{x}, t)$, $\mathfrak{z}^m(\mathbf{x}, t)$ are complete in spaces $\mathbf{L}_2(Q_T)$ and $L_2(Q_T)$, respectively.

OCEAN TIDES

Choose any $\mathfrak{W}^m$ from class (2.5.9) and $\mathfrak{Z}^m$ from class (2.5.10), bearing in mind that the functions $\mathbf{w}^n$, ζ^n at $n > m$ obey equalities (2.5.1), (2.5.2). Multiply each of the equalities from equation (2.5.1) by $c_\kappa(t)$ and those from (2.5.2) by $d_\kappa(t)$, sum them over κ from 1 to n, integrate them by t from zero to T and obtain

$$\int_{Q_T} \left[\frac{\partial \mathbf{w}^n}{\partial t} \mathfrak{W}^m + \kappa_h \frac{\partial \mathbf{w}^n}{\partial x} \frac{\partial \mathfrak{W}^m}{\partial x} + \frac{\iota}{D^2} |\mathbf{w}^n + \mathbf{w}^0| (\mathbf{w}^n + \mathbf{w}^0) \mathfrak{W}^m \right.$$

$$\left. + A_1 \mathbf{w}^n \cdot \mathfrak{W}^m \right] d\mathbf{x}\, dt + g \int_{Q_T} \zeta^n \cdot \operatorname{div}(D\mathfrak{Z}^m)\, d\mathbf{x}\, dt$$

$$= \int_{Q_T} \mathbf{f} \cdot \mathfrak{W}^m\, d\mathbf{x}\, dt \qquad (2.5.11)$$

$$\int_{Q_T} \mathfrak{Z}^m \left(\frac{\partial \zeta^n}{\partial t} + \operatorname{div} \mathbf{w}^n \right) d\mathbf{x}\, dt = 0. \qquad (2.5.12)$$

In equations (2.5.11), (2.5.12) let n ($n > m$) by the above subsequence n_k tend to infinity. We have already found the convergent character of $\mathbf{w}^n$ to $\mathbf{w}$ and ζ^n to ζ. It is quite sufficient to fulfil the limit transition of all the integrands into equations (2.5.11), (2.5.12). As a result equalities (2.5.11), (2.5.12) are also valid for the limit functions $\mathbf{w}$ and ζ, which proves that $\mathbf{w}$ and ζ are weak solutions of the problem (2.2.1)—(2.2.3) with the properties described in the theorem.

Let us finally prove that at a certain smoothness of the boundary and any $t \in [0, T]$ the function $\mathbf{w}$ belongs to $\overset{0}{\mathbf{W}}{}_2^2(S)$, and $\nabla \zeta$ to $\mathbf{L}_2(S)$. For this purpose let us see if the estimate (2.4.19) is valid at $\mathbf{w} = \mathbf{w}^n$. For $\{\mathbf{a}^\kappa(\mathbf{x})\}$ consider the following system of vectors:

$$\mathbf{a}^{2\kappa} = \begin{Bmatrix} u_\kappa \\ 0 \end{Bmatrix},$$

$$\mathbf{a}^{2\kappa-1} = \begin{Bmatrix} 0 \\ u_\kappa \end{Bmatrix}, \qquad (2.5.13)$$

where u_κ is the eigenfunction of the Laplace operator

$$\Delta u_\kappa = \lambda_\kappa u_\kappa; \quad u_\kappa|_\Gamma = 0, \quad \kappa = 1, 2, \ldots \qquad (2.5.14)$$

It is easy to see that the vector system obtained in this way is complete in the space $\overset{0}{\mathbf{W}}{}_2^2(S) \cap \mathbf{W}_2^1(S)$. Differentiate equation (2.5.1) by t followed by multiplication of each of these equalities by corresponding $c_{\kappa n}(t)$ and eigenvalue λ_κ.

Summarize the resulting expressions by k from 1 to n. Then, taking account of relation (2.5.2) written for the functions $\mathbf{w}^n$, ζ^n obtain the equality which makes possible the estimate (2.4.19). The same estimate is valid for the limit function $\mathbf{w}$ also.

Equation (2.2.1) and the properties of the weak solution $\mathbf{w}$ demonstrate that $\nabla \zeta$ is an element of $\mathbf{L}_2(S)$ for any $t \in [0, T]$.

2.6. On the Existence of a Periodic Solution of the Equations of Tidal Dynamics

Consider the case of a closed basin, in which the initial system of equations and boundary conditions (2.1.1)—(2.1.3) assumes the form

$$\frac{\partial \mathbf{w}}{\partial t} + \mathbf{A}_1 \mathbf{w} - \kappa_h \Delta \mathbf{w} + \frac{\iota}{D^2} |\mathbf{w}| \mathbf{w} = -gD \cdot \nabla \zeta + \mathbf{f}; \qquad (2.6.1)$$

$$\frac{\partial \zeta}{\partial t} + \operatorname{div} \mathbf{w} = 0; \qquad (2.6.2)$$

$$\mathbf{w}|_\Gamma = 0, \qquad (2.6.3)$$

and show the system (2.6.1)—(2.6.3) to have a periodic solution in the case when the free term is taken to be a periodic time function.

Introduce the Hilbert space $\mathbf{H}(Q_T) = \mathbf{H}_1(S) + \mathbf{H}_2(Q_T)$, where $\mathbf{H}_1(S)$ is a space of stationary vector functions from $\overset{0}{\mathbf{W}_2^1}(S)$, possessing solenoidal property, and $\mathbf{H}_2(Q_T)$ is a space of vector functions from $\overset{0}{\mathbf{W}_2^{1,1}}(Q_T)$, changing in time with period $T = 2\pi/\sigma$. Let the latter of the above vector functions be orthogonal in $\mathbf{L}_2(Q_T)$ to any vector from $\overset{0}{\mathbf{W}_2^1}(S)$.

In an analogous way introduce the Hilbert space

$$H'(Q_T) = H_1'(S) + H_2'(Q_T), \quad \text{where} \quad H_1'(S) \quad \text{and} \quad H_2'(Q_T)$$

are spaces of stationary and periodic scalar functions, belonging to the spaces $L_2(S)$ and $W_2^{0,1}(Q_T)$, respectively.

Assume further that $\mathbf{f} \in \mathbf{H}(Q_T)$,[†] and determine a weak solution as a pair

---

[†] A commonly used expression for the perturbing force $\mathbf{f} = \operatorname{real} \bar{\mathbf{f}} \exp i\sigma t$ belongs to this class.

of functions $\mathbf{w} \in H(Q_T)$, $\zeta \in H'(Q_T)$, obeying the identities

$$\mathfrak{G}_1(\mathbf{w}, \zeta, \mathfrak{W}) \equiv -\int_{Q_T} \mathbf{w}\,\frac{\partial \mathfrak{W}}{\partial t}\,d\mathbf{x}\,dt + \int_{Q_T} \mathbf{A}_1\mathbf{w}\cdot\mathfrak{W}\,d\mathbf{x}\,dt$$

$$+\varkappa_h \int_{Q_T}\frac{\partial \mathbf{w}}{\partial \mathbf{x}}\,\frac{\partial \mathfrak{W}}{\partial \mathbf{x}}\,d\mathbf{x}\,dt + \varepsilon \int_{Q_T}\frac{1}{D^2}|\mathbf{w}|\,\mathbf{w}\cdot\mathfrak{W}\,d\mathbf{x}\,dt$$

$$-g\int_{Q_T}\zeta\cdot\mathrm{div}\,(D\mathfrak{W})\,d\mathbf{x}\,dt = (\mathbf{f}, \mathfrak{W})_{Q_T}; \qquad (2.6.4)$$

$$\mathfrak{G}_2(\mathbf{w}, \zeta, \mathfrak{z}) \equiv \int_{Q_T}\left(\frac{\partial \zeta}{\partial t} + \mathrm{div}\,\mathbf{w}\right)\mathfrak{z}\,d\mathbf{x}\,dt = 0 \qquad (2.6.5)$$

at any $\mathfrak{W} \in \mathbf{H}(Q_T)$, $\mathfrak{z} \in H'(Q_T)$.

The approximate solution of problem (2.6.1)—(2.6.3) can be obtained from

$$\mathbf{w}^n(\mathbf{x}, t) = \sum_{\varkappa=1}^{n} c_{\varkappa n}\mathbf{a}_\varkappa^{(1)}(\mathbf{x}) + \sum_{i,j}^{n} \alpha_{ij}^n \mathbf{a}_i^{(2)}(\mathbf{x})\,\psi_j(t);$$

$$\zeta^n(\mathbf{x}, t) \equiv \zeta_1^n(\mathbf{x}) + \zeta_2^n(\mathbf{x}, t) = \sum_{\varkappa=1}^{n} d_{\varkappa n}b_\varkappa(\mathbf{x}) + \sum_{ij}^{n}\beta_{ij}^n b_i(\mathbf{x})\psi_j(t). \qquad (2.6.6)$$

Here $\{\mathbf{a}_\varkappa^{(1)}(\mathbf{x})\}$, $\{\mathbf{a}_\varkappa^{(2)}(\mathbf{x})\}$, $\{b_\varkappa\}$ are full systems in the spaces $\mathbf{H}_1(S)$, $\mathbf{W}_2^1(S)$ and $L_2(S)$, respectively, $\psi(t) = \{\cos j\sigma t, \sin j\sigma t, j = 1, 2, \ldots\}$.

The unknown coefficients $c_{\varkappa n}$, α_{ij}^n, $d_{\varkappa n}$ and β_{ij}^n obey the system of non-linear algebraic equations resulting from equations (2.6.4), (2.6.5), if one sets in them in succession that $\mathfrak{W} = \mathbf{a}_\varkappa^{(1)}(\mathbf{x})/D$, $\mathfrak{W} = \mathbf{a}_i^{(2)}(\mathbf{x})\psi_j(t)/D$, $\mathfrak{z} = b_\varkappa(\mathbf{x})$, $\mathfrak{z} = b_i(\mathbf{x})\psi_j(t), i, j, \varkappa = 1, 2, \ldots, n$.

This system has the form

$$\mathfrak{G}_1(\mathbf{w}^n, \zeta^n, \mathbf{a}_\varkappa^{(1)}/D) = (\mathbf{f}, \mathbf{a}_\varkappa^{(1)}/D)_{Q_T}, \qquad \varkappa = 1, 2, \ldots, n; \qquad (2.6.7)$$

$$\mathfrak{G}_1(\mathbf{w}^n, \zeta^n, \mathbf{a}_i^{(2)}\psi_j/D) = (\mathbf{f}, \mathbf{a}_i^{(2)}\psi_j/D)_{Q_T}, \qquad i, j = 1, 2, \ldots, n; \qquad (2.6.8)$$

$$\mathfrak{G}_2(\mathbf{w}^n, \zeta^n, b_\varkappa) = 0, \qquad \varkappa = 1, 2, \ldots, n; \qquad (2.6.9)$$

$$\mathfrak{G}_2(\mathbf{w}^n, \zeta^n, b_i\psi_j) = 0, \qquad i, j = 1, 2, \ldots, n. \qquad (2.6.10)$$

Note that by virtue of condition $\int_0^T \psi_j(t)\,dt = 0$, valid at any $j = 1, 2, \ldots$, equations (2.6.9) are identically fulfilled and, besides, all the coefficients at $d_{\varkappa n}$ in equations (2.6.7), (2.6.8), (2.6.10) become zero. Thus, in the system of equations (2.6.7), (2.6.8), (2.6.10) only $c_{\varkappa n}$, α_{ij}^n and β_{ij}^n remain unknown.

Rewrite the system in the operator form

$$\mathfrak{K}\mathfrak{r} = \mathfrak{F}, \qquad (2.6.11)$$

where $\mathfrak{K}$ is a non-linear matrix, $\mathfrak{r}$ is the vector of unknown values, $\mathfrak{F}$ is the vector of free terms, and prove the solvability of the system (2.6.11).

LEMMA 2. *Let*

$$S_1 = \left\{ \mathbf{x}, \frac{\partial^2 D}{\partial x^2}, \frac{\partial^2 D}{\partial y^2} > 0 \right\}, \quad S_2 = \{\mathbf{x}, D \neq \text{const}\}$$

and the condition

$$\min_{S} D - \min \{ M_1 (2\nu\nu_1)^{1/2}, M_2 (2\nu\nu_2)^{1/4} \} > 0$$

be fulfilled, where $\nu_1 = \text{mes } S_1$, $\nu_2 = \text{mes } S_2$, $\nu = \text{mes } S$, $M_1 = \max_{S_1} |\Delta D|$, $M_2 = \max_{S_2} |\nabla D|$ *Then the system (2.6.11) has at least one solution.*

A sufficient condition for the solvability of system (2.6.11) is the fulfilment of the estimation

$$\|\mathfrak{r}\| \leq C_1 \|\mathfrak{F}\|, \qquad (2.6.12)$$

where the symbol $\| \ \|$ denotes, as usual, the norm of the vector in the Euclidian space R^n.

To verify inequality (2.6.12) make up the quadratic form

$$(\mathfrak{K}\mathfrak{r}, \mathfrak{r}) = t \int_{Q_T} |\mathbf{u}^n|^3 \, d\mathbf{x} \, dt + \kappa_h \int_{Q_T} \frac{\partial \mathbf{w}^n}{\partial \mathbf{x}} \frac{\partial \mathbf{u}^n}{\partial \mathbf{x}} \, \partial \mathbf{x} \, dt. \qquad (2.6.13)$$

Here $\mathbf{u}^n = \mathbf{w}^n / D$.

The last term in the right-hand part can be rewritten as follows:

$$\kappa_h \int_{Q_T} \frac{\partial \mathbf{w}^n}{\partial \mathbf{x}} \frac{\partial \mathbf{u}^n}{\partial \mathbf{x}} \, d\mathbf{x} \, dt = \kappa_h \int_{Q_T} D \left(\frac{\partial \mathbf{u}^n}{\partial \mathbf{x}} \right)^2 d\mathbf{x} \, dt$$

$$+ \kappa_h \int_{Q_T} \mathbf{u}^n \frac{\partial D}{\partial \mathbf{x}} \frac{\partial \mathbf{u}^n}{\partial \mathbf{x}} \, d\mathbf{x} \, dt = \kappa_h \int_{Q_T} D \left| \frac{\partial \mathbf{u}^n}{\partial \mathbf{x}} \right|^2 d\mathbf{x} \, dt$$

$$- \frac{1}{2} \kappa_h \int_{Q_T} \Delta D \cdot |\mathbf{u}^n|^2 \, d\mathbf{x} \, dt. \qquad (2.6.14)$$

Substitute equation (2.6.14) into equation (2.6.13) and, employing the Hölder–Friedrichs inequality as well as the condition of Lemma 2, obtain the estimate

$$(\mathfrak{K}\mathfrak{r}, \mathfrak{r}) \geq C_2 \left\| \frac{\partial \mathbf{u}^n}{\partial \mathbf{x}} \right\|^2 \geq C_3 \|\mathbf{u}^n\|^2. \qquad (2.6.15)$$

By equation (2.6.15) and the relation

$$(\mathfrak{K}\mathfrak{r}, \mathfrak{r}) = (\mathfrak{F}, \mathfrak{r})$$

find

$$\left\| \frac{\partial \mathbf{u}^n}{\partial \mathbf{x}} \right\|, \quad \|\mathbf{u}^n\|_{2, Q_T} \leq C_4 \|\mathfrak{F}\|. \qquad (2.6.16)$$

Then the relation

$$\left\| \frac{\partial \zeta_2^n}{\partial t} \right\|^2 + \left(\operatorname{div} \mathbf{w}^n, \frac{\partial \zeta^n}{\partial t} \right) = 0$$

yields the estimate

$$\left\| \frac{\partial \zeta_2^n}{\partial t} \right\|_{2, Q_T} \leq C_5 \|\mathfrak{F}\|.$$

The latter inequality and the definition ζ^n makes it possible to estimate $\|\zeta_2^n\|$, i.e.

$$\|\zeta_2^n\| \leq C_6 \|\mathfrak{F}\|. \qquad (2.6.17)$$

Remember, all the constants in the above inequalities are n-independent.

Estimates (2.6.16) and (2.6.17) guarantee the fulfilment of condition (2.6.12). Thus, the solvability of system (2.6.11) is proved.

Note that from the relation

$$\mathfrak{G}_1 \left(\mathbf{w}^n, \zeta^n, \frac{1}{D} \frac{\partial \mathbf{w}^n}{\partial t} \right) = \left(\mathbf{f}, \frac{1}{D} \frac{\partial \mathbf{w}^n}{\partial t} \right),$$

which in fact results from equations (2.6.7), (2.6.8), as well as the uniform estimates for $\|\mathbf{w}^n\|$, $\|\partial \mathbf{w}^n/\partial \mathbf{x}\|$, $\|\partial \zeta_2^n/\partial t\|$ yield the inequality

$$\left\| \frac{\partial \mathbf{u}^n}{\partial t} \right\| \leq C_7 \|\mathfrak{F}\|, \qquad (2.6.18)$$

where C_7 is also n-independent.

From the above estimates for $\mathbf{w}^n$, ζ^n one can see the weak convergence of $\mathbf{w}^n$, ζ^n to certain functions of $\mathbf{w}$, ζ in $\mathbf{L}_2(Q_T)$ and $L_2(Q_T)$, respectively, $\partial \mathbf{w}^n/\partial \mathbf{x}$, $\partial \mathbf{w}^n/\partial t$ converging weakly in $\mathbf{L}_2(Q_T)$ and $\partial \zeta^n/\partial t$ in $L_2(Q_T)$ to $\partial \mathbf{w}/\partial \mathbf{x}$, $\partial \mathbf{w}/\partial t$,

$\partial \zeta / \partial t$. The established character of convergence is sufficient to fulfil a limiting transition by n under the integral sign. This transition results in equalities (2.6.4), (2.6.5) valid for any functions from the above-considered classes. It is the latter fact that means that **w**, ζ are the desired weak periodic solution of the problem (2.6.1)—(2.6.3).

Remark. In fact the definition of the weak solution contains only the function $\zeta_2(\mathbf{x}, t)$. Stationary addition $\zeta_1(\mathbf{x})$ from the Galerkin system is not defined since $\nabla \zeta_1$ belongs to the class of functions, orthogonal to the vectors from $\mathbf{H}(Q_T)$.

2.7. Conjugate Equations of Tidal Dynamics

While investigating oceanic tides one is concerned with some simplified model whose degree of validity is not obvious. The latter fact hampers as a rule quantitative assessment of calculations and interpretation of their results. Moreover, in this case the solution dependence upon the model parameters and input data of the problem is not always apparent. To help understand the above problem it is expedient to carry out the analysis of conjugate equations of tidal dynamics and to formulate the perturbation theory.

Rewrite the basic system of equations for tidal dynamics (2.1.1), (2.1.2) in a component-wise form

$$\frac{\partial u}{\partial t} - 2\omega_z v - \kappa_h \Delta u + \frac{t}{D^2} \sqrt{u^2 + v^2} u + gD \frac{\partial \zeta}{\partial x} = f_1;$$

$$\frac{\partial v}{\partial t} + 2\omega_z u - \kappa_h \Delta v + \frac{t}{D^2} \sqrt{u^2 + v^2} v + gD \frac{\partial \zeta}{\partial y} = f_2;$$

$$\frac{\partial \zeta}{\partial t} + \frac{\partial u}{\partial x} + \frac{\partial v}{\partial y} = 0. \qquad (2.7.1)$$

For simplicity, assume the depth and the Coriolis parameter to be constant within the domain in question, and the contour Γ to coincide with the coast line, so that, in accordance with equality (2.1.3), the no-slip condition

$$u, v = 0 \quad \text{on} \quad \Gamma \qquad (2.7.2)$$

is fulfilled.

As initial conditions assume [see relation (2.1.6)]

$$u = u_0(x, y),$$
$$v = v_0(x, y),$$
$$\zeta = \zeta_0(x, y) \quad \text{at} \quad t = 0. \qquad (2.7.3)$$

Now determine the operator

$$\mathfrak{P} = \begin{cases} \dfrac{\partial}{\partial t} - \kappa_h \Delta + \dfrac{t}{D^2}\sqrt{u^2+v^2} & -2\omega_z & gD\dfrac{\partial}{\partial x} \\ 2\omega_z & \dfrac{\partial}{\partial t} - \kappa_h \Delta + \dfrac{t}{D^2}\sqrt{u^2+v^2} & gD\dfrac{\partial}{\partial x} \\ \dfrac{\partial}{\partial x} & \dfrac{\partial}{\partial y} & \dfrac{\partial}{\partial t} \end{cases}$$

and vector-functions

$$\mathfrak{w} = \begin{vmatrix} u \\ v \\ \zeta \end{vmatrix};$$

$$\mathbf{f} = \begin{vmatrix} f_1 \\ f_2 \\ 0 \end{vmatrix}.$$

Then system (2.7.1)—(2.7.3) can be formally rewritten as follows:

$$\mathfrak{P}\mathfrak{w} = \mathbf{f};$$
$$\mathbf{I}\mathfrak{w} = 0 \quad \text{on} \quad \Gamma;$$
$$\mathfrak{w} = \mathfrak{m}_0 \quad \text{at} \quad t = 0, \tag{2.7.4}$$

where $\mathbf{I}$ is the matrix of the form

$$\mathbf{I} = \begin{bmatrix} 1 & 0 & 0 \\ 0 & 1 & 0 \\ 0 & 0 & 0 \end{bmatrix}$$

Note that in this problem the operator $\mathfrak{P}$ is quasi-linear and dependent on the solution itself.

Consider now the Hilbert space of functions Φ with each element being sufficiently smooth (for all the further transformations to have sense). Assume, then, that elements $\mathfrak{w} \in \Phi$ obey the boundary conditions from relations (2.7.4). The above conditions all together form the domain of definition of the operator $\mathfrak{P}$, which will be further referred to as $\Phi(\mathfrak{P})$. In the given space the scalar

EQUATIONS OF TIDAL DYNAMICS

product is defined as

$$(\mathbf{a}, \mathbf{b})_{S\times[0,T]} = \sum_{i=1}^{3} \int_0^T dt \int_S a_i b_i \, dS, \quad (2.7.5)$$

where a_i, b_i are components of vector functions $\mathbf{a}$, $\mathbf{b}$ in the domain $S\times[0,T]$.
Consider the functional

$$(\mathfrak{P}\mathfrak{m}, \mathfrak{w}^*)_{S\times[0,T]} = \int_0^T dt \int_S \left[\left(\frac{\partial u}{\partial t} - 2\omega_z v - \kappa_h \Delta u \right.\right.$$

$$\left. + \frac{\iota}{D^2}\sqrt{u^2+v^2}\,u + gD\,\frac{\partial \zeta}{\partial x}\right)u^* + \left(\frac{\partial v}{\partial t} + 2\omega_z u - \kappa_h \Delta u\right.$$

$$\left.\left. + \frac{\iota}{D^2}\sqrt{u^2+v^2}\,v + gD\,\frac{\partial \zeta}{\partial y}\right)_* + \left(\frac{\partial \zeta}{\partial t} + \frac{\partial u}{\partial x} + \frac{\partial v}{\partial y}\right)\zeta^*\right] dS.$$

On integrating by parts, and using the conditions (2.7.2), (2.7.3), it assumes the form

$$(\mathfrak{P}\mathfrak{w}, \mathfrak{w}^*)_{S\times[0,T]} = \int_0^T dt \int_S \left[\left(-\frac{\partial u^*}{\partial t} + 2\omega_z v^* - \kappa_h \Delta u^* \right.\right.$$

$$\left. + \frac{\iota}{D^2}\sqrt{u^2+v^2}\cdot u^* - \frac{\partial \zeta^*}{\partial x}\right)u + \left(-\frac{\partial v^*}{\partial t} - 2\omega_z u^* - \kappa_h \Delta v^*\right.$$

$$\left. + \frac{\iota}{D^2}\sqrt{u^2+v^2}\cdot v^* - \frac{\partial \zeta^*}{\partial y}\right)v + \left(-\frac{\partial \zeta^*}{\partial t} - gD\,\frac{\partial u^*}{\partial x}\right.$$

$$\left.\left. - gD\,\frac{\partial v^*}{\partial y}\right)\zeta\right] dS - \int_0^T dt \int_\Gamma \kappa_h \left(u^*\,\frac{\partial u}{\partial n} + v^*\,\frac{\partial v}{\partial n}\right) d\Gamma$$

$$+ \int_S [u_T u_T^* + v_T v_T^* + \zeta_T \zeta_T^* - (u_0 u_0^* + v_0 v_0^* + \zeta_0 \zeta_0^*)]\, dS. \quad (2.7.6)$$

Here indices 0 and T denote, as earlier, functions at $t=0$ and $t=T$.

Let the components of conjugate vector-functions $\mathfrak{w}^*$ on the domain Γ contour obey the conditions

$$u^*, v^* = 0 \quad \text{on} \quad \Gamma \quad (2.7.7)$$

and introduce the operator

$$\mathfrak{P}^* = \begin{Bmatrix} -\dfrac{\partial}{\partial t} - \kappa_h \Delta + \dfrac{\iota}{D^2}\sqrt{u^2+v^2} & 2\omega_z & -\dfrac{\partial}{\partial x} \\ -2\omega_z & -\dfrac{\partial}{\partial t} - \kappa_h \Delta + \dfrac{\iota}{D^2}\sqrt{u^2+v^2} & -\dfrac{\partial}{\partial y} \\ -gD\dfrac{\partial}{\partial x} & -gD\dfrac{\partial}{\partial y} & -\dfrac{\partial}{\partial t} \end{Bmatrix}$$

and the vector-function

$$\mathfrak{w}^* = \begin{vmatrix} u^* \\ v^* \\ \zeta^* \end{vmatrix}.$$

Then the expression (2.7.6), with allowance for equations (2.7.5), (2.7.7), can be presented as

$$(\mathfrak{P}\mathfrak{w}, \mathfrak{w}^*)_{S\times[0,T]} = (\mathfrak{P}^*\mathfrak{w}^*, \mathfrak{w})_{S\times[0,T]} + (\mathfrak{w}_T, \mathfrak{w}_T^*)_S - (\mathfrak{w}_0, \mathfrak{w}_0^*)_S, \quad (2.7.8)$$

where

$$(\mathbf{a}, \mathbf{b})_S = \sum_{i=1}^{3} \int_S a_i b_i \, dS.$$

Now consider the conjugate problem

$$\mathfrak{P}^*\mathfrak{w}^* = \mathbf{f}^*;$$
$$\mathbf{l}\mathfrak{w}^* = 0 \quad \text{on} \quad \Gamma;$$
$$\mathfrak{w}^* = \mathfrak{w}_T^* \quad \text{at} \quad t = T, \quad (2.7.9)$$

here $\mathbf{f}$ and $\mathfrak{w}_T^*$ are as yet arbitrary vector-functions (which will be dealt with later).

If $\mathfrak{w}$ and $\mathfrak{w}^*$ in expression (2.7.8) are assumed to be corresponding solutions of the problems (2.7.4) and (2.7.9), then substituting $\mathfrak{P}\mathfrak{w}$ for $\mathbf{f}$ and $\mathfrak{P}^*\mathfrak{w}^*$ for $\mathbf{f}^*$ in expression (2.7.8) yields a new identity

$$(\mathbf{f}, \mathfrak{w}^*)_{S\times[0,T]} = (\mathbf{f}^*, \mathfrak{w})_{S\times[0,T]} + (\mathfrak{w}_T, \mathfrak{w}_T^*)_S - (\mathfrak{w}_0, \mathfrak{w}_0^*)_S. \quad (2.7.10)$$

In the case when the perturbing force $\mathbf{f}$ is described by a simple harmonic, it is convenient to choose the time interval $[0, T]$ equal to its period. Then under asymptotic conditions (i.e. at $t \to \infty$) the solutions of basic and conjugate conditions will be harmonic functions of time and instead of relation (2.7.10) we shall have

$$(\mathbf{f}^*, \mathfrak{w})_{S \times [0, T]} = (\mathbf{f}, \mathfrak{w}^*)_{S \times [0, T]}. \qquad (2.7.11)$$

Set now

$$\mathbf{f}^* = \delta(t - t_i)\, \delta(\mathbf{x} - \mathbf{x}_\kappa)\, \mathbf{e}_j, \qquad (2.7.12)$$

where $\mathbf{e}_j$ is any of the vectors

$$\mathbf{e}_1 = \begin{vmatrix} 1 \\ 0 \\ 0 \end{vmatrix};$$

$$\mathbf{e}_2 = \begin{vmatrix} 0 \\ 1 \\ 0 \end{vmatrix};$$

$$\mathbf{e}_3 = \begin{vmatrix} 0 \\ 0 \\ 1 \end{vmatrix};$$

$\delta(\ )$ is a small variation of the corresponding functions, $\mathbf{x}_\kappa \in S$ and $0 < t_i < T$.

Then on the basis of relation (2.7.11) obtain the following expression for $\mathbf{m}_j(x_\kappa, t_i)$:

$$\mathbf{m}_j(\mathbf{x}_\kappa, t_i) = \int_0^T dt \int_S \mathbf{f} \cdot \mathfrak{w}^* \, dS, \qquad (2.7.13)$$

where $\mathbf{m}_j$ is the solution component corresponding to the element $\mathbf{e}_j$, and $\mathfrak{w}^*$ is the solution to the problem

$$\mathfrak{P}\mathfrak{w}^* = \delta(t - t_i)\, \delta(\mathbf{x} - \mathbf{x}_\kappa)\, \mathbf{e};$$
$$\mathbf{I}\mathfrak{w}^* = 0 \quad \text{on} \quad \Gamma;$$
$$\mathbf{w}^* = \mathbf{w}_T^* \quad \text{at} \quad t = T \qquad (2.7.14)$$

2.8. The Perturbation Theory

Assume that in the basic problem (2.7.4) the function $\mathbf{f}$ changes by a small quantity $\delta \mathbf{f}$, so that $\mathbf{f}' = \mathbf{f} + \delta \mathbf{f}$. Then the considerations in the previous section give the formula

$$\mathfrak{w}'_j(\mathbf{x}_\kappa, t_i) = \int_0^T dt \int_S \mathbf{f}' \mathfrak{w}^* \, dS. \qquad (2.8.1)$$

Subtracting (2.7.13) from (2.8.1) gives the formula of the theory of small perturbations

$$\delta w_j(\mathbf{x}_\kappa, t_i) = \int_0^T dt \int_S \delta \mathbf{f} \cdot \mathbf{w}^* \, dS. \tag{2.8.2}$$

This formula, in particular, makes it possible to evaluate the dependence of any solution component on the perturbation $\delta \mathbf{f}$. Indeed, consider three different "sources"

$$\mathbf{f}_1^* = \begin{vmatrix} \delta(t - t_i)\delta(\mathbf{x} - \mathbf{x}_\kappa) \\ 0 \\ 0 \end{vmatrix};$$

$$\mathbf{f}_2^* = \begin{vmatrix} 0 \\ \delta(t - t_i)\delta(\mathbf{x} - \mathbf{x}_\kappa) \\ 0 \end{vmatrix};$$

$$\mathbf{f}_3^* = \begin{vmatrix} 0 \\ 0 \\ \delta(t - t_i)\delta(\mathbf{x} - \mathbf{x}_\kappa) \end{vmatrix}$$

and solve three adjoint problems (2.7.9) with set vector-functions $\mathbf{f}_j^*$. As a result obtain three formulas analogous to relation (2.8.2)

$$\delta u(\mathbf{x}_\kappa, t_i) = \int_0^T dt \int_S (\delta \mathbf{f}_1 \cdot u_1^* + \delta \mathbf{f}_2 \cdot v_1^*) \, dS;$$

$$\delta v(\mathbf{x}_\kappa, t_i) = \int_0^T dt \int_S (\delta \mathbf{f}_1 \cdot u_2^* + \delta \mathbf{f}_2 \cdot v_2^*) \, dS;$$

$$\delta \zeta(\mathbf{x}_\kappa, t_i) = \int_0^T dt \int_S (\delta \mathbf{f}_1 \cdot u_3^* + \delta \mathbf{f}_2 \cdot v_3^*) \, dS, \tag{2.8.3}$$

where $\delta \mathbf{f}_1$ and $\delta \mathbf{f}_2$ are the first two components of the vector-function $\delta \mathbf{f}$, and u_j^*, v_j^* ($j = 1, 2, 3$) are the solutions of three conjugate problems with set "sources" $\mathbf{f}_1^*, \mathbf{f}_2^*$ and $\mathbf{f}_3^*$.

Formulae (2.8.3) allow one to evaluate the influence of variations of the assumed input data on the solution at any point $\mathbf{x}_\kappa$ of the domain S and at any time moment t_i.

Let us now discuss a more general perturbation theory when not only the external "sources" but also the problem operator itself are varied. To this end let us assume that we have the adjoint problem (2.7.9) with parameters and operators corresponding to the unperturbed problem.

EQUATIONS OF TIDAL DYNAMICS

Suppose further that a "perturbed" problem can be presented as

$$\mathfrak{P}'\mathfrak{w}' = \mathbf{f}';$$
$$\mathbf{I}\mathfrak{w}' = 0 \quad \text{on} \quad \Gamma;$$
$$\mathfrak{w}' = \mathfrak{w}_0 \quad \text{at} \quad t = 0 \qquad (2.8.4)$$

where

$$\mathfrak{P}' = \mathfrak{P} + \delta\mathfrak{P}$$
$$\mathbf{f}' = \mathbf{f} + \delta\mathbf{f};$$
$$\mathfrak{w}' = \mathfrak{w} + \delta\mathfrak{w}. \qquad (2.8.5)$$

Multiply scalarly the basic equation from problem (2.8.4) by $\mathfrak{w}^*$, and the conjugate equation from problem (2.7.9) by $\mathfrak{w}'$ and subtract their results from one another. The equation

$$(\mathfrak{w}^*, \mathfrak{P}'\mathfrak{w}') - (\mathfrak{w}', \mathfrak{P}^*\mathfrak{w}^*) = (\mathfrak{w}^*, \mathbf{f}') - (\mathfrak{w}', \mathbf{f}^*),$$

with account taken of relations (2.8.5) and the above identities

$$(\mathfrak{w}^*, \mathfrak{P}\mathfrak{w}) = (\mathfrak{w}, \mathfrak{P}^*\mathfrak{w}^*),$$
$$(\mathfrak{w}^*, \mathbf{f}) = (\mathfrak{w}, \mathbf{f}^*)$$

yields

$$(\mathfrak{w}^*, \delta\mathfrak{P}\mathfrak{w}') = (\mathfrak{w}^*, \delta\mathbf{f}) - (\delta\mathfrak{w}, \mathbf{f}^*). \qquad (2.8.6)$$

If $\mathbf{f}^*$ is chosen so that $\delta J = (\delta\mathfrak{w}, \mathbf{f}^*)$ is a considered variation of the functional of our problem, then we can finally present the perturbation theory formula as

$$\delta J = (\mathfrak{w}^*, \delta\mathbf{f}) - (\mathfrak{w}^*, \delta\mathfrak{P}\mathfrak{w}'). \qquad (2.8.7)$$

Let

$$\delta\mathfrak{P} = \left\{ \begin{array}{ccc} u'\dfrac{\partial}{\partial x} + v'\dfrac{\partial}{\partial y} - \delta\kappa_h \varDelta + & 0 & g\delta D \dfrac{\partial}{\partial x} \\ + \delta\left[\left(\dfrac{\iota}{D^2}\right)\sqrt{u'^2 + v'^2}\right] & & \\ 0 & u'\dfrac{\partial}{\partial x} + v'\dfrac{\partial}{\partial y} - \delta\kappa_h \varDelta + & g\delta D \dfrac{\partial}{\partial y} \\ & + \delta\left[\left(\dfrac{\iota}{D^2}\right)\sqrt{u'^2 + v'^2}\right] & \\ 0 & 0 & 0 \end{array} \right\},$$

where

$$\delta\left[\left(\dfrac{\iota}{D^2}\right)\sqrt{u'^2 + v'^2}\right] = \left(\dfrac{\iota}{D^2}\right)'\sqrt{u'^2 + v'^2} - \dfrac{\iota}{D^2}\sqrt{u^2 + v^2}.$$

With the operator $\delta \mathfrak{P}$ set in this way, equations for perturbations in a component-wise form are as follows:

$$\frac{\partial u'}{\partial t} + u' \frac{\partial u'}{\partial x} + v' \frac{\partial u'}{\partial y} - \kappa_h' \Delta u' + \left(\frac{\iota}{D^2}\right)' \sqrt{u'^2 + v'^2} u' + gD' \frac{\partial \zeta'}{\partial x} = f_1';$$

$$\frac{\partial v'}{\partial t} + u' \frac{\partial v'}{\partial x} + v' \frac{\partial v'}{\partial y} - \kappa_h' \Delta v' + \left(\frac{\iota}{D^2}\right)' \sqrt{u'^2 + v'^2} v' + gD' \frac{\partial \zeta'}{\partial y} = f_2';$$

$$\frac{\partial \zeta'}{\partial t} + \frac{\partial u'}{\partial x} + \frac{\partial v'}{\partial y} = 0. \qquad (2.8.8)$$

They are solved under the boundary conditions

$$u', v' = 0 \quad \text{on} \quad \Gamma;$$
$$u' = u_0, \quad v' = v_0, \quad \zeta' = \zeta_0, \quad \text{at} \quad t = 0. \qquad (2.8.9)$$

With the set type of the operator $\delta \mathfrak{P}$ allowed for, the formula (2.8.6) takes the form

$$\delta J = \int_0^T dt \int_S (\delta f_1 \cdot u^* + \delta f_2 \cdot v^*) \, dS - \int_0^T dt \int_S \left[u^* \left(u' \frac{\partial u'}{\partial x} + v' \frac{\partial u'}{\partial y} \right.\right.$$

$$\left. - \delta \kappa_h \Delta u' + \delta \left(\frac{\iota}{D^2}\right) \sqrt{u'^2 + v'^2} u' + g \delta D \frac{\partial \zeta'}{\partial x} \right)$$

$$+ v^* \left(u' \frac{\partial v'}{\partial x} + v' \frac{\partial v'}{\partial y} - \delta \kappa_h \Delta v' \right.$$

$$\left.\left. + \delta \left(\frac{\iota}{D^2}\right) \sqrt{u'^2 + v'^2} v' + g \delta D \frac{\partial \zeta'}{\partial y} \right) \right] dS. \qquad (2.8.10)$$

In the case of small perturbations the values u', v', ζ' in formula (2.8.10) can be substituted for by u, v, ζ, which results in the following formula of the theory of small perturbations:

$$\delta J = \int_0^T dt \int_S (\delta f_1 \cdot u^* + \delta f_2 \cdot v^*) \, dS - \int_0^T dt \int_S \left[u^* \left(u \frac{\partial u}{\partial x} + v \frac{\partial u}{\partial y} \right.\right.$$

$$\left. - \delta \kappa_h \Delta u + \delta \left(\frac{\iota}{D^2}\right) \sqrt{u^2 + v^2} u + g \delta D \frac{\partial \zeta}{\partial x} \right)$$

$$+ v^* \left(u \frac{\partial v}{\partial x} + v \frac{\partial v}{\partial y} - \delta \kappa_h \Delta v \right.$$

$$\left.\left. + \delta \left(\frac{\iota}{D^2}\right) \sqrt{u^2 + v^2} v + g \delta D \frac{\partial \zeta}{\partial y} \right) \right] dS. \qquad (2.8.10')$$

The perturbation theory can of course be posed in different ways, but each time in accordance with given input data and operator one may obtain a formula of type (2.8.10) or (2.8.10′), which allows one to estimate the discussed functionals.

By way of conclusion, note that far from the coast the conjugate solution changes only slightly from point to point. That is why it is expedient in formula (2.8.10′) to make use of the solution of the conjugate problem with the "source" at the point $(\mathbf{x}_\kappa, t_j)$ followed by its use to estimate δJ within a more or less considerable range in the vicinity of a given point.

2.9. The Spectral Problem

Linearize the equations of tidal dynamics (2.1.1), (2.1.2) and rewrite them in the form of a single vector equation

$$\frac{\partial \mathfrak{w}}{\partial t} + \mathfrak{B}\mathfrak{w} = \mathbf{f}, \qquad (2.9.1)$$

where

$$\mathfrak{w} = \begin{vmatrix} u \\ v \\ \zeta \end{vmatrix};$$

$$\mathbf{f} = \begin{vmatrix} f_1 \\ f_2 \\ 0 \end{vmatrix};$$

$$\mathfrak{B} = \begin{bmatrix} -\kappa_h \Delta + r & 2\omega_z & gD\dfrac{\partial}{\partial x} \\ 2\omega_z & -\kappa_h \Delta + r & gD\dfrac{\partial}{\partial y} \\ \dfrac{\partial}{\partial x} & \dfrac{\partial}{\partial y} & 0 \end{bmatrix}$$

Solve equation (2.9.1) under the boundary conditions

$$\mathbf{I}\mathfrak{w} = 0 \quad \text{on} \quad \Gamma;$$
$$\mathfrak{w} = \mathfrak{w}_0 \quad \text{at} \quad t = 0. \qquad (2.9.2)$$

Define the conjugate vectors

$$\mathfrak{w}^* = \begin{vmatrix} u^* \\ v^* \\ \zeta^* \end{vmatrix};$$

$$\mathbf{f}^* = \begin{vmatrix} f_1^* \\ f_2^* \\ 0 \end{vmatrix};$$

and correlate the problem (2.9.1), (2.9.2) with the conjugate problem

$$-\frac{\partial \mathfrak{w}^*}{\partial t} + \mathfrak{B}^* \mathfrak{w}^* = \mathbf{f}^*;$$

$$\mathbf{l}\mathfrak{w}^* = 0 \quad \text{on} \quad \Gamma;$$

$$\mathfrak{w}^* = \mathfrak{w}_T^* \quad \text{at} \quad t = T \qquad (2.9.3)$$

where

$$\mathfrak{B}^* = \begin{bmatrix} -\kappa_h \Delta + r & 2\omega_z & -\dfrac{\partial}{\partial x} \\ -2\omega_z & -\kappa_h \Delta + r & -\dfrac{\partial}{\partial y} \\ -gD \dfrac{\partial}{\partial x} & -gD \dfrac{\partial}{\partial y} & 0 \end{bmatrix}$$

Consider now two spectral problems

$$\mathfrak{B}\psi = \lambda\psi;$$

$$\mathfrak{B}^*\psi^* = \lambda\psi^*, \qquad (2.9.4)$$

with the components of vector-functions ψ and ψ^* obeying the conditions

$$\psi_1 = \psi_2 = 0; \quad \psi_1^* = \psi_2^* = 0 \quad \text{on} \quad \Gamma. \qquad (2.9.5)$$

Let problems (2.9.4), (2.9.5) from two biorthogonal bases $\{\psi_n\}$, $\{\psi_n^*\}$, with each of their eigenfunctions corresponding to eigennumbers λ_n. It results in the following conditions:

$$(\psi_n \psi_{n'})_* = \begin{cases} 1, & \text{if} \quad n' = n, \\ 0, & \text{if} \quad n' \neq n. \end{cases}$$

EQUATIONS OF TIDAL DYNAMICS

Seek the solution of problem (2.9.1), (2.9.2) in the form

$$\mathbf{w} = \sum_n \mathbf{w}_n \psi_n \qquad (2.9.6)$$

keeping in mind that

$$\mathbf{f} = \sum_n \mathbf{f}_n \psi_n;$$

$$\mathbf{w}_0 = \sum_n \mathbf{w}_{0n} \psi_n, \qquad (2.9.7)$$

where

$$\mathbf{f}_n = (\mathbf{f}, \psi_n^*),$$

$$\mathbf{w}_{0n} = (\mathbf{w}_0, \psi_n^*).$$

Substitute expressions (2.9.6) and (2.9.7) into relations (2.9.1), (2.9.2) and obtain the system of equations for the Fourier coefficients of the basic problem

$$\frac{d\mathbf{w}_n}{dt} + \lambda_n \mathbf{w}_n = \mathbf{f}_n;$$

$$\mathbf{w}_n = \mathbf{w}_{0n} \quad \text{at} \quad t = 0 \qquad (2.9.8)$$

$$(n = 1, 2, \ldots).$$

In an analogous way, expansion of

$$\mathbf{w}^* = \sum_n \mathbf{w}_n^* \psi_n^*;$$

$$\mathbf{f}^* = \sum_n \mathbf{f}_n^* \psi_n^*;$$

$$\mathbf{w}_T^* = \sum_n \mathbf{w}_{Tn}^* \psi_n^*, \qquad (2.9.9)$$

where

$$\mathbf{f}_n^* = (\mathbf{f}^*, \psi_n),$$

$$\mathbf{w}_{Tn}^* = (\mathbf{w}_T^*, \psi_n),$$

affords the conjugate problem

$$-\frac{d\mathbf{w}_n^*}{dt} + \lambda_n \mathbf{w}_n^* = \mathbf{f}_n^*;$$

$$\mathbf{w}_n^* = \mathbf{w}_{Tn}^* \quad \text{at} \quad t = T \qquad (2.9.10)$$

$$(n = 1, 2, \ldots).$$

In a special case of setting "conjugate sources" in expressions (2.9.3), (2.9.10) and $\mathbf{f}_n^* = 0$ the solution of problem (2.9.10) takes the form

$$\mathfrak{w}_n^* = \mathfrak{w}_{T_n}^* e^{-\lambda_n(T-t)}$$

and, hence,

$$\mathfrak{w}^* = \sum_n \mathfrak{w}_{T_n}^* e^{-\lambda_n(T-t)} \psi_n^*.$$

Multiply the equation from (2.9.8) by $\mathfrak{w}_n^*$, and the equation from (2.9.10) by $\mathfrak{w}_n$ and integrate the resulting expression by time from zero to T. Then we have for each Fourier component

$$\mathfrak{w}_{T_n}^* \mathfrak{w}_{T_n} = \mathfrak{w}_{0n}^* \mathfrak{w}_{0n} + \int_0^T f_n \mathfrak{w}_n^* \, dt. \qquad (2.9.11)$$

If $\mathfrak{w}_{T_n}^*$ is set as

$$\mathfrak{w}_{T_n}^* = \begin{vmatrix} 0 \\ 0 \\ 1 \end{vmatrix},$$

relation (2.9.11) reduces to the formula

$$\zeta_{T_n} = u_{0n} u_{0n}^* + v_{0n} v_{0n}^* + \zeta_{0n} \zeta_{0n}^* + \int_0^T (f_{1n} u_n^* + f_{2n} v_n^*) \, dt \qquad (2.9.12)$$

Summation of expression (2.9.12) results in the relation

$$\zeta = \sum_n \left[u_{0n} u_{0n}^* + v_{0n} v_{0n}^* + \zeta_{0n} \zeta_{0n}^* + \int_0^T (f_{1n} u^* + f_{2n} v_n^*) \, dt \right] \psi_n. \qquad (2.9.13)$$

Assume that together with the basic problem (2.9.1), (2.9.2) we have the "perturbation" problem

$$\frac{\partial \mathfrak{w}'}{\partial t} + \mathfrak{B}\mathfrak{w} = \mathbf{f}';$$

$$\mathbf{l}\mathfrak{w}' = 0 \quad \text{on} \quad \Gamma;$$

$$\mathfrak{w}' = \mathfrak{w}_0' \quad \text{at} \quad t = 0, \qquad (2.9.14)$$

where $\mathbf{f}' = \mathbf{f} + \delta\mathbf{f}$, $\mathfrak{w}_0' = \mathfrak{w}_0 + \delta\mathfrak{w}_0$.

In this case the formula of the perturbation theory assumes the form

$$\delta\zeta = \sum_n \left[\delta u_{0n} \cdot u_{0n}^* + \delta v_{0n} \cdot v_{0n}^* + \delta\zeta_{0n} \cdot \zeta_{0n}^* \right.$$

$$\left. + \int_0^T (\delta f_{1n} \cdot u_n^* + \delta f_{2n} \cdot v_n^*) \, dt \right] \psi_n. \qquad (2.9.15)$$

EQUATIONS OF TIDAL DYNAMICS

An analogous formula can be also obtained for δu, δv, provided the conjugate functions u^*, v^*, ζ^* are solutions to problem (2.9.3) and, also, $\mathbf{f}_n^* = 0$ and $\mathbf{w}_{Tn}^* = (1, 0, 0)$, $\mathbf{w}_{Tn} = (0, 1, 0)$, respectively.

If, among other things, the operator $\mathfrak{B}$ is also subjected to perturbation (i.e. $\mathfrak{B}' = \mathfrak{B} + \delta\mathfrak{B}'$) then instead of problem (2.9.14) we have

$$\frac{\partial \mathbf{w}'}{\partial t} + \mathfrak{B}\mathbf{w}' = \mathbf{f}';$$

$$\mathbf{l}\mathbf{w}' = 0 \quad \text{on} \quad \Gamma;$$

$$\mathbf{w}' = \mathbf{w}_0' \quad \text{at} \quad t = 0, \qquad (2.9.16)$$

in which case $\mathbf{f}' = \mathbf{f} + \delta\mathbf{f} - \delta\mathfrak{B}\mathbf{w}'$.

In this case one can approximately set

$$\mathbf{f}' = \mathbf{f} + \delta\mathbf{f}',$$

where

$$\delta\mathbf{f}' = \delta\mathbf{f} - \delta\mathfrak{B}\mathbf{w}.$$

Here in $\delta\mathfrak{B}\mathbf{w}'$ we substituted $\mathbf{w}'$ by $\mathbf{w}$ corresponding to the unperturbed state of the system.

As a result obtain the expression for $\delta\zeta$, analogous to formula (2.9.15), i.e.

$$\delta\zeta = \sum_n \left[\delta u_{0n} \cdot u_{0n}^* + \delta v_{0n} \cdot v_{0n}^* + \delta\zeta_{0n} \cdot \zeta_{0n}^* \right.$$
$$\left. + \int_0^T (\delta f_{1n}' \cdot u_n^* + \delta f_{2n}' \cdot v_n^* + \delta f_{3n}' \cdot \zeta_n^*) \, dt \right] \psi_n, \qquad (2.9.15')$$

where $\delta f_{1n}'$, $\delta f_{2n}'$, $\delta f_{3n}'$ are the components of the vector function $\delta\mathbf{f}_1'$.

2.10. References

In this chapter we mainly used the work by Marchuk *et al.* (1972). Detailed information on functional spaces and inequalities used in the present section are available in the monograph by Ladyzshenskaya (1970).

Arguments in favour of the validity of an arbitrary choice of initial conditions while solving the equations of tidal dynamics are borrowed from a paper by Kagan (1970).

The existence of a periodic solution of tidal dynamic equations is substantiated in the paper by Gordeev (1974). Conjugate equations of tidal dynamics and the perturbation theory are discussed in the monograph by Marchuk (1974).

CHAPTER 3

Numerical Methods for the Solution of the Equations of Tidal Dynamics

IN INTEGRATING the equations of tidal dynamics there are two different approaches in prescribing *a priori* information on the temporal character of the solution. One of them is associated with the solution of the so-called periodic boundary value problem of tidal dynamics in which the velocity components of the tidal flow and the level are represented by harmonic time functions. In the second approach, no assumptions about the nature of time variations of tidal characteristics are made. They are found as in a certain sense limiting solutions for the equations. In other words, an initial boundary value problem is being solved until the sought functions become periodic. This is ensured by the presence of tide-generating forces and periodic boundary conditions on the open boundary of the domain discussed.

There exist many numerical methods for the solution of the equation of tidal dynamics. The subject of the present chapter is restricted to those methods for which one can prove convergence of the finite-difference solution to the exact solution of the problem considered.

3.1. Method of Boundary Values

In this approach, the initial equations are linearized dynamic equations which, under the supposition of periodicity, and in the absence of macroturbulence effects, can be presented as

$$\mathbf{B}\bar{\mathbf{v}} = -g \cdot \nabla \bar{\zeta} + \bar{\mathbf{f}}; \tag{3.1.1}$$

$$i\sigma\bar{\zeta} + \mathrm{div}\,(\bar{\mathbf{v}}D) = 0. \tag{3.1.2}$$

Here, along with common notations, the following new ones are used: $\bar{\mathbf{v}}$ and $\bar{\zeta}$ are the complex amplitudes of the averaged vertical velocity of the tidal transport and elevation; $\bar{\mathbf{f}}$ is the perturbating force amplitude; $\mathbf{B}$ is the coefficient matrix equal to $\mathbf{B} = (i\sigma + r)\mathbf{E} + \mathbf{A}_1$; σ is the wave frequency; r is the dimensional coefficient of friction in a linear law of resistance; $\mathbf{E}$ is the unity matrix.

Let the solution of equations (3.1.1), (3.1.2) obey the condition of no transport through the contour Γ of the closed domain S

$$(\mathbf{v} \cdot \mathbf{n})_\Gamma = 0 \qquad (3.1.3)$$

and reduce the system of equations (3.1.1), (3.1.2) to a single equation with respect to the amplitudes of tidal oscillations of the elevation $\bar\zeta$. For this purpose substitute the expression for the amplitude of the tidal transport velocity

$$\bar{\mathbf{v}} = -g\mathbf{B}^{-1} \cdot \nabla \bar\zeta + \mathbf{B}^{-1}\bar{\mathbf{f}}, \qquad (3.1.4)$$

derived from equation (3.1.1) into the continuity equation (3.1.2) and obtain the following equation:

$$i\sigma\bar\zeta - g \cdot \operatorname{div}\left[D\mathbf{B}^{-1}\left(\nabla\bar\zeta - \frac{\bar{\mathbf{f}}}{g}\right)\right] = 0 \qquad (3.1.5)$$

or

$$\mathcal{L}_f\bar\zeta + i\sigma\bar\zeta = 0, \qquad (3.1.6)$$

where $\mathcal{L}_f\bar\zeta = -g \cdot \operatorname{div}[D\mathbf{B}^{-1}(\nabla\bar\zeta - \bar{\mathbf{f}}/g)]$, $\mathcal{L}_f$ is the elliptical operator.

Let us rewrite the boundary condition (3.1.3) in terms of $\bar\zeta$:

$$\left[\mathbf{B}^{-1}\left(\nabla\bar\zeta - \frac{\bar{\mathbf{f}}}{g}\right) \cdot n\right]_\Gamma = 0. \qquad (3.1.7)$$

The problem (3.1.1)–(3.1.3) is thus reduced to the solution of the operator equation (3.1.6) obeying the condition (3.1.7).

Solvability of the equations

Let us now investigate the solvability of equations (3.1.6), (3.1.7). For this purpose make up a quadratic form $(\mathcal{L}_f\bar\zeta, \bar\zeta)$ defining it in the following way:

$$(\mathcal{L}_f\bar\zeta, \bar\zeta) = \int_S \mathcal{L}_f\bar\zeta \cdot \bar\zeta^* \, dS = -g \int_S \operatorname{div}\left[D\mathbf{B}^{-1}\left(\nabla\bar\zeta - \frac{\bar{\mathbf{f}}}{g}\right)\right] \cdot \bar\zeta^* \, dS,$$

where the asterisk denotes the complex conjugate amplitude, $dS = dx\,dy$ is an element of the domain S.

Transform the integral once by parts and, with account taken of condition

(3.1.7), we obtain

$$(\mathfrak{L}_f \zeta, \zeta) = g \int_S D\mathbf{B}^{-1}\left(\nabla \zeta - \frac{\bar{\mathbf{f}}}{g}\right) \cdot \nabla \zeta^* \, dS.$$

Considering only the real part of the quadratic form, we get

$$\text{real } (\mathfrak{L}_f \zeta, \zeta) = gr \int_S D \, |\mathbf{B}^{-1} \cdot \nabla \zeta|^2 \, dS - \text{real} \int_S D\mathbf{B}^{-1} \frac{\bar{\mathbf{f}}}{g} \cdot \nabla \zeta^* \, dS,$$

and, hence

$$gr \int_S D \, |\mathbf{B}^{-1} \cdot \nabla \zeta|^2 \, dS \leqslant \text{real } (\mathfrak{L}_f \zeta, \zeta)$$

$$+ \left| \text{real} \int_S D\mathbf{B}^{-1} \frac{\bar{\mathbf{f}}}{g} \cdot \nabla \zeta^* \, dS \right|. \qquad (3.1.8)$$

Estimate now the bounds of the inequality obtained

$$gr \int_S D \, |\mathbf{B}^{-1} \cdot \nabla \zeta|^2 \, dS \geqslant gr \, ||\mathbf{B}||^2 \cdot \int_S D \, |\nabla \zeta|^2 \, dS \geqslant gr\mu \, ||\nabla \zeta||^2.$$

$$\left| \text{real} \int_S D\mathbf{B}^{-1} \frac{\bar{\mathbf{f}}}{g} \cdot \nabla \zeta^* \, dS \right| \leqslant \left\| D\mathbf{B}^{-1} \frac{\bar{\mathbf{f}}}{g} \right\| \cdot ||\nabla \zeta||$$

$$\leqslant \frac{\varepsilon^{-2}}{2} \left\| D\mathbf{B}^{-1} \frac{\bar{\mathbf{f}}}{g} \right\|^2 + \frac{\varepsilon^2}{2} ||\nabla \zeta||^2,$$

where $\mu = \min\limits_S D$, ε is any positive number.

Substituting these estimates into inequality (3.1.8) and setting $\varepsilon^2 = gr\mu$ affords the following inequality for the elliptical operator $\mathfrak{L}_f$:

$$||\nabla \zeta||^2 \leqslant \text{real } (\mathfrak{L}_f \zeta, \zeta) + C_1 \left\| D\mathbf{B}^{-1} \frac{\bar{\mathbf{f}}}{g} \right\|^2. \qquad (3.1.9)$$

Henceforth, if not specified, the constants depending upon known values are denoted by the symbol C with various indices.

This estimate is sufficient to state the solvability of the boundary value problem (3.1.6), (3.1.7) with any perturbing functions $\bar{\mathbf{f}}$ and at all σ not belonging to the spectrum of the problem under discussion.

Let us prove now that $i\sigma$ does not belong to the spectrum of problem (3.1.6), (3.1.7). For this purpose multiply equation (3.1.6) by ζ^* and then integrate it over the domain S. Distinguishing the real and imaginary parts of the resulting expression and setting $\bar{\mathbf{f}} = 0$ yields

$$\int_S D|\mathbf{B}^{-1} \cdot \nabla \zeta|^2 \, dS = 0; \tag{3.1.10}$$

$$\text{imag}\,(\mathfrak{L}_0 \zeta, \zeta^*) + \sigma \int_S |\zeta|^2 \, dS = 0, \tag{3.1.11}$$

where $\mathfrak{L}_0 = \mathfrak{L}_f|_{\bar{\mathbf{f}}=0} = g \cdot \text{div}\,(D\mathbf{B}^{-1} \cdot \nabla \zeta)$.

The above leads one to the conclusion that $\nabla \zeta$ and, hence, the first term in relation (3.1.11) is equal to zero. In this case the second term in relation (3.1.11) is also equal to zero and, thus, the function ζ identically equals zero within the whole domain S.

Hence, the initial boundary value problem (3.1.6), (3.1.7) is uniquely solvable at any $\bar{\mathbf{f}}$ if the bottom friction coefficient r is not zero.

In the case of the first boundary value problem the initial equation (3.1.6) is reduced to the equation for a new unknown function ζ_1 with the help of the substitution $\zeta_1 = \zeta - \zeta_\Gamma$ (here ζ_Γ are ζ values extended inside the basin through out the contour). The resulting equation will include a modified perturbing function $\bar{\mathbf{f}}$ and obey the homogeneous boundary conditions on the basin contour. The fact that this problem is uniquely solvable can be proved using the above considerations.

Finite-difference scheme

There exist many finite-difference schemes to solve the first boundary problem for equation (3.1.5). Let us consider only one of them, for which one can prove the unique solvability at any mesh length h and convergence of the numerical solution to the exact solution of equation (3.1.5) at $h \to 0$.

Let us cover the domain S by a uniform grid with a mesh length h and approximate the derivatives under the sign of gradient and divergence by difference relations forward and backward, respectively. Then the difference analogue of equation (3.1.5) can be represented as follows:

$$g(q_{\bar{x}}^{(1)} + q_{\bar{y}}^{(2)}) + i\sigma \zeta = 0. \tag{3.1.12}$$

Here the same notations are employed for the grid functions

$$q = \begin{matrix} q^{(1)} \\ q^{(2)} \end{matrix} = \left\{ -D\mathbf{B}^{-1} \left[\begin{pmatrix} \zeta_x \\ \zeta_y \end{pmatrix} - \frac{\bar{\mathbf{f}}}{g} \right] \right\},$$

the indices denoting the correspondingly directed difference relations (without a bar — forward, with a bar — backward).

According to the problem conditions the values of the function ζ on the grid boundary Γ_h of the domain investigated are set, which allows one to assume that

$$\zeta|_{\Gamma_h} = \zeta_\Gamma. \qquad (3.1.13)$$

Let us prove convergence of the grid solution of system (3.1.12), (3.1.13) to the solution of the first boundary value problem for equation (3.1.5).

Equation (3.1.12) is seen to approximate equation (3.1.5) with an error of an order of h. Denote the deviation of the grid solution of the problem considered from the precise one by e_ζ and find the system for its determination with the use made of equations (3.1.12), (3.1.13). In this case, under the assumption that the grid boundary Γ_h coincides with Γ (this condition may be replaced by assuming Γ to be smooth), we get

$$g\big(q_{\bar{x}}^{(1)} + q_{\bar{y}}^{(2)}\big) + i\sigma e_\zeta = \mathfrak{C}_\zeta, \qquad (3.1.14)$$

$$e_\zeta|_{\Gamma_h} = 0, \qquad (3.1.15)$$

where $\mathfrak{C}_\zeta$ is the discrepancy due to the approximation of the initial problem by the system of equations (3.1.12), (3.1.13),

$$\begin{pmatrix} q^{(1)} \\ q^{(2)} \end{pmatrix} = \left\{ -D\mathbf{B}^{-1} \begin{matrix} (e_\zeta)_x \\ (e_\zeta)_y \end{matrix} \right\}.$$

The substitution $\bar{\mathbf{v}} = gD^{-1}\mathbf{q}$ reduces equation (3.1.14) to

$$(i\sigma + r)\bar{\mathbf{v}} + \mathbf{A}_1 \bar{\mathbf{v}} = -gD^{-1} \cdot \mathbf{B}q; \qquad (3.1.16)$$

$$i\sigma e_\zeta + (\bar{u}D)_x + (\bar{v}D)_y = \mathfrak{C}_\zeta, \qquad (3.1.17)$$

where $\bar{u}, \bar{v}$ are components of the vector $\bar{\mathbf{v}}$.

Let us take the scalar products of equation (3.1.16) with $\bar{\mathbf{v}}^*$ and equation (3.1.17) with e_ζ^* (here the asterisk denotes complex conjugate values of the corresponding characteristics), summing the resulting expressions over the whole domain and adding them up. Employing the formula of summation by parts and distinguishing real and imaginary parts in the resulting expression, we obtain the following equalities:

$$r\|\bar{\mathbf{v}}\|^2 = \text{real}\,(\mathfrak{C}_\zeta, e_\zeta^*); \qquad (3.1.18)$$

$$\sigma\|\bar{\mathbf{v}}\|^2 + \text{imag}\,(\mathbf{A}_1\bar{\mathbf{v}} \cdot \bar{\mathbf{v}}^*) + g\sigma\|e_\zeta\|^2 = \text{imag}\,(\mathfrak{C}_\zeta, e_\zeta^*), \qquad (3.1.19)$$

where symbols $\|\ \|$ and $(\ ,\ )$ denote, respectively, the grid norm and the grid scalar product in the space $L_2(S_n)$, which can be defined as

$$||\mathbf{a}||^2 = h^2 \sum_{S_h} |\mathbf{a}_h|^2,$$

$$(\mathbf{a}, \mathbf{b}) = h^2 \sum_{S_h} \mathbf{a}_h \mathbf{b}_n,$$

S_h is a grid division of the domain considered.

Estimate the right side of relation (3.1.18) by means of the Cauchy inequality. Then

$$||\bar{\mathbf{v}}||^2 \leq C_2 ||\mathfrak{C}_\zeta|| \cdot ||e_\zeta||. \qquad (3.1.20)$$

Substituting inequality (3.1.20) into relation (3.1.19) gives

$$||e_\zeta|| \leq C_3 \cdot ||\mathfrak{C}_\zeta||.$$

However, the fact that $||\mathfrak{C}_\zeta|| \to 0$ at $h \to 0$ yields

$$||e_\zeta|| \to 0, \qquad (3.1.21)$$

which proves convergence of the difference scheme (3.1.12) at $h \to 0$.

Note that inequality (3.1.21) guarantees unique solvability of system (3.1.12), (3.1.13) since for the difference of two solutions $(\zeta_1 - \zeta_2)$ this inequality turns into the identity

$$||\zeta_1 - \zeta_2|| \equiv 0.$$

Thus, the unique solvability of system (3.1.12), (3.1.13) at $r \neq 0$ is proved.

Let us consider equation (3.1.5) with boundary condition (3.1.7) brought to conformity with it. Problem (3.1.5), (3.1.7) is the classic Poincaré problem, whose numerical solution is hampered by a number of known difficulties associated with the realization of boundary condition (3.1.7). Indeed, approximation of the derivatives contained in problem (3.1.5), (3.1.7) in the above way results in a difference equation with the solution obeying the condition

$$\zeta_x + 2\omega_z(i\sigma + r)^{-1}\zeta_y = f_1 \qquad (3.1.22a)$$

on the part of the contour Γ_h parallel to the axis y, and the condition

$$\zeta_y - 2\omega_z(i\sigma + r)^{-1}\zeta_x = \bar{f}_2 \qquad (3.1.22b)$$

on the part of the contour Γ_h parallel to the axis x. Here $\bar{f}_1$ and f_2 are free terms regarded as combinations of the perturbating force components along axes x, y.

Thus, it appears that for the solution of the problem considered one has to know values of the function ζ in the fictitious set of grid points located at a mesh-length distance from the right and upper parts of the solid contour Γ_h. The most trivial method is to set the function $\bar{\zeta}$ in fictitious grid points. For instance, these points may be considered to be located on the mainland where the condition $\bar{\zeta} = 0$ is fulfilled. However, one then has to pay special attention to the smooth extention of the function $\bar{\zeta}$ outside the limits of the region of determination of the solution. Otherwise both on the boundary itself and in its vicinity an appreciable distortion of the field of tidal elevation can occur.

The following method of solving the difference problem is apparently more expedient. Instead of using an additional "computing" boundary condition in grid points extrapolated outside the boundary of the domain investigated, let us ascribe the difference equation and boundary condition (3.1.22) to all the points of the right and upper parts of the boundary. Then solving relation (3.1.22) with respect to the value of the function $\bar{\zeta}$ at the fictitious grid points and substituting the resulting expressions into the difference equation, we get as a result a closed system of algebraic equations (their number will coincide with that of the unknown values). An analogous method can be used when considering angular points. In this case we add to the difference equation the following conditions

$$\bar{\zeta}_x = f_1; \quad \bar{\zeta}_y = f_2; \qquad (3.1.23)$$

meaning that in the vicinity of the angular points both components of the velocity are equal to zero.

An important fact is that, with such an algorithm for solving the Poincaré problem, the condition that the coefficient matrix of the system of linear algebraic equations obtained is positive (see difference equation (3.1.12)) is not violated. This enables one to use a number of effective methods when solving it, including the method of minimal discrepancies.

Remark 1. Instead of the variant of the method of boundary values described above, it often appears to be more convenient to use another method based on solving the following problem for complex amplitudes of the vertically averaged velocity of the tidal flow

$$\mathbf{B}\bar{\mathbf{v}} = \frac{ig}{\sigma} \cdot \nabla \operatorname{div} \bar{\mathbf{v}} D + \bar{\mathbf{f}}; \qquad (3.1.24)$$

$$(\bar{\mathbf{v}} \cdot \mathbf{n})_\Gamma = 0. \qquad (3.1.25)$$

An indisputable advantage of this variant is the simplicity of realization of boundary condition (3.1.25). Of importance also is the fact that when numerically integrating the system (3.1.24), (3.1.25) there is no necessity to extend the solution outside the domain of its definition.

However, this method is also hampered by a number of difficulties, which are, incidentally, also typical of the first variant of the method of boundary values. They are due to energy accumulation in short waves induced in the process of tidal wave reflection from convex and concave boundary angles. The latter fact, in its turn, is a source of analytical peculiarities in the tidal characteristics.

Remark 2. The requirement of the bottom friction coefficient non-zero throughout the domain under investigation seems to be important from the standpoint of both the existence of the solution and the possibility of realizing the solution by difference methods. The possible difficulties encountered in the absence of bottom friction can be demonstrated by calculations of tides in the World Ocean carried out with the use of the central-difference scheme.

In this case the system of linear algebraic equations corresponding to the first boundary problem for equation (3.1.5) can be represented as

$$\zeta + \Lambda_1 \zeta = \varphi_1, \qquad (3.1.26)$$

where ζ is the matrix of unknown values, Λ_1 is the matrix of coefficients having zero main diagonal and φ_1 is the matrix of free terms.

If system (3.1.26) is solved within a sufficiently large domain through the use of the Zeidel-Nekrasov iteration method, then at $r = 0$ the maximum eigenvalue of the matrix Λ_1 can exceed unity. In this case the iteration process diverges and system (3.1.26) has to be replaced by a similar one

$$(1 - i\varepsilon)\zeta + \Lambda_1 \zeta = \varphi_1 \qquad (3.1.27)$$

for which the Zeidel-Nekrasov method converges. Here ε is a small positive number.

In this case the question whether the solution of system (3.1.27) converges to that of system (3.1.26) at $\varepsilon \to 0$ remains, however, unsolved, since Λ_1 does not seem to be a fixed sign matrix and the solution of system (3.1.27) at $\varepsilon = 0$ is, generally speaking, possibly non-existent. Besides, the addition of the term $-i\varepsilon\zeta$ into equation (3.1.27) appears to be equivalent to introducing a computed viscosity which, for instance, at $\varepsilon = 0.01$ is certain to exceed the physical viscosity.

To elucidate the above point, consider system (3.1.27) setting the Coriolis parameter equal to zero. In this case the matrix Λ_1 can be easily calculated and represented as

$$\Lambda_1 = (C_1 + iC_2 r)^{-1} \Lambda_2,$$

where $C_1 = C_2\sigma - 4/h^2$; $C_2 = a^2\sigma/gD$; h is the angular mesh length; a is the Earth's radius; Λ_2 is the r-independent matrix.

At $r = 0$, instead of equation (3.1.27) we have

$$(1-i\varepsilon)\bar{\zeta}+C_1^{-1}\Lambda_2\bar{\zeta} = \varphi_1. \tag{3.1.28}$$

Multiplying equation (3.1.28) by $(1-i\varepsilon)^{-1}$ and comparing it with equation (3.1.26) gives

$$\varepsilon = -\frac{C_2}{C_1}r.$$

Substituting expressions for the coefficients C_1 and C_2 in the above relation yields the following equation,

$$\varepsilon = \frac{\sigma(ah)^2}{4gD-(\sigma ah)^2}r. \tag{3.1.29}$$

Formula (3.1.29) offers a means of determining the order of magnitude of the bottom friction coefficient r corresponding to the chosen value of the parameter ε. In particular, if we assume $h = 5°$, $D = 4000$ m, $\sigma = 1.405\times 10^{-4}$ c^{-1}, $a = 6.4\times 10^6$ m and $\varepsilon = 0.01$ then $0\,(r) = 10^{-5}$ c^{-1}, i.e. the artificially introduced computed viscosity exceeds the corresponding physical viscosity obtained by means of the quadratic resistance law by two or three orders.

3.2. HN-method

The initial equations of the HN (Hydrodynamical-Numerical)-method are vertically averaged tidal dynamics equations, which with the use of the quadratic resistance law can be written as follows:

$$\frac{\partial \mathbf{v}}{\partial t}+A_1\mathbf{v} = -g\cdot\nabla\zeta-\frac{\iota}{D}|\mathbf{v}|\mathbf{v}+\mathbf{f}; \tag{3.2.1}$$

$$\frac{\partial \zeta}{\partial t}+\text{div}\,(\mathbf{v}D) = 0. \tag{3.2.2}$$

In addition to equations (3.2.1), (3.2.2), which can be regarded as a quasi-linear hyperbolic system, we set corresponding boundary and initial conditions. For instance, if the contour Γ of the basin under investigation S consists of two parts, i.e. a coastline (or a shelf edge) and a boundary Γ_2 between a given and a neighbouring water basins, then, as in the cases discussed above, on Γ_1 we set a condition of no transport

$$(\mathbf{v},\mathbf{n})_{\Gamma_1} = 0, \tag{3.2.3}$$

while on Γ_2 oscillations in elevation are defined by

$$\zeta|_{\Gamma_2} = \zeta_\Gamma. \tag{3.2.4}$$

Initial values of the functions

$$\mathbf{v}|_{t=0} = \mathbf{v}_0, \quad \zeta|_{t=0} = \zeta_0 \tag{3.2.5}$$

are, to a large extent, arbitrarily chosen. Specifically, as has been shown in the preceding chapter, $\mathbf{v}_0$ and ζ_0 can be set equal to zero, which is equivalent to the assumption that at the initial moment the liquid in the basin is in a state of rest.

Difference scheme

System (3.2.1), (3.2.2) is solved by the grid method using a uniform mesh-length $2h$ with respect to spatial variables and a mesh-length 2τ with respect to time.

Let the plane (x, y) be subdivided by straight lines parallel to coordinate axes into elementary squares with mesh-length h and let $2m$, $2n$ be a number of a grid nodes, and $x = mh$, $y = nh$ ($m, n = 0, \pm 1, \pm 2, \ldots$). Then the coordinates of the points at which values of u, v and ζ are calculated (henceforth referred to as u-, v- and ζ-points) will be equal to $x = (2m+1)h$, $y = 2nh$; $x = 2mh$, $y = (2n+1)h$ and $x = 2mh$, $y = 2nh$, respectively. In this case moments of time, at which the components of the vertically averaged tidal velocity and the level are determined, are shifted with respect to each other by half a time step τ and fall on $t = (2\kappa+1)\tau$ and $t = 2\kappa\tau$ $\kappa = 0, 1, 2, \ldots$, respectively.

This grid covers the contour of the basin in such a way that either u- or v-points (depending upon the meridional or latitudinal boundary orientation) are located on the solid part of the contour, while ζ-points are located on the liquid contour. Grid point location of this kind provides a means for good approximation to the boundary conditions; including the condition of the normal velocity component being zero on the coast. However, since when calculating u and v by the equations of motion at u- and v-grid-points one has to know both velocity components simultaneously (only under this condition can the Coriolis forces and bottom friction effects be taken into account), various interpolating formulae have to be used to determine v at u-nodes and u at v-nodes.

The method employs the simplest of interpolating formulae, i.e. velocity components at corresponding nodes are calculated as arithmetic means from four values in the neighbouring nodal points

$$\tilde{u}_{2m,\,2n+1} = \tfrac{1}{4}(u_{2m+1,\,2n+2} + u_{2m-1,\,2n+2} + u_{2m+1,\,2n} + u_{2m-1,\,2n});$$

$$\tilde{v}_{2m+1,\,2n} = \tfrac{1}{4}(v_{2m,\,2n+1} + v_{2m,\,2n-1} + v_{2m+2,\,2n+1} + v_{2m+2,\,2n-1}).$$

If spatial derivatives in system (3.2.1) (3.2.2) are approximated by central differences and time derivatives by forward differences, the system of difference equations can be presented as follows

$$\frac{u^{2\kappa+1}_{2m+1,\,2n} - u^{2\kappa-1}_{2m+1,\,2n}}{2\tau} - 2\omega_z \tilde{v}^{2\kappa-1}_{2m+1,\,2n} = -\frac{g}{2h}(\zeta^{2\kappa}_{2m+2,\,2n} - \zeta^{2\kappa}_{2m,\,2n})$$

$$- \frac{r_u^{2\kappa-1}}{2}(u^{2\kappa-1}_{2m+1,\,2n} + u^{2\kappa+1}_{2m+1,\,2n}) + f_1^{2\kappa+1}; \quad (3.2.6)$$

$$\frac{v^{2\kappa+1}_{2m,\,2n+1} - v^{2\kappa-1}_{2m,\,2n+1}}{2\tau} + 2\omega_z \tilde{u}^{2\kappa-1}_{2m,\,2n+1} = -\frac{g}{2h}(\zeta^{2\kappa}_{2m,\,2n+2} - \zeta^{2\kappa}_{2m,\,2n})$$

$$- \frac{r_v^{2\kappa-1}}{2}(v^{2\kappa-1}_{2m,\,2n+2} + v^{2\kappa+1}_{2m,\,2n+1}) + f_2^{2\kappa+1}; \quad (3.2.7)$$

$$\frac{\zeta^{2\kappa+2}_{2m,\,2n} - \zeta^{2\kappa}_{2m,\,2n}}{2\tau} + [((Du^{2\kappa+1})_x)_{2m,\,2n} + ((Dv^{2\kappa+1})_y)_{2m,\,2n}] = 0. \quad (3.2.8)$$

Here

$$r_u = \frac{t}{D}\sqrt{u^2 + \tilde{v}^2}; \qquad r_v = \frac{t}{D}\sqrt{\tilde{u}^2 + v^2};$$

$$((Du)_x)_{2m,\,2n} = \frac{1}{2h}[(Du)_{2m+1,\,2n} - (Du)_{2m-1,\,2n}];$$

$$((Dv)_y)_{2m,\,2n} = \frac{1}{2h}[(Dv)_{2m,\,2n+1} - (Dv)_{2m,\,2n-1}],$$

f_1 and f_2 are perturbing force components along the axes x and y, the upper indices 2κ, $2\kappa+1$, etc., denote functions at time layers $t = 2\kappa\tau$, $t = (2\kappa+1)\tau$, ..., in which case $\kappa = 0, 1, 2, \ldots$.

Boundary conditions (3.2.3)—(3.2.5) for difference equations (3.2.6)—(3.2.8) can be written as follows:

$$u|_{\Gamma^M_{1h}} = 0; \qquad v|_{\Gamma^Z_{1h}} = 0; \qquad \zeta|_{\Gamma_{2h}} = \zeta_{\Gamma_h}(t); \quad (3.2.9)$$

$$u|_{t=0} = u_0; \qquad v|_{t=0} = v_0, \qquad \zeta|_{t=0} = \zeta_0, \quad (3.2.10)$$

where Γ^M_{1h} and Γ^Z_{1h} are sets of the meridional and zonal parts of the grid contour Γ_{1h}, respectively.

A priori *estimates*

Let us investigate *a priori* the time boundedness of the solution for system (3.2.6)—(3.2.9) by the method of energy inequalities. Let us confine ourselves to the case when both the depth and the Coriolis parameter remain constant throughout the basin, and the elevations on its liquid contour are equal to zero. Note that while setting homogeneous boundary conditions on the liquid contour there arises a necessity to add to the right side of equations (3.2.6), (3.2.7) the terms which together with f_1 and f_2 will represent the perturbing force. However, since in this case we are interested in stability with respect to the initial data, we do not intend to write these terms out explicitly, assuming that f_1 and f_2 characterize the combined effect of tide generating forces and periodic boundary conditions on the liquid contour.

Introduce the following notations: let S_u, S_v and S_ζ be the set of u-, v- and ζ-points and

$$(a, b)_i = h^2 \sum_{S_i} a_h b_h, \quad \|a\|_i^2 = (a, a)_i$$

be the grid scalar product and the norm within the domain S_i, respectively, where i is any of symbols u, v and ζ; a and b are any of the functions in question set on the i-grid.

Assume the grid functions of u, v, ζ to have zero extension beyond the limits of the corresponding domains S_u, S_v, S_ζ, so that formally

$$\|a\|_{S_h}^2 \equiv h^2 \sum_{S_h} a^2 = h^2 \sum_{S_i} a^2 = \|a\|_i^2,$$

where $S_h = S_u \cup S_v \cup S_\zeta$.

Multiply equation (3.2.6) by $2\tau h^2(u_{2m+1,\,2n}^{2\kappa-1} + u_{2m+1,\,2n}^{2\kappa+1})$, and equation (3.2.7) by $2\tau h^2(v_{2m,\,2n+1}^{2\kappa-1} + v_{2m,\,2n+1}^{2\kappa+1})$, sum them over the domains S_u, S_v, respectively, and sum the resulting expressions. Then, employing the formula for summation by parts and boundary conditions (3.2.9) we obtain

$$(\|v^{2\kappa+1}\|_{S_h}^2 - \|v^{2\kappa-1}\|_{S_h}^2) + \tau h^2 \left[\sum_{S_u} r_u^{2\kappa-1}(u^{2\kappa-1}+u^{2\kappa+1})^2 \right.$$
$$\left. + \sum_{S_v} r_v^{2\kappa-1}(v^{2\kappa-1}+v^{2\kappa+1})^2 \right] - 4\tau\omega_z[(u^{2\kappa-1}+u^{2\kappa+1},\, \tilde{v}^{2\kappa-1})_u$$
$$- (v^{2\kappa-1}+v^{2\kappa+1},\, \tilde{u}^{2\kappa+1})_v] - 2\tau g\, (\zeta^{2\kappa},\, u_x^{2\kappa-1}+u_x^{2\kappa+1}+v_y^{2\kappa-1}$$
$$+ v_y^{2\kappa+1})_\zeta = 2\tau(\mathbf{f},\, \mathbf{v}^{2\kappa-1}+\mathbf{v}^{2\kappa+1}). \qquad (3.2.11)$$

Transform now the latter term on the left side of equation (3.2.11), making use of relation (3.2.8). Rewrite it at time layers $t = (2\kappa-2)\tau$, $t = 2\kappa\tau$, with the following multiplication by $2\tau h^2 \zeta^{2\kappa}$ and sum over the domain S_ζ. In this case

$$-2\tau g(\zeta^{2\kappa}, u_x^{2\kappa-1}+u_x^{2\kappa+1}+v_y^{2\kappa-1}+v_y^{2\kappa+1})$$
$$= [(\zeta^{2\kappa}, \zeta^{2\kappa+2})_\zeta - (\zeta^{2\kappa}, \zeta^{2\kappa-2})_\zeta]. \quad (3.2.12)$$

Nevertheless, since

$$[(\zeta^{2\kappa}, \zeta^{2\kappa+2})_\zeta - (\zeta^{2\kappa}, \zeta^{2\kappa-2})_\zeta] = \tfrac{1}{4}(\|\zeta^{2\kappa}-\zeta^{2\kappa-2}\|_\zeta^2$$
$$-\|\zeta^{2\kappa+2}-\zeta^{2\kappa}\|_\zeta^2 + \|\zeta^{2\kappa+2}+\zeta^{2\kappa}\|_\zeta^2 - \|\zeta^{2\kappa}+\zeta^{2\kappa-2}\|_\zeta^2),$$

then, employing equation (3.2.8), we get

$$-2\tau g(\zeta^{2\kappa}, u_x^{2\kappa-1} - u_x^{2\kappa+1}+v_y^{2\kappa-1}+v_y^{2\kappa+1})_\zeta = gD\tau^2(\|u_x^{2\kappa-1}$$
$$+v_y^{2\kappa-1}\|_\zeta^2 - \|u_x^{2\kappa+1}+v_y^{2\kappa+1}\|_\zeta^2) + \frac{g}{4D}(\|\zeta^{2\kappa+2}+\zeta^{2\kappa}\|_\zeta^2$$
$$+\|\zeta^{2\kappa}+\zeta^{2\kappa-2}\|_\zeta^2). \quad (3.2.13)$$

Note that

$$(u, \tilde{v})_u = (v, \tilde{u})_v.$$

Hence, the third term on the left side of relation (3.2.11) can be presented as follows:

$$-4\tau\omega_z[(u^{2\kappa-1}+u^{2\kappa+1}, \tilde{v}^{2\kappa-1})_u - (v^{2\kappa-1}+v^{2\kappa+1}, \tilde{u}^{2\kappa-1})_v]$$
$$= 4\tau\omega_z(\tilde{u}^{2\kappa+1}, v^{2\kappa+1})_v - (\tilde{u}^{2\kappa-1}, v^{2\kappa-1})_v]$$
$$= 2\tau\omega_z(\|\tilde{u}^{2\kappa+1}+v^{2\kappa+1}\|_v^2 - \|\tilde{u}^{2\kappa-1}+v^{2\kappa-1}\|_v^2$$
$$-\|\tilde{u}^{2\kappa+1}\|_v^2 - \|v^{2\kappa+1}\|_v^2 + \|\tilde{u}^{2\kappa-1}\|_v^2 + \|v^{2\kappa-1}\|_v^2). \quad (3.2.14)$$

Substituting (3.2.13), (3.2.14) into relation (3.2.11) and neglecting positive terms affords the following energy estimate;

$$(\mathfrak{Q}^{2\kappa+1})^2 - (\mathfrak{Q}^{2\kappa-1})^2 \leq 2\tau \|\mathbf{f}^{2\kappa+1}\|_{S_h}(\|\mathbf{v}^{2\kappa+1}\|_{S_h} + \|\mathbf{v}^{2\kappa-1}\|_{S_h}), \quad (3.2.15)$$

where

$$(\mathfrak{Q}^{2\nu+1})^2 = \|\mathbf{v}^{2\kappa+1}\|_{S_h}^2 - 2\tau\omega_z(\|\tilde{u}^{2\kappa+1}\|_{S_h}^2 + \|v^{2\kappa+1}\|_{S_h}^2)$$
$$-gD\tau^2\|u_x^{2\kappa+1}+v_y^{2\kappa+1}\|_{S_h}^2 + \frac{g}{4D}\|\zeta^{2\kappa+2}+\zeta^{2\kappa}\|_{S_h}^2,$$
$$\kappa = 0, 1, \ldots.$$

Now take advantage of inequalities

$$\|\tilde{u}\|_{S_h}^2 \leq \|u\|_{S_h}^2$$

and obtain

$$\|u_x+v_y\|^2_{S_h} \leq 2(\|u_x\|^2_{S_h}+\|v_y\|^2_{S_h}) \leq \frac{2}{h^2}\|v\|^2_{S_h}.$$

In this at any κ for $\mathfrak{Q}^2$ the estimate

$$\mathfrak{Q}^2 \geq \left(1-2\tau\omega_z - \frac{2gD\tau^2}{h^2}\right)\|v\|^2_{S_h} \quad (3.2.16)$$

is valid.

Thus, $(\mathfrak{Q}^{2\kappa+1})^2$ and $(\mathfrak{Q}^{2\kappa-1})^2$ will be positive if

$$1-2\tau\omega_z - \frac{2gD\tau^2}{h^2} > 0. \quad (3.2.17)$$

Denote the left side of inequality (3.2.17) by $\mathfrak{h}$ and rewrite inequality (3.2.15) as follows:

$$\sqrt{\mathfrak{h}}(\mathfrak{Q}^{2\kappa+1}-\mathfrak{Q}^{2\kappa-1})(\mathfrak{Q}^{2\kappa+1}+\mathfrak{Q}^{2\kappa-1})$$
$$\leq 2\tau\|\mathbf{f}^{2\kappa+1}\|_{S_h}(\sqrt{\mathfrak{h}}\|\mathbf{v}^{2\kappa+1}\|_{S_h}+\sqrt{\mathfrak{h}}\|\mathbf{v}^{2\kappa-1}\|_{S_h}),$$

which in accordance with inequality (3.2.16) yields

$$\sqrt{\mathfrak{h}}(\mathfrak{Q}^{2\kappa+1}-\mathfrak{Q}^{2\kappa-1})(\mathfrak{Q}^{2\kappa+1}+\mathfrak{Q}^{2\kappa-1}) \leq 2\tau\|\mathbf{f}^{2\kappa+1}\|_{S_h}(\mathfrak{Q}^{2\kappa+1}+\mathfrak{Q}^{2\kappa-1}).$$

Thus, the final inequality can be presented as

$$\mathfrak{Q}^{2\kappa+1} \leq \mathfrak{Q}^{2\kappa-1} + C_1\tau\|\mathbf{f}^{2\kappa+1}\|_{S_h}, \quad (3.2.18)$$

Having been summed over j from 0 to κ it takes the form

$$\mathfrak{Q}^{2\kappa+1} \leq \mathfrak{Q}_0 + C_1\tau\sum_{j=0}^{\kappa}\|\mathbf{f}^{2j+1}\|_{S_h}, \quad (3.2.19)$$

where $\mathfrak{Q}_0$ is the value of the function $\mathfrak{Q}$ at the initial moment of time; C_1 is the κ-independent constant.

The latter estimate indicates the difference scheme will remain stable provided condition (3.2.17) is fulfilled.

3.3. Modified Variant of the HN-method

The modified HN-method consists of adding to the equation of motion (3.2.1) the term $\kappa_h \Delta \mathbf{v}$ describing macroturbulence effects. The increase by one order of the equation of motion makes it necessary to add to system (3.2.3),

(3.2.4) one more condition, i.e.

$$\frac{\partial v_\Gamma}{\partial n}\bigg|_\Gamma = 0, \qquad (3.3.1)$$

where v_Γ is a vertically averaged velocity component tangential to the contour Γ; ∂n is an element of the normal to the contour Γ.

Approximation of the dynamic equations within the modified variant of the HN-method is carried out in the same way as for equations (3.2.1), (3.2.2). Hence, the system of difference equations (3.2.6)—(3.2.8) remains the same but now the right sides of equations (3.2.6), (3.2.7) will have the additional terms

$$\frac{\kappa_h}{h^2} (u^{2\kappa-1}_{2m+3,\,2n} + u^{2\kappa-1}_{2m+1,\,2n+2} + u^{2\kappa-1}_{2m-1,\,2n} + u^{2\kappa-1}_{2m+1,\,2n-2} - 4u^{2\kappa-1}_{2m+1,\,2n});$$

$$\frac{\kappa_h}{h^2} (v^{2\kappa-1}_{2m+2,\,2n+1} + v^{2\kappa-1}_{2m,\,2n+3} + v^{2\kappa-1}_{2m-2,\,2n+1} + v^{2\kappa-1}_{2m,\,2n-1} - 4v^{2\kappa-1}_{2m,\,2n+1}),$$

respectively, and condition (3.3.1) will be replaced by its difference analogue $\mathfrak{L}v_\Gamma|_{\Gamma_h} = 0$, where $\mathfrak{L}$ is the difference approximation of the second order of precision of the differential operator $\partial/\partial n$.

A priori estimates

Let us now find *a priori* estimates for the modified variant of the HN-method, setting again, for the sake of simplicity that both the depth and the Coriolis parameter remain unchanged throughout the basin, and that its elevations are equal to zero on the open boundary.

Transformations analogous to those carried out in the preceding section result in the equality differing from relation (3.2.11) only in an additional term on the left side

$$\mathcal{J} = 2\tau\kappa_h[(u_x^{2\kappa-1} + u_x^{2\kappa+1}, u_x^{2\kappa-1})_u + (u_y^{2\kappa-1} + u_y^{2\kappa+1}, u_y^{2\kappa-1})_u$$
$$+ (v_x^{2\kappa-1} + v_x^{2\kappa+1}, v_x^{2\kappa-1})_v + (v_y^{2\kappa-1} + v_y^{2\kappa+1}, v_y^{2\kappa-1})_v].$$

Take into account the identity $a(a+b) = \frac{1}{2}[(a+b)^2 + a^2 - b^2]$ and transform the expression for $\mathcal{J}$ to the following form:

$$\mathcal{J} = \tau\kappa_h(\|u_x^{2\kappa-1} + u_x^{2\kappa+1}\|_u^2 + \|u_y^{2\kappa-1} + u_y^{2\kappa+1}\|_u^2 + \|v_x^{2\kappa-1} + v_x^{2\kappa+1}\|_v^2$$
$$+ \|v_y^{2\kappa-1} + v_y^{2\kappa+1}\|_v^2 + \|u_x^{2\kappa-1}\|_u^2 + \|u_y^{2\kappa-1}\|_u^2 + \|v_x^{2\kappa-1}\|_v^2$$
$$+ \|v_y^{2\kappa-1}\|_v^2 - \|u_x^{2\kappa+1}\|_u^2 - \|u_y^{2\kappa+1}\|_u^2 - \|v_x^{2\kappa+1}\|_v^2 - \|v_y^{2\kappa+1}\|_v^2). \quad (3.3.2)$$

Substituting formula (3.3.2) into relation (3.2.11) and neglecting the positive terms, we again obtain inequality (3.2.15) in a somewhat modified form

$$(\mathfrak{Q}^{2\kappa+1})^2 = \|\mathbf{v}^{2\kappa+1}\|_{S_h}^2 - 2\tau\omega_z(\|\tilde{u}^{2\kappa+1}\|_{S_h}^2 + \|v^{2\kappa+1}\|_{S_h}^2)$$

$$- gD\tau^2 \|u_x^{2\kappa+1} + v_y^{2\kappa+1}\|_{S_h}^2 + \frac{g}{4D}\|\zeta^{2\kappa+2} + \zeta^{2\kappa}\|_{S_h}^2$$

$$- \tau\kappa_h(\|u_x^{2\kappa+1}\|_{S_h}^2 + \|u_y^{2\kappa+1}\|_{S_h}^2 + \|v_x^{2\kappa+1}\|_{S_h}^2 + \|v_y^{2\kappa+1}\|_{S_h}^2).$$

With allowance made for the estimate

$$(\|u_x\|_{S_h}^2 + \|u_y\|_{S_h}^2 + \|v_x\|_{S_h}^2 + \|v_y\|_{S_h}^2) \leq \frac{2}{h^2}\|\mathbf{v}\|_{S_h}^2$$

we find that for any κ the inequality

$$\mathfrak{Q}^2 \geq \left(1 - 2\tau\omega_z - \frac{2gD\tau^2}{h^2} - \frac{2\tau\kappa_h}{h^2}\right)\|\mathbf{v}\|_{S_h}^2 \qquad (3.3.3)$$

is valid.

It can be easily proved that in this case estimate (3.2.19) holds. Thus, the difference scheme of the modified variant of the HN-method will remain stable, provided

$$1 - 2\tau\omega_z - \frac{2gD\tau^2}{h^2} - \frac{2\tau\kappa_h}{h^2} > 0. \qquad (3.3.4)$$

Convergence

Let us now prove convergence of the solution for the difference system (3.2.6)—(3.2.9). If u', v', ζ' denote the precise solution of the initial problem, u_h, v_h, ζ_h denote its grid solution, then for errors of the corresponding tidal characteristic calculations $e_u = u_h - u'$, $e_v = v_h - v'$, $e_\zeta = \zeta_h - \zeta'$ the following system of difference equations will be valid:

$$\frac{e_u^{2\kappa+1} - e_u^{2\kappa+1}}{2\tau} - 2\omega_z \tilde{e}_v^{2\kappa-1} + \frac{r_u^{2\kappa-1}}{2}(e_u^{2\kappa-1} + e_u^{2\kappa+1})$$

$$= -g(e_\zeta^{2\kappa})_x + \frac{1}{2}(r_u^{2\kappa-1} - r_u'^{2\kappa-1})(u'^{2\kappa-1} + u'^{2\kappa+1}) + \mathfrak{G}_u^{2\kappa+1}; \qquad (3.3.5)$$

$$\frac{e_v^{2\kappa+1} - e_v^{2\kappa-1}}{2\tau} + 2\omega_z \tilde{e}_u^{2\kappa-1} + \frac{r_v^{2\kappa-1}}{2}(e_v^{2\kappa-1} + e_v^{2\kappa+1})$$

$$= -g(e_\zeta^{2\kappa})_y + \frac{1}{2}(r_v^{2\kappa+1} - r_v'^{2\kappa-1})(v'^{2\kappa-1} + v'^{2\kappa+1}) + \mathfrak{G}_v^{2\kappa+1}; \qquad (3.3.6)$$

$$\frac{e_\zeta^{2\kappa+2} - e_\zeta^{2\kappa}}{2\tau} + (De_u^{2\kappa+1})_x + (De_v^{2\kappa+1})_y = \mathfrak{G}_\zeta^{2\kappa+1}. \qquad (3.3.7)$$

Here

$$(e_\zeta)_x = \frac{1}{2h}[(e_\zeta)_{2m+2,\,2n}-(e_\zeta)_{2m,\,2n}]; \qquad (e_\zeta)_y = \frac{1}{2h}[(e_\zeta)_{2m,\,2n+2}-(e_\zeta)_{2m,\,2n}];$$

$$\mathfrak{C} = (\mathfrak{C}_u, \mathfrak{C}_v, \mathfrak{C}_\zeta)$$

is the error of approximation determined by the formula

$$\mathfrak{C} = \mathfrak{L}(u', v', \zeta') - \mathfrak{L}_{h\tau}(u', v', \zeta'), \tag{3.3.8}$$

where $\mathfrak{L}_{h\tau}$ is the difference approximation of the initial equations of the problem.

Let u', v', ζ' be the functions, sufficiently differentiated. Then for $\|\mathfrak{C}^{2\kappa+1}\|_{S_h}$ the estimate

$$\|\mathfrak{C}^{2\kappa+1}\|_{S_h} \leq C_2(\tau+h), \tag{3.3.9}$$

will be valid, where C_2 is the κ-independent constant.

Indeed, presenting all the difference functions contained in formula (3.3.8) as a Taylor series, we have

$$\frac{u'^{2\kappa+1}-u'^{2\kappa-1}}{2\tau} = \left(\frac{\partial u'}{\partial t}\right)^{2\kappa-1} + \tau\frac{\partial^2 u'}{\partial t^2} + O(\tau^2);$$

$$\tilde{u}'^{2\kappa-1} = u'^{2\kappa-1} + \frac{h^2}{2}\Delta u'^{2\kappa-1} + O(h^4);$$

$$\sqrt{u'^2+\tilde{v}'^2}u' = \sqrt{u'^2+v'^2}u' + \frac{h^2}{2}u'\sqrt{u'^2+v'^2}v'(\Delta v')^2 + O(h^4).$$

etc.

These expansions hold only for internal grid points. In fact due to the transition of the boundary conditions from the real contour to the grid boundary the error of approximation is of order h. Hence for $\mathfrak{C}_u, \mathfrak{C}_v, \mathfrak{C}_\zeta$, respectively, in each of u-, v-, ζ-points the estimate

$$\{|\mathfrak{C}_u|, \ |\mathfrak{C}_v|, \ |\mathfrak{C}_\zeta|\} \leq C_3(\tau+h), \tag{3.3.10}$$

is valid, where C_3 is the constant independent of the choice of points.

Summing each of the inequalities (3.3.10) over its domain S_i and introducing the notation

$$\|\mathfrak{C}\|_{S_h}^2 = \|\mathfrak{C}_u\|_{S_h}^2 + \|\mathfrak{C}_v\|_{S_h}^2 + \|\mathfrak{C}_\zeta\|_{S_h}^2,$$

gives inequality (3.3.9), which means that the difference scheme is in good agreement with the initial system of differential equations and is of the first order of precision with respect to time and horizontal coordinates.

Multiply now equality (3.3.5) by $2\tau h^2(e_u^{2\kappa-1}+e_u^{2\kappa+1})$, equality (3.3.6) by $2\tau h^2(e_v^{2\kappa-1}+e_v^{2\kappa+1})$ with the following summation of the resulting expressions over corresponding domains and over all time layers $t = (2\kappa+1)\tau$, $\kappa = 0, 1, 2, \ldots$. Then, using condition (3.2.17) and the estimate

$$\left|\sum_{S_u} (r_u^{2\kappa-1} - r_u'^{2\kappa-1})(u'^{2\kappa-1}+u'^{2\kappa+1})(e_u^{2\kappa-1}+e^{\kappa+1}) \right.$$
$$\left. + \sum_{S_v}(r_v^{2\kappa-1}-r_v'^{2\kappa-1})(v'^{2\kappa-1}+v'^{2\kappa+1})(e_v^{2\kappa-1}+e_v^{2\kappa+1})\right|$$
$$\leq C_4(\|e_u^{2\kappa-1}\|_{S_h}^2 + \|e_u^{2\kappa+1}\|_{S_h}^2 + \|e_v^{2\kappa-1}\|_{S_h}^2 + \|e_v^{2\kappa+1}\|_{S_h}^2),$$

and after a number of transformations analogous to those carried out while proving the stability, we obtain the inequality

$$\|\mathbf{e}^{2\kappa+1}\|_{S_h}^2 \leq C_5\tau \sum_{j=0}^{\kappa} \|\mathbf{e}^{2j+1}\|_{S_h}^2 + C_6\tau \sum_{j=0}^{\kappa} \|\mathbf{C}^{2j+1}\|_{S_h}^2.$$

Here

$$\|\mathbf{e}^{2\kappa+1}\|_{S_h}^2 = \|e_u^{2\kappa+1}\|_{S_h}^2 + \|e_v^{2\kappa+1}\|_{S_h}^2 + \|e_\zeta^{2\kappa}\|_{S_h}^2.$$

The latter inequality yields the estimate

$$\|\mathbf{e}^{2\kappa+1}\|_{S_h}^2 \leq C_7 \mathfrak{q}, \qquad (3.3.11)$$

where

$$\mathfrak{q} = C_6\tau \sum_{j=0}^{\kappa} \|\mathbf{C}^{2j+1}\|_{S_h}^2.$$

However, since at $\tau, h \to 0$ the quantity q also tends to zero [see inequality (3.3.9)], on the basis of inequality (3.3.11) we have $\|\mathbf{e}^{2\kappa+1}\|_{S_h} \to 0$ at $\tau, h \to 0$. Thus, the convergence of the HN-method difference scheme is proven.

Remark 1. Condition (3.3.1) and relation (3.3.2) demonstrate that liquid is slipping along the basin boundary. Obviously, such a motion can occur only in the case of a non-viscous liquid. In the present case, when viscosity is taken into account by introducing the term $\kappa_h \Delta \mathbf{v}$ into the equation of motion, it would be natural to require the no-slip condition to be fulfilled on one of the parts of the contour Γ (namely, on its solid part Γ_1). The no-slip condition results in the appearance in the coastal zones of the basin of boundary layers in which the velocity gradient in the direction perpendicular to the coast not only does not vanish, as required by condition (3.3.1), but conversely reaches its maximum value.

Replacing the slip condition by one of non-slip must bring about energy redistribution within the limits of the basin, which inevitably influences the regime of tidal motions throughout the basin, especially in its coastal zones. However, condition (3.3.1) cannot be replaced an exact no-slip condition because of peculiarities of the grid structure used by the HN-method, which allows one to calculate only one velocity component at each point.

Remark 2. In presenting the difference schemes of the HN-method, we have somewhat improved them, having approximated the terms describing bottom friction effects, for instance, by $\frac{1}{2} r_u^{2\kappa-1}(u_{2m+1,2n}^{2\kappa-1} + u_{m+1,2n}^{\kappa+2})$ instead of $r_u^{2\kappa-1} u_{2m+1,2n}^{2\kappa-1}$ in equation (3.2.6) and by $\frac{1}{2} r_v^{2\kappa-1}(v_{2m,2n+1}^{2\kappa-1} + v_{2m,2n+1}^{2\kappa+1})$ instead of $r_v^{2\kappa-1} v_{2m,2n+1}^{2\kappa-1}$ in equation (3.2.7). Besides, approximation of the terms describing the Coriolis force effect is also somewhat different from the initial variant. Remember that it is common practice to ascribe these terms to the layer $2\kappa-1$, while in our case the values of $\tilde{v}$ in equation (3.2.6) and the values of $\tilde{u}$ in equation (3.2.7) are ascribed to the $(2\kappa-1)$-th and to the $(2\kappa+1)$-th time layers, respectively.

The first of the above peculiarities results in the fact that the condition of stability (3.2.17), unlike analogous conditions

$$\tau \leq \frac{r}{r^2 + 4\omega_z^2}, \quad \sqrt{2gD}\, \frac{\tau}{h} \leq 1, \tag{3.3.12}$$

obtained by the spectral method, appeared to be independent of the value of the bottom friction coefficient r. Incidentally, it is due to this fact that condition (3.2.17) is weaker than inequalities (3.3.12).

Remark 3. In common tidal calculations besides the difference schemes of the HN-method, some other explicit schemes of integration are employed. The simplest of these is the central-difference scheme in which the velocity components and the level are determined at all points of the spatial grid

$$\frac{u^{\kappa+1} - u^{\kappa-1}}{\tau} - 2\omega_z v^{\kappa-1} = -g(\zeta_x^{\kappa-1/2} + \zeta_{\bar{x}}^{\kappa-1/2}) - r^{\kappa-1} u^{\kappa-1} + f^{\kappa+1}; \tag{3.3.13}$$

$$\frac{v^{\kappa+1} - v^{\kappa-1}}{\tau} + 2\omega_z u^{\kappa-1} = -g(\zeta_y^{\kappa-1/2} + \zeta_{\bar{y}}^{\kappa-1/2}) - r^{\kappa-1} v^{\kappa-1} + f_2^{\kappa+1}; \tag{3.3.14}$$

$$\frac{\zeta^{\kappa-1/2} - \zeta^{\kappa-3/2}}{\tau} + [(uD)_x^{\kappa-1} + (uD)_{\bar{x}}^{\kappa-1} + (vD)_y^{\kappa-1} + (vD)_{\bar{y}}^{\kappa-1}] = 0; \tag{3.3.15}$$

$$u|_{\Gamma_{1h}^{\kappa}} = 0, \quad v|_{\Gamma_{1h}^{y}} = 0, \quad \zeta|_{\Gamma_{2h}} = \zeta_{\Gamma_h}(t); \tag{3.3.16}$$

$$u|_{t=0} = u_0, \quad v|_{t=0} = v_0, \quad \zeta|_{t=0} = \zeta_0. \tag{3.3.17}$$

Unfortunately, while using the above scheme one has to set an additional boundary condition on the basin contour. Indeed, in order to determine u and v at points in the vicinity of the boundary it is necessary to know the level and the tangential velocity component at the boundary itself. It can be achieved in two ways: (1) by setting the elevation value on the contour, which is expedient in many respects, and (2) by determining the value of the tangential velocity component on the contour by the equations of motion. For instance, the tangential velocity component v for the meridional part of the boundary can be found by the relations

$$-2\omega_z v^{\kappa-1} = -g(\zeta_x^{\kappa-1/2}+\zeta_{\bar{x}}^{\kappa-1/2})+f_1^{\kappa+1}; \qquad (3.3.18)$$

$$\frac{v^{\kappa-1}-v^{\kappa-2}}{\tau} = -g(\zeta_y^{\kappa-3/2}+\zeta_{\bar{y}}^{\kappa-3/2})+f_2^{\kappa-1}. \qquad (3.3.19)$$

However, since the system (3.3.18), (3.3.19) is overdetermined with respect to $v^{\kappa-1}$, we have to deal here with the consequences of ambiguity in the tangential velocity components on the boundary.

The same undesirable properties are typical of the other explicit schemes using, for instance, grids of the chess type in which both velocity components and the level are calculated at the points spaced at a half-step distance.

The only way to overcome the above difficulties is apparently to retain viscous terms in the equations of motion.

3.4. The Method of Fractional Steps

This method is based on equations (2.1.1)—(2.1.6) presented in terms of total transports. They have been studied in detail in Chapter 2. It has been shown, in particular, that these equations are uniquely solved within the finite interval of time variation. Some *a priori* estimates of the precise solution of problem (2.1.1)—(2.1.6) or, equivalently, of problem (2.2.1)—(2.2.3), have been obtained. Later they were used as the basis for the proof of the existence-theorem. And finally, we elucidated the fact that the initial equations have at least one periodic solution for any value of the bottom-friction coefficient.

Difference scheme

When seeking the numerical solution of system (2.2.1)—(2.2.3) let us use the following difference scheme. First, divide our time interval into subintervals τ. Denote the number of a step by κ and introduce an auxiliary step with a frictional index $\kappa+\frac{1}{2}$. Then, splitting the equations of motion, taking into consideration the possibility of further employment of the implicit scheme of variable directions, and approximating derivatives by differences, we get

$$\frac{u^{\kappa+1/2}-u^\kappa}{\tau}+\frac{r_1^\kappa}{2}u^{\kappa+1/2}=-\frac{1}{2}gD\frac{\partial\zeta^\kappa}{\partial x}+\omega_z v^\kappa-\frac{r_1^\kappa}{2}(u^0)^{\kappa+1/2}$$

$$+\frac{K_h}{2}(u_{x\bar{x}}^\kappa+u_{y\bar{y}}^{\kappa+1/2})+\frac{f_1^{\kappa+1/2}}{2}\ ; \qquad (3.4.1)$$

$$\frac{v^{\kappa+1/2}-v^\kappa}{\tau}+\frac{r_1^\kappa}{2}v^{\kappa+1/2}=-\frac{1}{2}gD\frac{\partial\zeta^\kappa}{\partial y}-\omega_z u^{\kappa+1/2}-\frac{r_1^\kappa}{2}(v^0)^{\kappa+1/2}$$

$$+\frac{K_h}{2}(v_{x\bar{x}}^{\kappa+1/2}+v_{y\bar{y}}^\kappa)+\frac{f_2^{\kappa+1/2}}{2}\ ; \qquad (3.4.2)$$

$$\frac{v^{\kappa+1}-v^{\kappa+1/2}}{\tau}+\frac{r_1^{\kappa+1/2}}{2}v^{\kappa+1}=-\frac{1}{2}gD\frac{\partial\zeta^{\kappa+1}}{\partial y}-\omega_z u^{\kappa+1/2}$$

$$-\frac{r_1^{\kappa+1/2}}{2}(v^0)^{\kappa+1}+\frac{K_h}{2}(v_{x\bar{x}}^{\kappa+1/2}+v_{y\bar{y}}^{\kappa+1})+\frac{f_2^{\kappa+1}}{2}\ ; \qquad (3.4.3)$$

$$\frac{u^{\kappa+1}-u^{\kappa+1/2}}{\tau}+\frac{r_1^{\kappa+1/2}}{2}u^{\kappa+1}=-\frac{1}{2}gD\frac{\partial\zeta^{\kappa+1}}{\partial x}+\omega_z v^{\kappa+1}$$

$$-\frac{r_1^{\kappa+1/2}}{2}(u^0)^{\kappa+1}+\frac{K_h}{2}(u_{x\bar{x}}^{\kappa+1}+u_{y\bar{y}}^{\kappa+1/2})+\frac{f_1^{\kappa+1}}{2}. \qquad (3.4.4)$$

To determine horizontal level gradients at the interval $\kappa+1$ let us employ the continuity equation (2.2.2) differentiated with respect to x and y. Present the operator $\partial/\partial t$ as its difference analogue and correlate the derivatives of the total transport divergence to the intermediate layer $\kappa+\frac{1}{2}$. Then, substituting $\partial^2 u^{\kappa+1/2}/\partial x^2$, $\partial^2 u^{\kappa+1/2}/\partial y\,\partial x$, $\partial^2 v^{\kappa+1/2}/\partial x\,\partial y$ and $\partial^2 v^{\kappa+1/2}/\partial y^2$ by

$$\tfrac{1}{2}(u_{x\bar{x}}^{\kappa+1}+u_{x\bar{x}}^{\kappa+1/2}),\quad \tfrac{1}{2}(u_{y\bar{x}}^{\kappa+1/2}+u_{y\bar{x}}^\kappa);$$

$$\tfrac{1}{2}(v_{x\bar{y}}^{\kappa+1/2}+v_{x\bar{y}}^\kappa),\quad \tfrac{1}{2}(v_{y\bar{y}}^{\kappa+1}+v_{y\bar{y}}^{\kappa+1/2}),$$

respectively, we have

$$\frac{\partial\zeta^{\kappa+1}}{\partial x}=\frac{\partial\zeta^\kappa}{\partial x}-\frac{\tau}{2}(u_{x\bar{x}}^{\kappa+1}+u_{x\bar{x}}^{\kappa+1/2}+v_{y\bar{x}}^{\kappa+1/2}+v_{y\bar{x}}^\kappa); \qquad (3.4.5)$$

$$\frac{\partial\zeta^{\kappa+1}}{\partial y}=\frac{\partial\zeta^\kappa}{\partial y}-\frac{\tau}{2}(u_{x\bar{y}}^{\kappa+1/2}+u_{x\bar{y}}^\kappa+v_{y\bar{y}}^{\kappa+1}+v_{y\bar{y}}^{\kappa+1/2}). \qquad (3.4.6)$$

Here and in all the above expressions for any of the complete flow components u, v and coordinates x, y

$$u_x(x, y, t) = \frac{1}{h}[u(x+h, y, t) - u(x, y, t)];$$

$$u_{\bar{x}}(x, y, t) = \frac{1}{h}[u(x, y, t) - u(x-h, y, t)];$$

$$u_{x\bar{x}} = (u_x)_{\bar{x}};$$

$$r_1 = \frac{t}{D^2}\sqrt{(u+u^0)^2 + (v+v^0)^2};$$

u^0, v^0 are components of the vector $\mathbf{w}^0$; f_1 and f_2 are the components of the free term $\mathbf{f}$, which, as has been pointed out in Chapter 2, comprises not only the tide-generating force but also the values of the functions u^0, v^0 and their difference ratios with respect to x, y, t. The latter are limited with respect to the modulus, because u^0 and v^0 are considered to be sufficiently smooth functions.

The boundary conditions u, $v|_{\Gamma_h} = 0$ and arbitrary initial conditions for u, v, $\partial \zeta/\partial x$, $\partial \zeta/\partial y$ are added to the system (3.4.1)—(3.4.6).

Let S_h be, as before, the set of the grid points belonging to the domain S and let Γ_h be its boundary. Introduce the following notations for the total transport vector $\mathbf{w}_h = (u_h, v_h)$:

$$w_h^2 = u_h^2 + v_h^2;$$

$$w_{hx}^2 = u_{hx}^2 + v_{hx}^2;$$

$$w_{h\bar{x}}^2 = u_{h\bar{x}}^2 + v_{h\bar{x}}^2,$$

and notations for the grid norm and grid scalar product

$$\|\mathbf{w}_h^\kappa\|^2 = h^2 \sum_{S_h} (\mathbf{w}_h^\kappa)^2;$$

$$\|\mathbf{w}_{hx}^\kappa\|^2 = h^2 \sum_{S_h} (\mathbf{w}_{hx}^\kappa)^2;$$

$$(\mathbf{f}_h^\kappa, \mathbf{w}_h^\kappa) = h^2 \sum_{S_h} \mathbf{f}_h^\kappa \mathbf{w}_h^\kappa,$$

where $\bar{S}_h = S_h \cup \Gamma_h$, the indices h and κ implying that the discussed function is considered at the grid points at the layer $t = \kappa\tau$, $\kappa = 0, 1/2, 1, \ldots$.

Later on we shall frequently use the well-known relations

$$2\tau u_{\bar{t}}^\kappa u^\kappa = (u^\kappa)^2 - (u^{\kappa-1})^2 + \tau^2(u_{\bar{t}}^\kappa)^2, \tag{3.4.7}$$

where
$$u_{\bar{t}}^\kappa = \frac{1}{\tau}(u^\kappa - u^{\kappa-1/2}),$$

and
$$h^2 \sum_{S_h} u_{hx} v_h = -h^2 \sum_{S_h} u_h v_{h\bar{x}}, \tag{3.4.8}$$

the latter being valid for arbitrary functions u_h, v_h set on the grid, provided $\mathbf{w}_h|_{\Gamma_h} = 0$.

To prove unique solvability for the system (3.4.1)—(3.4.6) it is sufficient to show that its corresponding homogeneous system has but zero solution at every time layer.

At the layers $t = (\kappa + 1/2)\tau$ and $t = (\kappa + 1)\tau$ homogeneous systems can be written as follows:

$$u^{\kappa+1/2} + \frac{\tau}{2} r_1^\kappa u^{\kappa+1/2} = \frac{\tau K_h}{2} u_{y\bar{y}}^{\kappa+1/2}; \tag{3.4.9}$$

$$v^{\kappa+1/2} + \frac{\tau}{2} r_1^\kappa v^{\kappa+1/2} = \frac{\tau K_h}{2} v_{x\bar{x}}^{\kappa+1/2} - \tau \omega_z u^{\kappa+1/2} \tag{3.4.10}$$

and

$$v^{\kappa+1} + \frac{\tau}{2} r_1^{\kappa+1/2} v^{\kappa+1} = \left(\frac{gD\tau^2}{4} + \frac{\tau K_h}{2}\right) v_{y\bar{y}}^{\kappa+1}; \tag{3.4.11}$$

$$u^{\kappa+1} + \frac{\tau}{2} r^{\kappa+1/2} u^{\kappa+1} = \left(\frac{gD\tau^2}{4} + \frac{\tau K_h}{2}\right) u_{x\bar{x}}^{\kappa+1} + \tau \omega_z v^{\kappa+1}. \tag{3.4.12}$$

Multiply equations (3.4.9) and (3.4.10) by $h^2 u^{\kappa+1/2}$ and $h^2 v^{\kappa+1/2}$, respectively. Sum the resulting expressions over all grid points and employ relation (3.4.8). Then, with allowance made for the assumption that the Coriolis parameter is constant, obtain throughout the domain

$$\|u^{\kappa+1/2}\|^2 + \frac{\tau h^2}{2} \sum_{S_h} r_1^\kappa (u^{\kappa+1/2})^2 + \frac{\tau K_h}{2} \|u_y^{\kappa+1/2}\|^2 = 0;$$

$$\|v^{\kappa+1/2}\|^2 + \frac{\tau h^2}{2} \sum_{S_h} r_1^\kappa (v^{\kappa+1/2})^2 + \frac{\tau K_h}{2} \|v^{\kappa+1/2}\|^2$$

$$= -\tau \omega_z (u^{\kappa+1/2}, v^{\kappa+1/2}),$$

which yields $u^{\kappa+1/2}$, $v^{\kappa+1/2} \equiv 0$ throughout the domain $\bar{S}_h$.

At the layer $t = (\kappa+1)\tau$ as a result of similar transformations we have

$$\|v^{\kappa+1}\|^2 + \frac{\tau h^2}{2}\sum_{S_h} r_1^{\kappa+1/2}(v^{\kappa+1})^2 + \frac{\tau K_h}{2}\|v_y^{\kappa+1}\|^2$$

$$+ \frac{g\tau^2}{4}\sum_{S_h} D(v_y^{\kappa+1})^2 = -\frac{g\tau^2}{4}\sum_{S_h} v_y^{\kappa+1}v^{\kappa+1}D_y; \quad (3.4.13)$$

$$\|u^{\kappa+1}\|^2 + \frac{\tau h^2}{2}\sum_{S_h} r_1^{\kappa+1/2}(u^{\kappa+1})^2 + \frac{\tau K_h}{2}\|u_x^{\kappa+1}\|^2 + \frac{g\tau^2}{4}\sum_{S_h} D(u_x^{\kappa+1})^2$$

$$- \tau \omega_z(v^{\kappa+1}, u^{\kappa+1}) = -\frac{g\tau^2}{2}\sum_{S_h} u_x^{\kappa+1}u^{\kappa+1}D_x. \quad (3.4.14)$$

Relations (3.4.13), (3.4.14) yield the following inequalities:

$$\|v^{\kappa+1}\|^2 + \frac{\tau K_h}{2}\|v^{\kappa+1}\|^2 \leq \frac{g\tau^2 M_h^2}{16\mu h}\|v^{\kappa+1}\|^2; \quad (3.4.15)$$

$$\|u^{\kappa+1}\|^2 + \frac{\tau K_h}{2}\|u_x^{\kappa+1}\|^2 \leq \frac{g\tau^2 M_h^2}{16\mu h}\|u^{\kappa+1}\|^2$$

$$+ \tau \omega_z \|v^{\kappa+1}\| \cdot \|u^{\kappa+1}\|, \quad (3.4.16)$$

where $M_h = \max\{D_{hx}, D_{hy}\}, \mu_h = \min D_h$.

At sufficiently small τ (for instance, at $\tau < 4\mu_h^{1/2}/M_h g^{1/2}$) inequalities (3.4.15), (3.4.16) afford $\|u^{\kappa+1}\| = \|v^{\kappa+1}\| = 0$. In this case through the use of equations (3.4.5), (3.4.6) we get that $\partial \zeta^{\kappa+1}/\partial x$, $\partial \zeta^{\kappa+1}/\partial y$ are also equal to zero throughout the domain $\overline{S}_h$.

Thus, we have proved that the homogeneous system has only a trivial solution, and hence, the inhomogeneous system (3.4.1)—(3.4.6) is uniquely solvable for any right-hand side.

Remark. In the case of constant depth D when $M_h = 0$ system (3.4.1)—(3.4.6) is uniquely solvable for any τ.

Stability

The subject of the present section is the stability of the difference scheme (3.4.1)—(3.4.6). To simplify the case, set initially the total transport components u, v and the horizontal level gradients in the direction of the axes x, y equal to zero everywhere.

Multiply equation (3.4.1) by $2\tau h^2 u^{\kappa+1/2}$, equation (3.4.2) by $2\tau h^2 v^{\kappa+1/2}$, equation (3.4.3) by $2\tau h^2 v^{\kappa+1}$ and equation (3.4.4) by $2\tau h^2 u^{\kappa+1}$. Sum the resulting expressions over all $\overline{S}_h$ points and add expressions 1 and 2 and expressions 3 and 4. Employing formulae (3.4.7), (3.4.8) and an easily verifiable relation

$$\tau K_h(u_x^\kappa, u_x^{\kappa+1/2}) = \tau K_h \|u_x^{\kappa+1/2}\|^2 - \tau^2 K_h(u_{x\bar{t}}^{\kappa+1/2}, u_x^{\kappa+1/2}),$$

$$\kappa = 0, 1/2, 1, \ldots, \qquad (3.4.17)$$

we get

$$\|\mathbf{w}^{\kappa+1/2}\|^2 - \|\mathbf{w}^\kappa\|^2 + \tau^2\|\mathbf{w}_{\bar{t}}^{\kappa+1/2}\|^2 + \mathfrak{U}^{\kappa+1/2} + \tau K_h(\|\mathbf{w}_x^{\kappa+1/2}\|^2$$
$$+ \|\mathbf{w}_y^{\kappa+1/2}\|^2) = -g\tau\left[\left(D\frac{\partial \zeta^\kappa}{\partial x}, u^{\kappa+1/2}\right)\right.$$
$$+ \left.\left(D\frac{\partial \zeta^\kappa}{\partial y}, v^{\kappa+1/2}\right)\right] + \tau^2 K_h[(u_{y\bar{t}}^{\kappa+1/2}, u_y^{\kappa+1/2})$$
$$+ (v_{x\bar{t}}^{\kappa+1/2}, v_x^{\kappa+1/2})] + \mathfrak{U}_1^{\kappa+1/2}; \qquad (3.4.18)$$

$$\|\mathbf{w}^{\kappa+1}\|^2 - \|\mathbf{w}^{\kappa+1/2}\|^2 + \tau^2\|\mathbf{w}_{\bar{t}}^{\kappa+1}\|^2 + \mathfrak{U}^{\kappa+1} + \tau K_h(\|\mathbf{w}_x^{\kappa+1}\|^2$$
$$+ \mathbf{w}_y^{\kappa+1}\|^2) = -g\tau\left[\left(D\frac{\partial \zeta^{\kappa+1}}{\partial x}, u^{\kappa+1}\right)\right.$$
$$+ \left.\left(D\frac{\partial \zeta^{\kappa+1}}{\partial y}, v^{\kappa+1}\right)\right] + \tau^2 K_h[(u_{x\bar{t}}^{\kappa+1}, u_x^{\kappa+1})$$
$$+ (v_{y\bar{t}}^{\kappa+1}, v_y^{\kappa+1})] + \mathfrak{U}_1^{\kappa+1}. \qquad (3.4.19)$$

Here

$$\mathfrak{U}^{\kappa+1/2} = \tau h^2 \sum_{\overline{S}_h} r_1^\kappa [(u^{\kappa+1/2})^2 + (v^{\kappa+1/2})^2];$$

$$\mathfrak{U}_1^{\kappa+1/2} = \tau[(f_1^{\kappa+1/2}, u^{\kappa+1/2}) + (f_2^{\kappa+1/2}, v^{\kappa+1/2})]$$
$$- 2\omega_z\tau[(u^{\kappa+1/2}, v^{\kappa+1/2}) - (u^{\kappa+1/2}, v^\kappa)]$$
$$- \tau[(r_1^\kappa(u^0)^{\kappa+1/2}, u^{\kappa+1/2}) + (r_1^\kappa(v^0)^{\kappa+1/2}, v^{\kappa+1/2})],$$

$$\kappa = 0, 1/2, 1, \ldots.$$

Estimate the second term on the right-hand side of equation (3.4.17) and the last terms in the right-hand side of equations (3.4.18), (3.4.19) in the following way:

$$\tau^2 \kappa_h |(u_{x\bar{t}}^{\kappa+1/2}, u_x^{\kappa+1/2})| \leq \frac{2\tau\kappa_h}{h^2}\left(||u_x^{\kappa+1/2}||^2 + \frac{\tau^2}{2}||u_{\bar{t}}^{\kappa+1/2}||^2\right); \quad (3.4.20)$$

$$|\mathfrak{U}_1^{\kappa+1/2}| \leq \tau \left\{\frac{\varepsilon_1^2}{2}||\mathbf{w}^{\kappa+1/2}||^2 + \frac{1}{2\varepsilon_1^2}||\mathbf{f}^{\kappa+1/2}||^2\right.$$

$$+ |2\omega_z|(||\mathbf{w}^{\kappa+1/2}||^2 + ||\mathbf{w}^{\kappa}||^2)$$

$$+ \tau \mu_h^{-2}\left[\max_{\bar{S}_h}|(\mathbf{w}^0)^{\kappa+1/2}| \cdot \left(||\mathbf{w}^{\kappa+1/2}||^2 + ||\mathbf{w}^{\kappa}||^2\right.\right.$$

$$\left.\left.\left. + \frac{\varepsilon_2^2}{2}||\mathbf{w}^{\kappa+1/2}||^2 + \frac{1}{2\varepsilon_2^2}||(\mathbf{w}^0)^{\kappa+1/2}||^2\right)\right]\right\}$$

$$\equiv C_1\tau(||\mathbf{w}^{\kappa+1/2}||^2 + ||\mathbf{w}^{\kappa}||^2) + C_2, \quad (3.4.21)$$

where ε_1 and ε_2 are any positive numbers; C_1 and C_2 are constants depending on the choice of ε_1 and ε_2 and the initial data.

Adding equalities (3.4.18), (3.4.19) affords the relation containing in its right-hand side the expression of type

$$\mathfrak{T} = -g\tau\left[\left(D\frac{\partial \zeta^{\kappa}}{\partial x}, u^{\kappa+1/2}\right) + \left(D\frac{\partial \zeta^{\kappa}}{\partial y}, v^{\kappa+1/2}\right)\right.$$

$$\left. + \left(D\frac{\partial \zeta^{\kappa+1}}{\partial x}, u^{\kappa+1}\right) + \left(D\frac{\partial \zeta^{\kappa+1}}{\partial y}, v^{\kappa+1}\right)\right].$$

Let us extend the expression for $\mathfrak{T}$. In line with relations (3.4.5), (3.4.6) we have

$$\frac{\partial \zeta^{\kappa+1}}{\partial x} = -\frac{\tau}{2}\sum_{j=0}^{\kappa}(u_{x\bar{x}}^{j+1} + u_{x\bar{x}}^{j+1/2} + v_{y\bar{x}}^{j+1/2} + v_{y\bar{x}}^{j}): \quad (3.4.22)$$

$$\frac{\partial \zeta^{\kappa+1}}{\partial y} = -\frac{\tau}{2}\sum_{j=0}^{\kappa}(u_{x\bar{y}}^{j+1/2} + u_{x\bar{y}}^{j} + v_{y\bar{y}}^{j+1} + v_{y\bar{y}}^{j+1/2}). \quad (3.4.23)$$

Then introduce the following definitions:

$$\mathfrak{d} = u_x^{\kappa} + v_y^{\kappa};$$

$$\mathfrak{d}^{\kappa+1/2} = u_x^{\kappa+1/2} + v_y^{\kappa+1/2};$$

$$\mathfrak{D}^{\kappa} = \sum_{j=0}^{\kappa}\mathfrak{d}^j + \sum_{j=0}^{\kappa-1}\mathfrak{d}^{j+1/2};$$

$$\mathfrak{D}^{\kappa+1/2} = \sum_{j=0}^{\kappa}\mathfrak{d}^j + \sum_{j=0}^{\kappa}\mathfrak{d}^{j+1/2}$$

and indicate any vector $\{\mathbf{a}_h \sqrt{D_h}\}$ by symbol D. Using relations (3.4.22), (3.4.23) and carrying out summation by parts gives

$$\mathfrak{T} = -\frac{g\tau^2}{2}[(\mathfrak{D}_D^{\kappa+1}, \mathfrak{d}_D^{\kappa+1} + \mathfrak{d}_D^{\kappa+1/2}) + \mathfrak{D}_1 + \mathfrak{D}_2], \qquad (3.4.24)$$

where

$$\mathfrak{D}_1 = -(\mathfrak{d}_D^{\kappa+1}, \mathfrak{d}_D^{\kappa+1/2}) - \|\mathfrak{d}_D^{\kappa+1/2}\|^2 - (v_{yD}^{\kappa+1}, u_{xD}^{\kappa+1/2})$$
$$\quad - (u_{xD}^{\kappa+1}, v_{yD}^{\kappa+1/2}) - 2(v_{yD}^{\kappa+1}, u_{xD}^{\kappa+1});$$
$$\mathfrak{D}_2 = (\mathfrak{D}^{\kappa+1}, uD_x^{\kappa+1/2}) + (\mathfrak{D}^{\kappa+1}, uD_x^{\kappa+1}) + (\mathfrak{D}^{\kappa+1}, vD_y^{\kappa+1/2})$$
$$\quad + (\mathfrak{D}^{\kappa+1}, vD_y^{\kappa+1}) - (\mathfrak{d}^{\kappa+1} + v_y^{\kappa} + \mathfrak{d}^{\kappa+1/2}, uD_x^{\kappa+1/2})$$
$$\quad - (\mathfrak{d}^{\kappa+1} + u_x^{\kappa} + \mathfrak{d}^{\kappa+1/2}, vD_y^{\kappa+1/2}) - (v_y^{\kappa+1}, uD_x^{\kappa+1})$$
$$\quad - (u_x^{\kappa+1}, vD_y^{\kappa+1}).$$

Rewrite the first two brackets in formula (3.4.24) in the following way:

$$\frac{g\tau^2}{2}[(\mathfrak{D}_D^{\kappa+1}, \mathfrak{d}_D^{\kappa+1} + \mathfrak{d}_D^{\kappa+1/2}) + \mathfrak{D}_1] = \frac{g\tau^2}{4}\{\|\mathfrak{D}_D^{\kappa+1}\|^2 - \|\mathfrak{D}_D^{\kappa}\|^2$$
$$\quad + \|u_{xD}^{\kappa+1} + v_{yD}^{\kappa+1}\|^2 - \|\mathfrak{d}_D^{\kappa+1/2}\|^2 - 2(v_{yD}^{\kappa+1}, u_{xD}^{\kappa+1/2})$$
$$\quad - 2(u_{xD}^{\kappa+1}, v_{yD}^{\kappa+1/2})| \qquad (3.4.25)$$

and sum this relation over j from 0 to κ. As a result

$$\frac{g\tau^2}{2}\sum_{j=0}^{\kappa}[(\mathfrak{D}_D^{j+1}, \mathfrak{d}_D^{j+1} + \mathfrak{d}_D^{j+1/2}) + \mathfrak{D}_1] = \frac{g\tau^2}{4}\{\|\mathfrak{D}_D^{\kappa+1}\|^2$$
$$\quad + \sum_{j=0}^{\kappa}[\|u_{xD}^{j+1} - v_{yD}^{j+1}\|^2 - \|\mathfrak{d}^{j+1/2}\|^2$$
$$\quad - 2(v_{yD}^{j+1}, u_{xD}^{j+1/2}) - 2(u_{xD}^{j+1}, v_{yD}^{j+1/2})]\} \qquad (3.4.26)$$

Estimate the third term in formula (3.4.24) using the Cauchy inequality and inequality $\|u_x^{\kappa}\| \leq 2/h \|u^{\kappa}\|$. In this case

$$\frac{g\tau^2}{2}|\mathfrak{D}_2| \leq M_h\tau^3\sqrt{\frac{g^3}{\mu_h}}\|\mathfrak{D}_D^{\kappa+1}\|^2 + M_h\tau\sqrt{\frac{g}{\mu_h}}(\|\mathbf{w}^{\kappa+1}\|^2$$
$$\quad + \|\mathbf{w}^{\kappa+1/2}\|^2) + \frac{5g\tau^2 M_h}{h}(\|\mathbf{w}^{\kappa+1}\|^2 + \|\mathbf{w}^{\kappa+1/2}\|^2). \qquad (3.4.27)$$

Suppose that

$$\frac{\tau\kappa_h}{h^2} < 1. \qquad (3.4.28)$$

In this case after adding equalities (3.4.18), (3.4.19), summing the resulting expression over j from 0 to κ and considering expressions (3.4.20), (3.4.21), (3.4.27) as well as estimates of the difference relations contained in relation (3.4.26) we get

$$\|\Im^{\kappa+1}\|^2 \leq \left[\frac{2g\tau^2 \max D_h}{h^2} + C_3(\varepsilon_1, \varepsilon_2, \varepsilon, 2\omega_z)\tau \right.$$

$$\left. + M_h\tau \sqrt{\frac{g}{\mu_h}} + \frac{5g\tau^2 M_h}{h}\right] \sum_{j=0}^{\kappa} (\|\Im^{j+1}\|^2$$

$$+ \|\Im^{j+1/2}\|^2) + C_4. \qquad (3.4.29)$$

Here $\|\Im\|^2 = \frac{1}{2}(\|w\|^2 + (g\tau^2/4)\|\mathfrak{D}_D\|^2)$; C_3, C_4 are constants uniting all the constants in corresponding estimates.

A similar inequality is obviously valid at the layer $t = (\kappa + \frac{1}{2})\tau$ also.

Inequality (3.4.29) gives the second condition of stability of the difference scheme (3.4.1)—(3.4.6):

$$\frac{2g\tau^2 \max D_h}{h^2} + C_3\tau + M_h\tau\sqrt{\frac{g}{\mu_h}} + \frac{5g\tau^2 M_h}{h} < 1. \qquad (3.4.30)$$

Using inequality (3.4.29) we derive on the basis of inequality (3.4.30) the following estimate:

$$\|\Im^\kappa\| \leq C(T), \quad \kappa = 0, 1/2, 1, \ldots, \qquad (3.4.31)$$

where C is a κ-independent constant.

Remark 1. This estimate is valid for the case when eigenvalues of the spectral matrix are equal to $1 + 0(\tau)$.

Remember that in the HN-method the modulus of eigenvalues of the spectral matrix did not exceed unity. The above difference is associated with the fact that when analysing stability of the method of fractional steps we did not exclude the layers with fractional index, whereas in the HN-method these layers were not taken into account.

Remark 2. With $D = D_0$ (const), condition (3.4.30) can be reduced to

$$\frac{2gD_0\tau^2}{h^2} + C_3\tau < 1,$$

whence, at ε_1, ε_2, τ, $2\omega \to 0$, (which is equivalent to $C_3 \to 0$), follows the Courant–Friederichs–Levy criterion

$$\frac{\tau}{h}\sqrt{2gD_0} < 1. \qquad (3.4.32)$$

Remark 3. Fulfilment of condition (3.4.28) with $\tau, h \to 0$, naturally, involves fulfilment of conditions (3.4.30) or (3.4.32). Therefore, one can employ only one condition, i.e. (3.4.28), as a stability criterion for the difference scheme (3.4.1)—(3.4.6). However, in actual calculations, when τ and h take fixed values, both conditions (3.4.28) and (3.4.30) must be checked.

Convergence

Let the precise solution of the initial problem (2.2.1)—(2.2.3) coincide on the grid with the difference solution for system (3.4.1)—(3.4.6) with discrepancy $\mathfrak{C} = (\mathfrak{C}_u, \mathfrak{C}_v, \mathfrak{C}_{\zeta_x}, \mathfrak{C}_{\zeta_y})$. This discrepancy vanishes if $\tau, h \to 0$ when the solution of problem (2.2.1)—(2.2.3) is sufficiently smooth. The latter condition is fulfilled when all functions and their derivatives appearing in the initial system are continuous. In this case to calculate discrepancies of the components of total transport e_u, e_v and those of the level gradient e_{ζ_x}, e_{ζ_y} one can write the following difference system:

$$\frac{e_u^{\kappa+1/2} - e_u^{\kappa}}{\tau} + \frac{\iota}{2D^2}\,|\,\mathbf{e}^{\kappa} + \mathbf{w}'^{\kappa}\,|\,e_u^{\kappa+1/2} = -\frac{gD}{2}\,e_{\zeta_x}^{\kappa}$$

$$+ \frac{\kappa_h}{2}[(e_u^{\kappa})_{\bar{x}} + (e_u^{\kappa+1/2})_{y\bar{y}}] + \omega_z e_v^{\kappa}$$

$$- \frac{\iota}{2D^2}(|\,\mathbf{e}^{\kappa} + \mathbf{w}'^{\kappa}\,| - |\,\mathbf{w}'^{\kappa}\,|)\,u'^{\kappa+1/2} + \mathfrak{C}_u^{\kappa+1/2}; \qquad (3.4.33)$$

$$\frac{e_v^{\kappa+1/2} - e_v^{\kappa}}{\tau} + \frac{\iota}{2D^2}\,|\,\mathbf{e}^{\kappa} + \mathbf{w}'^{\kappa}\,|\,e_v^{\kappa+1/2} = -\frac{gD}{2}\,e_{\zeta_y}^{\kappa}$$

$$+ \frac{\kappa_h}{2}[(e_v^{\kappa+1/2})_{x\bar{x}} + (e_v^{\kappa})_{y\bar{y}}] - \omega_z e_u^{\kappa+1/2}$$

$$- \frac{\iota}{2D^2}(|\,\mathbf{e}^{\kappa} + \mathbf{w}'^{\kappa}\,| - |\,\mathbf{w}'^{\kappa}\,|)\,v'^{\kappa+1/2} + \mathfrak{C}_v^{\kappa+1/2}; \qquad (3.4.34)$$

$$\frac{e_v^{\kappa+1}-e_v^{\kappa+1/2}}{\tau}+\frac{t}{2D^2}|\mathbf{e}^{\kappa+1/2}+\mathbf{w}'^{\kappa+1/2}|e_v^{\kappa+1}=-\frac{gD}{2}e_{\zeta_y}^{\kappa+1}$$

$$+\frac{\varkappa_h}{2}[(e_v^{\kappa+1/2})_{x\bar{x}}+(e_v^{\kappa+1})_{y\bar{y}}]-\omega_z e_u^{\kappa+1/2}$$

$$-\frac{t}{2D^2}(|\mathbf{e}^{\kappa+1/2}+\mathbf{w}'^{\kappa+1/2}|$$

$$-|\mathbf{w}'^{\kappa+1/2}|)v'^{\kappa+1}+\mathfrak{E}_v^{\kappa+1}; \qquad (3.4.35)$$

$$\frac{e_u^{\kappa+1}-e_u^{\kappa+1/2}}{\tau}+\frac{t}{2D^2}|\mathbf{e}^{\kappa+1/2}+\mathbf{w}'^{\kappa+1/2}|e_u^{\kappa+1}=-\frac{gD}{2}e_{\zeta_x}^{\kappa+1}$$

$$+\frac{\varkappa_h}{2}[(e_u^{\kappa+1})_{x\bar{x}}+(e_u^{\kappa+1/2})_{y\bar{y}}]+\omega_z e_v^{\kappa+1}$$

$$-\frac{t}{2D^2}(|\mathbf{e}^{\kappa+1/2}+\mathbf{w}'^{\kappa+1/2}|$$

$$-|\mathbf{w}'^{\kappa+1/2}|)u'^{\kappa+1}+\mathfrak{E}_u^{\kappa+1}; \qquad (3.4.36)$$

$$e_{\zeta_x}^{\kappa+1}=e_{\zeta_x}^{\kappa}-\frac{\tau}{2}[(e_u^{\kappa+1})_{x\bar{x}}+(e_u^{\kappa+1/2})_{x\bar{x}}+(e_v^{\kappa+1/2})_{y\bar{x}}$$

$$+(e_v^{\kappa})_{y\bar{x}}]+\tau\mathfrak{E}_{\zeta_x}^{\kappa+1}; \qquad (3.4.37)$$

$$e_{\zeta_y}^{\kappa+1}=e_{\zeta_y}^{\kappa}-\frac{\tau}{2}[(e_u^{\kappa+1/2})_{x\bar{y}}+(e_u^{\kappa})_{x\bar{y}}+(e_v^{\kappa+1})_{y\bar{y}}$$

$$+(e_v^{\kappa+1})_{y\bar{y}}]+\tau\mathfrak{E}_{\zeta_y}^{\kappa+1}. \qquad (3.4.38)$$

Here **e** is the difference between the grid $\mathbf{w}_h$ and precise $\mathbf{w}'$ values of the total transport vector.

In order to reduce calculations, assume that $\Gamma_h = \Gamma$. Then boundary conditions for **e** can be presented in a form $\mathbf{e}|_{\Gamma_h} = 0$.

Let us multiply each equation of system (3.4.33)—(3.4.36) by $2\tau h^2$ and by a corresponding function and then sum the resulting expressions over the whole domain S_h. After transformations identical to those made in proving stability we get an inequality which differs from (3.4.29) only in some additional terms, i.e.

$$\|\vartheta^{\kappa+1}\|^2 \leq C_5\tau \sum_{j=0}^{\kappa}(\|\vartheta^{j+1}\|^2+\|\vartheta^{j+1/2}\|^2)+Y^{\kappa+1},$$

$$\kappa = 0, 1/2, 1, \ldots, \qquad (3.4.39)$$

where $\|\mathfrak{I}^{\kappa+1}\|$ is found in a similar way to inequality (3.4.29), by substituting $\mathbf{e}$ for $\mathbf{w}$; C_5 is a κ-dependent constant;

$$Y^{\kappa+1} = \tau \left[\sum_{j=0}^{\kappa} (\mathfrak{E}_\mathbf{w}^j, \mathbf{e}^j) + \sum_{j=0}^{\kappa} (\mathfrak{E}_\mathbf{w}^{j+1/2}, \mathbf{e}^{j+1/2}) \right]$$
$$+ g\tau^2 \max D_h \sum_{j=0}^{\kappa} \sum_{i=0}^{j} (\mathfrak{E}_{\nabla\zeta}^i, \mathbf{e}^i),$$

$\mathfrak{E}_\mathbf{w}$ and $\mathfrak{E}_{\nabla\zeta}$ being vectors with components

$$\begin{pmatrix} \mathfrak{E}_u \\ \mathfrak{E}_v \end{pmatrix} \quad \text{and} \quad \begin{pmatrix} \mathfrak{E}_{\zeta_x} \\ \mathfrak{E}_{\zeta_y} \end{pmatrix}, \quad \text{respectively.}$$

We now write

$$|Y^{\kappa+1}| \leq C_6 \tau \sum_{j=0}^{\kappa} (\|\mathbf{e}^{j+1}\|^2 + \|\mathbf{e}^{j+1/2}\|^2) + \mathcal{Y}^{\kappa+1},$$

where

$$\mathcal{Y}^{\kappa+1} = C_7 \tau \sum_{j=0}^{\kappa} (\|\mathfrak{E}_\mathbf{w}^{j+1}\|^2 + \|\mathfrak{E}_\mathbf{w}^{j+1/2}\|^2 + \|\mathfrak{E}_{\nabla\zeta}^j\|^2).$$

Then by means of inequalities (3.2.39), if $\tau (\tau < \frac{1}{2}(\max C_5 + C_6))$ is sufficiently small, we find that

$$\|\mathfrak{I}^\kappa\| \leq C_8 \mathcal{Y}^\kappa, \quad \kappa = 0, 1/2, 1, \ldots. \tag{3.4.40}$$

Assuming the smoothness of a solution, then $\mathcal{Y}^\kappa \to 0$ if $\tau, h \to 0$. Hence follows the convergence of the grid solution of problem (3.4.1)—(3.4.6) to the precise one in the grid norm. Thus, we have proved the convergence of the difference scheme (3.4.1)—(3.4.6).

Remark 1. The assumption of coincidence of Γ_h and Γ can be changed for a requirement of smoothness of Γ.

Remark 2. A more subtle analysis involving a consideration of polylinear and simplicial corrections of a grid solution in the domain S_h allows one to relax the requirement of smoothness of the solution and, in particular, to prove convergence of the grid solution to the precise one

$$w' \in \overset{0}{\mathbf{W}}_2^{1,1}(Q_T), \quad \zeta' \in W_2^{0,1}(Q_T).$$

3.5. A Modified Variant of the Method of Fractional Steps

The subject of the preceding section was the difference scheme in which the components of the horizontal level gradient are determined by the continuity equation differentiated with respect to x, y. Let us modify this scheme, approximating the horizontal gradient of elevation in the initial equations of tidal dynamics by its difference-analogue.

In the difference scheme obtained, all the terms except those containing $\nabla \zeta$ are written in the same way as in system (3.4.1)—(3.4.4). In this case, however, equations (3.4.1)—(3.4.4) instead of

$$-\frac{1}{2} gD \frac{\partial \zeta^\kappa}{\partial x}, \quad -\frac{1}{2} gD \frac{\partial \zeta^\kappa}{\partial y}, \quad -\frac{1}{2} gD \frac{\partial \zeta^{\kappa+1}}{\partial y}, \quad -\frac{1}{2} gD \frac{\partial \zeta^{\kappa+1}}{\partial x}$$

will contain

$$-\tfrac{1}{2} gD\zeta^\kappa_{\bar{x}}, \quad -\tfrac{1}{2} gD\zeta^\kappa_{\bar{y}}, \quad -\tfrac{1}{2} gD\zeta^{\kappa+1/2}_{\bar{y}}, \quad -\tfrac{1}{2} gD\zeta^{\kappa+1/2}_{\bar{x}},$$

respectively. Besides, relations (3.4.5), (3.4.6) are substituted for the equalities

$$\frac{\zeta^{\kappa+1/2} - \zeta^\kappa}{\tau} + \frac{1}{2}(u_x^{\kappa+1/2} + v_y^{\kappa+1/2}) = 0; \tag{3.5.1}$$

$$\frac{\zeta^{\kappa+1} - \zeta^{\kappa+1/2}}{\tau} + \frac{1}{2}(u_x^{\kappa+1} + v_y^{\kappa+1}) = 0. \tag{3.5.2}$$

Using similar transformations obtain the equality in which the term containing $\zeta_{\bar{x}}$ and $\zeta_{\bar{y}}$ will have the form

$$\mathfrak{T} = \frac{1}{2} g\tau \left[\sum_{S_h} D(\zeta^\kappa_{\bar{x}} u^{\kappa+1/2} + \zeta^{\kappa+1/2}_{\bar{x}} u^{\kappa+1} + \zeta^\kappa_{\bar{y}} v^{\kappa+1/2} + \zeta^{\kappa+1/2}_{\bar{y}} v^{\kappa+1}) \right].$$

After summation by parts and allowing for equations (3.5.1), (3.5.2), we obtain

$$\mathfrak{T} = -g\tau[(\zeta^\kappa_D, \zeta^{\kappa+1/2}_{iD}) + (\zeta^{\kappa+1/2}_D, \zeta^{\kappa+1}_{iD})] - g\tau[(\zeta^\kappa, u^{\kappa+1/2}D_x + v^{\kappa+1/2}D_y) + (\zeta^{\kappa+1/2}, u^{\kappa+1}D_x + v^{\kappa+1}D_y)]. \tag{3.5.3}$$

Estimate the first term in relation (3.5.3) by the use of formula (3.4.7) and inequality $\|u_x\| \leq 2/h \|u\|$. We then have

$$-g\tau[(\zeta^\kappa_D, \zeta^{\kappa+1/2}_{iD}) + (\zeta^{\kappa+1/2}_D, \zeta^{\kappa+1}_{iD})] = -g[\|\zeta^{\kappa+1}_D\|^2 - \|\zeta^\kappa_D\|^2$$
$$-\tau^2(\|\zeta^{\kappa+1/2}_{iD}\|^2 + \|\zeta^{\kappa+1}_{iD}\|^2)] \leq -g(\|\zeta^{\kappa+1}_D\|^2 - \|\zeta^\kappa_D\|^2)$$
$$+\frac{2g\tau^2}{h^2}(\|\mathbf{w}^{\kappa+1/2}_D\|^2 + \|\mathbf{w}^{\kappa+1}_D\|^2). \tag{3.5.4}$$

The second bracket in relation (3.5.3) can be estimated as follows:

$$-g\tau[(\zeta^\kappa, u^{\kappa+1/2}D_x + v^{\kappa+1/2}D_y) + (\zeta^{\kappa+1/2}, u^{\kappa+1}D_x + v^{\kappa+1}D_y)]$$
$$\leq M_h\tau \sqrt{\frac{g}{\mu_h}} [g(||\zeta_D^\kappa||^2 + ||\zeta_D^{\kappa+1/2}||^2) + ||\mathbf{w}^{\kappa+1/2}||^2 + ||\mathbf{w}^{\kappa+1}||^2]. \quad (3.5.5)$$

Apply condition (3.4.28) to τ and h, sum all equalities (3.5.3) over j from 0 to κ and employ estimates (3.5.4), (3.5.5). As a result,

$$||\Im^{\kappa+1}||^2 \leq \left(\frac{2g\tau^2 \max D_h}{h^2} + C_1\tau + M_h\tau \sqrt{\frac{g}{\mu_h}}\right) \sum_{j=0}^{\kappa} (||\Im^{j+1}||^2$$
$$+ ||\Im^{j+1/2}||^2) + C_2; \quad \kappa = 0, 1/2, 1, \ldots, \quad (3.5.6)$$

where

$$||\Im^{\kappa+1}||^2 = \tfrac{1}{2}(||\mathbf{w}^{\kappa+1}||^2 + g||\zeta_D^{\kappa+1}||^2)$$

which shows that the considered difference scheme is stable if the condition

$$\frac{2g\tau^2 \max D_h}{h^2} + C_1\tau + M_h \sqrt{\frac{g}{\mu_h}} < 1 \quad (3.5.7)$$

is fulfilled.

Its convergence can be easily proved as in the preceding section.

3.6. References

The reader interested in the original description of the method of boundary values can be referred to the article by Hansen (1952). A review of some problems associated with the application of this method can be found in the monograph by Kagan (1968), which also contains a detailed bibliography of works by different authors who used this method to describe tides in marginal seas.

The above method served as an apparatus to study tides in the World Ocean. For a detailed investigation of this application of the method of boundary values the reader is referred to the review by Hendershott and Munk (1970), as well as to the papers by Tiron, Sergeev and Michurin (1967), Bogdanov and Magarick (1967).

A detailed discussion of numerous methods of solution for the system of linear algebraic equations, to which the equation for complex amplitudes of tidal elevations (3.1.5) is reduced, is given in the monograph by Marchuk (1973).

The failure of the Zeidel–Nekrasov iteration method to converge when applied to the solution of equation (3.1.5) within sufficiently large domains has been mentioned in a paper by Bogdanov, Kim and Magarick (1964). Substitution of equation (3.1.2) by equation (3.1.3) was suggested in the paper by Bogdanov and Magarick (1967), already mentioned.

Description of the procedure for numerical solution of the Poincaré problem follows the paper by Marchuk (1969). The application of this procedure to solve tidal problems under concrete physico-geographic conditions is presented in papers by Zalesny and Tamsalu (1972), Zalesny *et al.* (1972), Zalesny and Pospelov (1972).

Another variant of the boundary values method based on equations for the complex amplitudes of vertically averaged velocity of the tidal flow was considered by Pekeris and Accad who also take the credit for a bold attempt to apply this method to the World Ocean (see Pekeris and Accad, 1969). A delayed convergence of the numerical solution for problem (3.1.5), (3.1.7) in the case of convex and concave boundary angles, was studied by Pnueli and Pekeris (1968).

The HN-method as well as its modified variant was suggested by Hansen (1956). A detailed analysis of the HN-method properties is presented in papers by Hansen (1962), Kempff (1968) and Kagan (1970). Basic results obtained by this method are summarized in a monograph by Hansen (1966), where the reader can also find references to the application of the HN-method to calculating problems of tides, surges and wind currents. Here a more recent paper by Zahel (1970) should be mentioned, its subject also being the application of the HN-method to tide calculations in the World Ocean.

Some other explicit schemes of integrating tidal dynamics equations are suggested in papers by Ueno (1964) and Heaps (1969a); see also Laska (1971).

The method of fractional steps as applied to tidal calculations was elaborated by Marchuk, Kagan and Tamsalu (1969), the computing aspects of this method being discussed in a paper by Marchuk *et al.* (1973). Results of tidal calculations in marginal seas carried out by the method of fractional steps can be found in a paper by Dvorkin, Kagan and Kleschova (1972) as well as in the above-mentioned paper by Marchuk *et al.* (1973). A modified variant of the method is given in Marchuk (1972).

CHAPTER 4

Tides in the World Ocean

THUS, we have considered the most popular methods of numerical integration of the equations of tidal dynamics. The task now is to employ them quantitatively to describe and explain the nature of tides in the World Ocean. The question arises, however, whether it is in principle possible to solve this problem and, if so, what will be the degree of approximation to the real picture of the phenomenon in question.

It is apparent that opinions on this question may differ drastically. We do not indend to thrust our personal opinion on the reader but let him draw his own conclusions, having described the complete situation.

4.1. Empirical Cotidal Charts

As is clear from the title of the section we are going to dwell here on cotidal charts made from direct measurements of elevation in the World Ocean.

However, since these data are available mainly on the coastline of the mainlands and at a small number of oceanic islands, and, moreover, their distribution is far from being uniform, shortage of data can be compensated by some ideas on the nature of the phenomenon under consideration.

The above comments can serve as a logical proof, though it may seem a paradox, of the fact that practically all current hypotheses on the mechanism of tide generation in the World Ocean were put forward at the end of the last and in the beginning of the present centuries, when the observational data were much more scarce than nowadays, and when investigators had no idea of employing computers.

The Whewell chart

Since Newton it had been thought that tides in the World Ocean are induced by forced tidal waves following the disturbing body from east to west. Imagine, for instance, that the only data available are the above conclusions, some

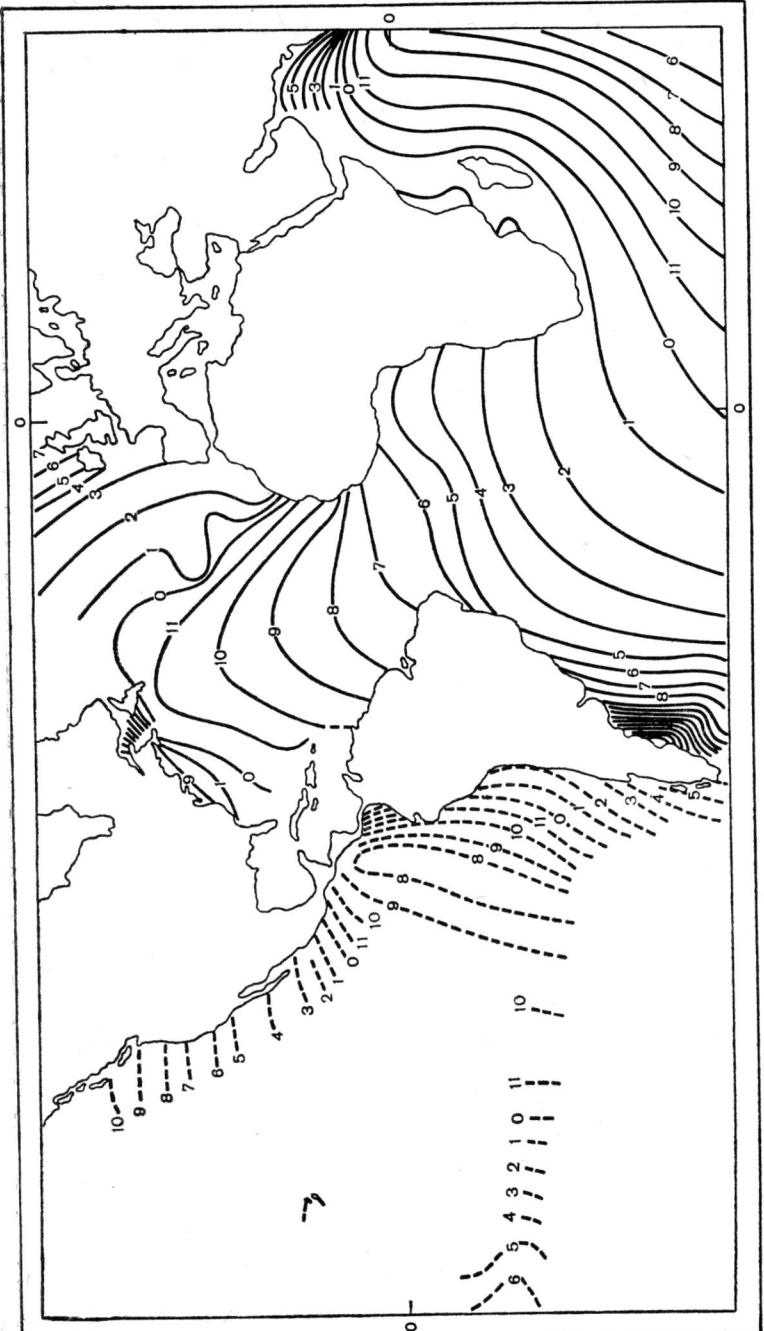

FIG. 4.1. Cotidal chart of a semidiurnal tide in the World Ocean on the days of new and full moons (according to Whewell, 1833). Numbers on the isolines indicate the time of high water occurrence (in solar hours) with respect to the moment of the Sun's transit over the Greenwich meridian.

occasional level measurements on the coastline and oceanic islands as well as two obvious facts, namely that the effect of tide-generating forces is of a global character and that the Southern ocean is the only place where the mainlands do not hinder zonal wave propagation.

Under this supposition one can readily justify the concept by Whewell (1833) who, when making the first cotidal chart of the World Ocean in the history of tidal investigations (Fig. 4.1), assumed that oceanic tides originated in the southern ring girdling the Antarctic Regions. Forced tidal waves originating in this region in the course of their westward propagation diffract at the southern extremities of Australia, Africa and South America and, in their turn, generate free tidal waves. The latter propagate this time in the meridional direction with velocity dependent on the local oceanic depth.

This is the way tides in the Atlantic and Indian Oceans were thought to be generated. As regards the Pacific Ocean, taking into account its huge size and the extremely scarce observational data available, Whewell did not take the risk of making the cotidal chart for the whole ocean, but limited himself to approximate indications of the time of high water along the western coast of America only.

The Harris chart

This chart is based on the supposition that semidiurnal tides in the World Ocean generate as a result of resonance interaction of free oscillations existing within closed regions with the tide-generating forces of semidiurnal period. Dimensions of such regions delineated by coastlines, island or underwater reef chains, were chosen in such a way that the period of inducing forces coincided precisely with that of free oscillations at the mean oceanic depth within each region. The period of free oscillations was calculated in the same manner as in the case of a closed basin under the condition that the Earth does not rotate. Free modes were considered to be of the character of standing oscillations with one or several nodal lines, the cotidal hour changing by half the tidal period in the course of its passing over the nodal lines (Fig. 4.2). Superposition of the standing oscillatory systems gives rise to amphidromic systems.

In oceanic regions with dimensions and depth characterized by the fact that the period of free oscillations induced in them greatly exceeds that of the perturbing forces, tides are mainly generated by the "resonance" waves, i.e. by progressive waves propagating from the resonance regions. This is the case, for example, for the regions with unknown time of high water located far from the nodal lines or on the boundaries with other water basins.

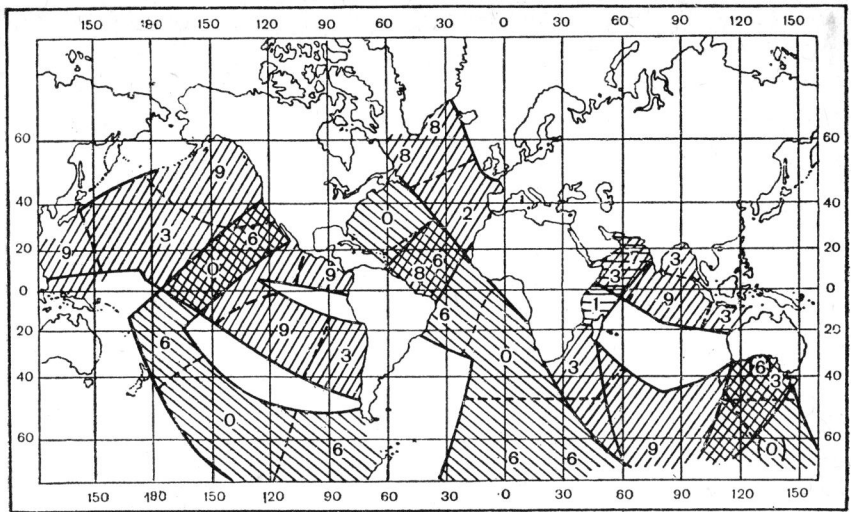

FIG. 4.2. Resonance regions for the oscillations of semidiurnal period (according to Harris, 1904). Dashed line denotes nodal lines. Numbers indicate the time of high water (in lunar hours) with respect to the moment of the Moon's transit over the Greenwich meridian.

When this region is located between two neighbouring "resonance" regions, the latter having different cotidal hours, the nodal line in the intermediate region is replaced by the nodal zone with the characteristic concentration of cotidals. However, if the two neighbouring "resonance" regions have the same cotidal hour, the same hour is ascribed to the intermediate region.

The above considerations allowed Harris (1904) to draw the cotidal lines even in oceanic regions lacking the necessary tidal data and to make a new cotidal chart of the World Ocean (Fig. 4.3).

The Sterneck chart

It is a tradition with modern literature to emphasize the formalism of the Sterneck method of making up the cotidal chart of the World Ocean. However, we consider it to be as good as any other uniform method of observational data systematization.

Initial assumptions made by Sterneck (1920) were as follows. Semidiurnal tides in the World Ocean are represented as the result of the superposition of two global systems of synchronous oscillations. These systems are standing waves with phases equidistant from one another by a quarter of the tidal period. Nodal lines are formed by the totality of points at which the correspond-

FIG. 4.3. Cotidal chart of a semidiurnal tide in the World Ocean (according to Harris, 1904). Numbers on the isolines indicate the time of high water (in lunar hours) with respect to the moment of the Moon's transit over the Greenwich meridian.

ing values of the tidal amplitudes are equal to zero. Thus, to establish the location of the nodal lines it is sufficient to connect the points of elevation measurements with the tidal phases equal to 3 and 9 hours for the first, and 0 and 6 hours for the second of these oscillations, respectively.

Sterneck believed that the very nature of the nodal lines of the same oscillation prevents them from intersecting. They intersect only with the nodal lines of another system of oscillations which results in the appearance of amphidromic systems with the direction of rotation dependent only on the location of the nodal lines of both standing oscillations. Specifically, in cases when the centres of the two neighbouring amphidromic systems are located on one and the same nodal line, the rotation of these amphidromes will always be opposite in direction.

The scheme of location of the nodal lines and amphidromic points of the semidiurnal tide in the World Ocean presented by Sterneck is given in Fig. 4.4. The reader's attention can be drawn to three peculiarities of the pattern of nodal lines characteristic of the system of standing oscillations:

(i) The distance between nodal lines of each oscillation is conditioned by the bottom contour and corresponds approximately to the value obtained for standing waves by the well-known Merian formula. Remember that Sterneck determined the location of the nodal lines only by the elevation data on islands and the coastline of the World Ocean;

(ii) The intersection points of two nodal lines belonging to different oscillatory systems is the location of the amphidrome centres, the number of these amphidromes being two in the Atlantic Ocean, four in the Indian Ocean and six in the Pacific Ocean. In the North Atlantic both amphidromes have a counter-clockwise rotation, in the Indian ocean two north amphidromes have clockwise and two south ones a counter-clockwise rotation. In the Pacific Ocean the direction of the cotidal lines joining amphidromic centres located south and north of the equator does not reveal any strict regularity. Remember in this connection that the Coriolis force effect should have resulted in the amphidromes' counter-clockwise rotation of amphidromes in the north and clockwise rotation in the south hemispheres;

(iii) in the Southern ocean, instead of a theoretically predicted tidal wave travelling from east to west, the motion in the opposite direction is observed.

The scheme of nodal line location presented in Fig. 4.4 served as the basis for constructing a cotidal chart of the semidiurnal tide in the World Ocean

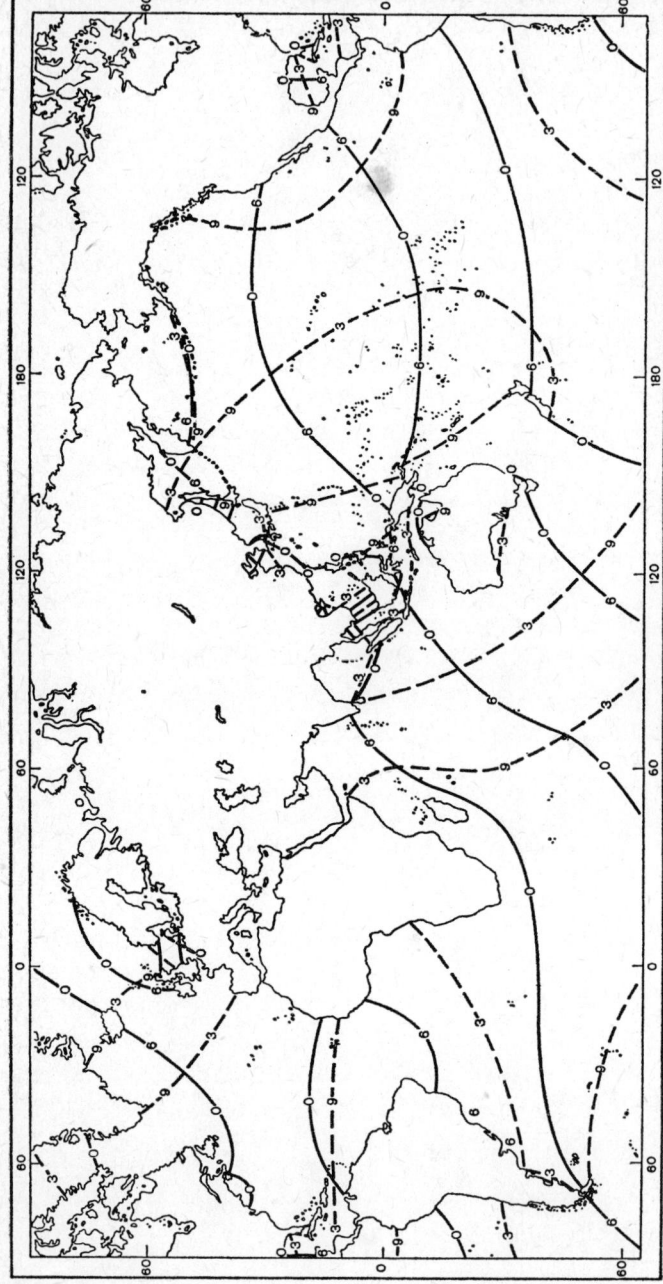

FIG. 4.4. Nodal lines of the two systems of standing oscillations of a semidiurnal period (by Sterneck, 1920). Solid and dashed curves denote the nodal lines of different systems of standing oscillations. Numbers indicate the time of high water (in lunar hours) with respect to the moment of the Moon's transit over the Greenwich meridian.

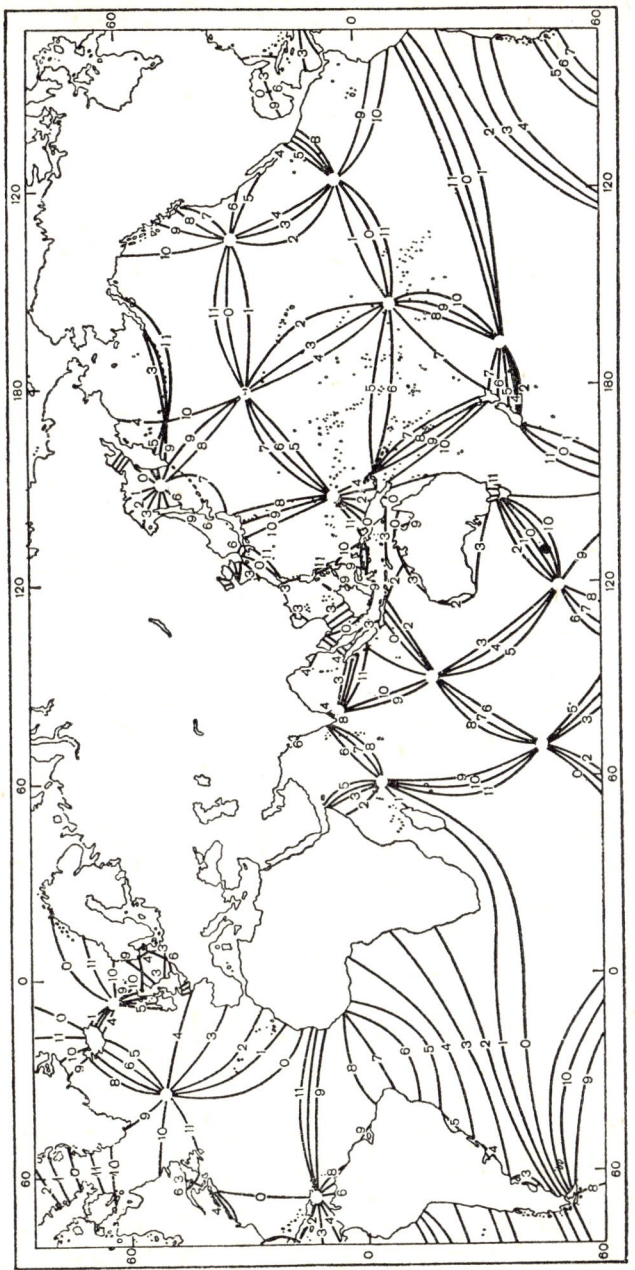

FIG. 4.5. Cotidal chart of a semidiurnal tide in the World Ocean on the days of new and full Moons (according to Sterneck, 1920). Numbers on the isolines indicate the time of high water (in lunar hours) with respect to the moment of the Moon's transit over the Greenwich meridian.

(Fig. 4.5). Additional cotidal lines in the chart were drawn according to the observational data, taking account of the amphidromic centres already obtained.

The Dietrich and Villain charts

It would be tempting to create a cotidal chart of the World Ocean based solely on the observational data, taking no advantage of any hypotheses on the nature of the phenomenon. The question arises, however, whether these data alone are sufficient?

This question could be easily answered in Harris's and Sterneck's times when the general number of points with known values of harmonic constants of elevation was negligibly small as compared to modern data, Harris having at his disposal only 183, and Sterneck 204 points. It is evident that in those conditions it was impossible to manage with solely observational data.

By the middle of the present century the situation had changed dramatically. Dietrich (1944) made use of harmonic constants from 1665, and Villain (1952) with approximately 2800 points of recorded elevations. At first sight, such a network of stations might seem to be quite sufficient to make up a reliable cotidal chart of the World Ocean. Note, however, that the majority of those stations was located on mainland coasts, although by no means all coasts were represented, e.g. the Antarctic coast. As to elevation records in the open sea, they were extremely few in number. Besides, the accuracy of determination of harmonic constants on islands, especially on the islands of the South hemisphere, left much to be desired.

That is why due credit must be given to the inventiveness and courage of the researchers to whom we owe today our knowledge of the general spatial distribution of tides in the World Ocean. Their attempts to create a global picture solely from observational data concluded an important stage in tidal research, the stage of systematization of empirical data.

All such attempts are known to be more or less subjective, the cotidal charts of the M_2 wave constructed by Dietrich and Villain (Figs. 4.6, 4.7) being typical examples. The first thing that we notice in comparing the two charts is the huge spaces of the Pacific and South Oceans which are, by Villain, "white spots" in the cotidal chart of the World Ocean. In other words, despite the fact that Villain had at his disposal all the data the earlier investigators had made use of, he considered them either insufficient or unreliable.

But it is the striking similarity between the charts and with the Sterneck chart that impresses one most of all. Of course, it does not necessarily mean that the increased volume of information has in any way influenced the pattern

FIG. 4.6. Cotidal chart of the M_2 wave in the World Ocean (by Dietrich, 1944). Numbers on the isolines indicate the time of high water (in mean lunar hours) with respect to the moment of the Moon's transit over the Greenwich meridian.

FIG. 4.7. Cotidal chart of the M_2 wave in the World Ocean (by Villain, 1952). For details see Fig. 4.6.

of the cotidal lines. In particular, the quantity of amphidromes and their location have been defined more exactly. The process, however, can be very much delayed unless the problem of increasing the network of recording stations in the open ocean and that of their optimal location is solved. Note that the question of optimizing the location of the stations has not yet even been mentioned. It is still a debatable question what minimal density of the stations in different parts of the World Ocean can ensure the required degree of accuracy of tidal mapping.

4.2. Basic Features of the Spatial Distribution of Tides in the World Ocean

Let us comment on Figs. 4.6 an 4.7, paying special attention to the validity of some peculiarities of the cotidal chart of the M_2 wave discovered by Dietrich and Villain.

The Pacific Ocean

The picture of tides in this ocean is still poorly understood though some of its peculiarities can be considered quite reliably established. One of them is the nodal zone in the north-west part of the ocean. It is the observational data obtained on the Marshall and Caroline islands, the Palau and Ogasawara (Bonin) islands and, finally, on the western and eastern coasts of Japan and the north coast of New Guinea that served as the basis for singling out this nodal zone.

The pattern of the cotidal lines in the north-east part of the Pacific Ocean is determined by observational data obtained along the western coast of North America as well as on the Aleutian and Hawaiian islands. Here one first of all notices an increase in cotidal hours of the M_2 wave observed throughout the region from California to the Gulf of Alaska. An especially rapid increase is, however, observed on the coast between Los Angeles and San Francisco which results in a concentration of the cotidal lines. The latter fact as well as an increase of the tidal amplitudes northward and southward from this region indicates the existence of a nodal zone.

It was Harris (Fig. 4.2) who first discovered this zone, representing it as an arc of the circumference passing through observational data on the Hawaiian and Aleutian islands. An indication that this nodal zone could be transformed into an amphidromic system was also first represented in the Harris chart (Fig. 4.3).

The central part of the Pacific Ocean is one of the least studied, as regards tides. Even by the end of the first half of the present century the harmonic constants had been obtained from only a few points elevation measurements on the Society, Fiji, Samoa, Ellis, Phoenix and Fanning islands, as well as on the Marshall and Tuamotu islands. The bulk of these data greatly exceeds that which Harris and Sterneck had at their disposal, which made it possible to confirm on the basis of new material the existence of the amphidrome of a counter-clockwise rotation between the Tahiti and Fanning islands and to refine the location of the cotidal lines radiating from the amphidrome centre in north-west and north-east directions.

At the same time, the lack of observational data south of the Society islands and west of the Marshall and Tuamotu islands suggests that the distribution of the cotidal lines presented in Figs. 4.6 and 4.7 is hypothetical rather than true.

A peculiar feature of the cotidal chart of the southwest part of the Pacific Ocean is a change of the cotidal hours in the region of New Zealand. Here, according to the observational data, the time of high water runs through the complete tidal cycle. Among other things, comparatively large amplitudes have been observed on the coast of New Zealand and the Oakland islands. Both latter facts allow one to state that New Zealand is the place of a false amphidrome.

One more amphidrome can be detected in the Solomon sea, its centre being fairly precisely fixed by observations at the Nususong point, where the elevation amplitudes are equal to zero. However, due to lack of reliable data on the south coast of the New Britain islands and on the northern islands of the Luisiado archipelago, the distribution of the cotidal hours in the sea itself remains obscure.

Approximately the same (if not a worse) situation is found in the south-west portion of the Pacific Ocean. Here sufficiently reliable data are available only as far as the coast of South America and the Easter, Sala-y-Gomez, San Felix and Juan-Fernandez islands are concerned. Thus, the problem of location and even of the existence of the amphidrome located by Dietrich to the south of Easter island still remains unresolved.

Lack of observational data is characteristic of the western part of the Pacific Ocean too. The current cotidal charts of this region have been constructed by observational data obtained at a few points of the Central and South American coast, as well as on the Galapagos, Clipperton and Revilla Gigedo islands. Unfortunately, not all of the harmonic constant values available are equally reliable. This was only one of the difficulties hampering the analysis, the basic ones being associated with the presence of the nodal zone stretching from Mexico to Peru.

Observational data enable one precisely to determine the position of the north end of this nodal zone. The time of high water appears to change from 3 to 9 hours within only a 100-miles stretch of the Mexico coast. However, as is often the case, the main fact, i.e. the character of the phase change within this stretch of the coast, remains obscure. The supposition that the phases increase from Acapulco to Civitavecchia results in the Villain pattern of the cotidal lines; the supposition that the phases between these points decrease yields the Dietrich pattern.

In the former case, Harris's conclusion on the existence of a degenerate amphidrome centred in the vicinity of Mexico is invalidated, while the latter case is a proof of Sterneck's conclusion on the appearance of the amphidrome having a clockwise rotation to the west of the Galapagos islands. Thus, observational data give rise to ambiguous interpretation of the tidal picture in the western part of the Pacific Ocean.

The Indian Ocean

The Indian Ocean is hardly more studied than the Pacific Ocean. Luckily, however, the whole of its central part is occupied by an antiamphidromic area, the time of high water remaining practically constant within its limits. Therefore the lack of data here is not as critical as that in the Pacific Ocean.

Besides the central region there are four more regions in the Indian Ocean with small variations of the cotidal hour:

(i) The western region. It stretches along the whole eastern coast of Africa from the Cape of Good Hope to the Somalia Peninsula including the Madagascar, Amirante, Seychelles and Comoro islands. The time of high water varies here from 0 to 1 hour.

(ii) The Arabian Sea and the adjacent part of the Indian Ocean limited in the south by the parallel 12°N. The time of high water here equals 4–5 hours according to Dietrich and 5–6 hours according to Villain,

(iii) The Bay of Bengal and the Andaman Sea, in this region the time of high water does not exceed 2–3 hours,

(iv) The eastern part. It adjoins the western coast of Australia and the Java islands. The time of high water here is 1–2 hours.

The boundary between the above four regions and the central part of the Indian Ocean runs through the nodal zones, amphidromes being formed at the points of their crossing. One of them is located in the north-west part of the Indian Ocean on the line connecting the Seychelles and Maldive islands. The other two are to be found in the south-east and south-west parts of the Ocean.

Note that the location of none of them can be considered to be reliably determined, the same being true as far as the amphidrome in the north-west part of the Indian Ocean is concerned. This can be accounted for by the absence and, in a number of cases, by the unreliability of observational data in the open ocean.

The amphidrome in the south-west part of the Indian Ocean can serve as an illustration of the last statements. It was discovered by Sterneck who located it in the vicinity of Kerguelen island (Fig. 4.5). In Dietrich's chart (Fig. 4.6) this amphidrome appeared to be shifted to the north-west and located in the vicinity of Madagascar. Finally, Villain, having analysed the observational data obtained on Kerguelen, Saint Paul, Mauritius, Reunion and Madagascar decided that the very existence of this amphidrome was dubious.

The Atlantic Ocean

Tides in the Atlantic Ocean seem to have been studied so thoroughly that any new observational data are unable to change the common concepts, which have been formed under the influence of investigations by Harris, Sterneck, Dietrich and Villain.

They all suggest that the tides in the South Atlantic are generated by a progressive tidal wave travelling throughout the entire width of the basin from south to north. Progressive movement of the tidal wave is violated only on the Patagonia shelf, where three amphidromes are formed not far from the Saint Torge and Saint Matias Bays, the existence of two of these having been confirmed by direct measurements.

However, somewhat northward of the La Plata Bay the tidal wave acquires its progressive character again. Marked concentration of the cotidal lines as well as small tidal amplitudes indicate the fact that it is the beginning of a nodal zone, its cotidal lines radiating in the direction of the open sea in a fan-like way.

One more well-defined nodal zone crosses the Ocean in the zonal direction from the Antilles Islands to Africa. At the western end of this nodal zone (in the western part of the Caribbean Sea) there is an amphidrome of a counter-clockwise rotation and somewhat northward there is an antiamphidromic region embracing the western part of the North Atlantic as far as Newfoundland. The north-west and the whole eastern parts of the North Atlantic lie within the region of influence of the second amphidrome of a counter-clockwise rotation.

The above picture of the cotidal lines of the M_2 wave in the Atlantic Ocean was constructed from all available observational data at islands, and there have been no appreciable changes in recent times. However, Villain called attention to the fact that the accuracy of the harmonic constants from the islands located, in particular, in the South Atlantic, is far from exact. Their verification has become necessary since publication of the results of numerical calculations of tides in the World Ocean (see below), which suggested the possibility of the existence of an amphidrome of a counter-clockwise rotation in the South Atlantic.

The history of the discovery of this amphidrome is rather instructive. The first indications of its existence in the South Atlantic at the $\sim 25°S$ latitude can be found in the paper by Defant (1924), who presented tidal calculations for the Atlantic/Arctic Ocean system by the method of a "narrow sea". The resulting picture of the cotidal lines, however, was so inconsistent with the observational data that Defant had to abandon it. He suggested that only longitudinal oscillations are developed in the South Atlantic while lateral oscillations vanish due to compensation of the Coriolis force effect and the lateral component of the tide-generating force.

Twenty years later the amphidrome in the South Atlantic was rediscovered by Proudman (1944). Traditional concepts prevailing on him as well, he brought the calculated results into accord with the observational data by a special tune of the parameters of Kelvin and Poincaré free waves.

Finally, 20 years later at the Fourteenth General Assembly of the IUGG a tidal chart of the M_2 wave constructed on the basis of numerical solution for the dynamics equations of tides in the World Ocean without use of any empirical data was presented (Pekeris and Accad, 1969). The chart again presented the amphidrome in the South Atlantic, with its centre located at the 20° latitude among the Ascension, Trinidad, Tristan da Cunha and Saint Helena islands. Observational data on all the islands were available, only those obtained on the Ascension and Trinidad islands being reliable. Harmonic constants for the other two islands in the Admiralty tidal tables (1938) were marked "presumable".

It was Cartwright (1971) who verified the island constants. For this purpose he carried out 39-day bottom pressure registrations in the region of the Saint Helena island and a fortnightly series of standard level measurements on Tristan da Cunha island. The resulting values of the harmonic constants of elevation at these islands along with the refined observational data on the Ascension and Trinidad islands served as the basis for making up a new cotidal chart of the central part of the South Atlantic (Fig. 4.8). The theoretically predicted fact of the existence of an amphidrome in the South Atlantic was thus verified.

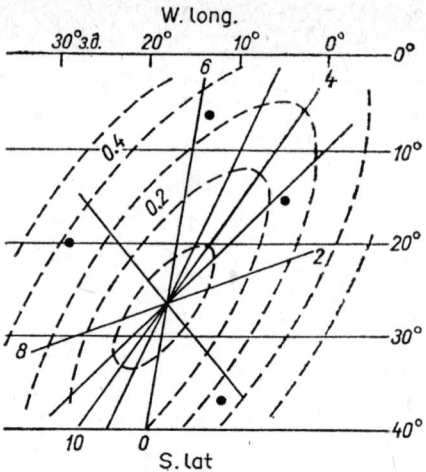

FIG. 4.8. Tidal chart of the M_2 wave in the central part of the South Atlantic (according to Cartwright, 1971). Solid lines denote cotidals, dashed lines denote isoamplitudes. The time of high water with respect to the moment of the Moon's transit over the Greenwich meridian is given in mean lunar hours, the tidal amplitude in metres. Ascension, Saint Helena, Tristan da Cunha and Trinidad islands are denoted by points.

The Southern Ocean

Observational data on tides in the Southern Ocean are still more scarce than those on any other ocean. This can be mainly accounted for by the extremely scarce elevation data at Southern Ocean islands and on the coast of the Antarctic Continent. The latter circumstance is aggravated by a seasonal variation of harmonic constants associated with the fact that the sea surface is covered with ice.

It is clear that under such conditions any attempts to interpret observational data in the South Ocean can be regarded as a preliminary and extremely rough approximation to the real picture. Nevertheless, they are worth considering as they are testimony to, first, the necessity to abandon the conclusions of the statical theory concerning the nature and the direction of the tidal wave travelling in the South Ocean, and, second, the probable existence of an amphidromic system between the Cape of Good Hope and Enderbury Land (the Antarctic Continent). The coordinates of the centres of the above and all the other amphidromic systems presented in the charts by Harris, Sterneck, Dietrich and Villain are given in Table 4.1.

It is thus clear that there is not enough basic data to elucidate the overall features and the mechanism of tide generation in the World Ocean. The simplest

TIDES IN THE WORLD OCEAN 127

TABLE 4.1. *Location, coordinates of centres and direction of rotation of the amphidromes presented in empirical charts of different investigators*

Ocean	Region	Coordinates of centres and direction of rotation			
		Harris (1904)	Sterneck (1920)	Dietrich (1944)	Villain (1952)
Atlantic	Central region of North Atlantics	40°N, 40°W (+)	53°N, 33°W (+)	52°N, 33°W (+)	44°N, 34°W (+)
	Caribbean Sea	–	16°N, 66°W (+)	15°N, 67°W (+)	17°N, 63°W (+)
Indian	Arabian Sea	1°S, 65°E (–)	5°S, 60°E (–)	10°N, 63°E (–)	0°, 62°E (–)
	Western region	–	50°S, 70°E (+)	30°S, 53°E (+)	–
	Eastern region	–	23°S, 92°E (–)	33°S, 103°E (–)	29°S, 97°E (–)
Pacific	North region	–	52°N, 169°E (+)	–	54°N, 168°W (+)
	North-east region	30°N, 142°W (+)	38°N, 135°W (+)	40°N, 148°W (+)	40°N, 150°W (+)
	North-west region	–	33°N, 178°E (–)	–	–
	West region	–	7°N, 146°E (+)	–	–
		–	5°S, 155°E (–)	8°S, 157°E (–)	12°S, 157°E (–)
	Central region	14°S, 153°W (+)	10°S, 154°W (+)	5°S, 152°W (+)	15°S, 150°W (+)
	East region	–	6°N, 115°W (–)	2°S, 103°W (–)	–
	South-east region	–	–	40°S, 105°W (–)	–
	South-west region	51°S, 172°W (–)	41°S, 166°W (–)	51°S, 160°W (–)	49°S, 160°W (–)
South	Afro-Antarctic sector	–	–	55°S, 15°E (–)	–
	Australo-Antarctic sector	–	52°S, 120°E (+)	–	–

Remark. Sign (+) corresponds to amphidromes of counterclockwise rotation, sign (–) to those of clockwise rotation; dash denotes the absence of amphidromes.

procedure would be, of course, to postpone the solution of the problem for the time being. The question arises whether it is worth trying to replace the missing data with numerical modelling. The answer to this question will be the subject of the next section.

4.3. An Example of Numerical Modelling of Tides in the World Ocean

Formulation and peculiarities of its solution

Consider a "classical" formulation of the tidal problem discussed in Chapter 2 and rewrite it as applied to oceanic conditions in a spherical system of coordinates, yielding the following system of equations

$$\frac{\partial \mathbf{w}}{\partial t} - \mathbf{A}_1 \mathbf{w} = -gD \cdot \nabla \zeta - \frac{\iota}{D^2} |\mathbf{w}| \mathbf{w}$$
$$+ k_h \left[\Delta \mathbf{w} - \frac{\cos 2\Theta}{a^2 \sin^2 \Theta} \mathbf{w} + \frac{1}{a^2} \mathbf{A}_2 \frac{\partial \mathbf{w}}{\partial \lambda} \right] + \mathbf{f}; \qquad (4.3.1)$$

$$\frac{\partial \zeta}{\partial t} + \operatorname{div} \mathbf{w} = 0. \qquad (4.3.2)$$

where, as earlier, $\mathbf{w}$ is the vector-function with the components u and v, which denote here the total transport components in the direction of axes λ and Θ (λ is the longitude reckoned eastward from the Greenwich meridian; Θ is the colatitude); ζ is the vertical displacement of the surface; D is the ocean depth; ι and k_h are the coefficients of bottom and horizontal turbulent friction; a is the Earth's mean radius; $\mathbf{A}_1$ and $\mathbf{A}_2$ are the coefficient matrices equal, respectively, to

$$A_1 = \begin{bmatrix} 0 & -2\omega \cos \Theta \\ 2\omega \cos \Theta & 0 \end{bmatrix}$$

$$A_2 = \begin{bmatrix} 0 & -\dfrac{2 \cos \Theta}{\sin^2 \Theta} \\ \dfrac{2 \cos \Theta}{\sin^2 \Theta} & 0 \end{bmatrix}$$

the rest of the notation is standard.

Complement the system of equations (4.3.1), (4.3.2) by an expression for a free term $\mathbf{f}$ characterizing the effect of tide-generating forces

$$f = gD \cdot \nabla \zeta^+ \qquad (4.3.3)$$

and by the relation for the static tide ζ^+ which, for instance, for the M_2 harmonic of the tide-generating forces and for mean astronomic conditions can be

presented as
$$\zeta^+(\text{cm}) = 24.25 \sin^2 \Theta \cos(\sigma t + 2\lambda), \qquad (4.3.4)$$
where σ is the M_2 wave frequency.

As boundary conditions use the no-slip condition on the contour Γ
$$\mathbf{w}|_\Gamma = 0 \qquad (4.3.5)$$
and the condition
$$\mathbf{w}, \zeta|_{t=0} = 0, \qquad (4.3.6)$$
which means that at the initial moment the ocean is in a state of rest.

Demonstration of unique solvability of problem (4.3.1)—(4.3.6) is carried out in exactly the same manner as for the case of the Cartesian system of coordinates (see Chapter 2). It will be briefly copied here with account taken of the Earth's sphericity.

Let us first prove the uniqueness theorem.

THEOREM. *If S is an arbitrary domain of Θ and λ lying in a spherical rectangle $\{0 < \Theta_0 \leqslant \Theta \leqslant \Theta_1 < \pi; 0 \leqslant \lambda \leqslant 2\pi\}$, then the solution of problem (4.3.1)—(4.3.6) does exist and it is unique.*

Assume that there are two solutions $(\mathbf{w}_1, \zeta_1)$, $(\mathbf{w}_2, \zeta_2)$ for problem (4.3.1)—(4.3.6). Subtract then equations (4.3.1), (4.3.2) as written for the functions $(\mathbf{w}_1, \zeta_1)$ and $(\mathbf{w}_2, \zeta_2)$ term by term. Multiply the resulting expressions by $\mathfrak{W} = \mathbf{w}_1 - \mathbf{w}_2$ and $\mathfrak{z} = \zeta_1 - \zeta_2$, respectively, and integrate them over the domain S. After summation and allowances made for boundary condition (4.3.5) we obtain

$$\begin{aligned}
\frac{d}{dt} \int_S \frac{1}{2}(\mathfrak{W}^2 + gD\mathfrak{z}^2)\, dS &+ \varepsilon \int_S \frac{|\mathbf{w}_1|}{D^2} \mathfrak{W}^2 \, dS \\
&+ \frac{k_h}{a^2} \int_S \left(\left| \frac{\partial \mathfrak{W}}{\partial \Theta} \right|^2 + \left| \frac{\partial \mathfrak{W}}{\sin \Theta \, \partial \lambda} \right|^2 \right) dS \\
&= g \int_S \mathfrak{z}\mathfrak{W} \cdot \nabla D \, dS + \varepsilon \int_S \frac{\mathbf{w}_2}{D^2}(|\mathbf{w}_1| - |\mathbf{w}_2|)\mathfrak{W} \, dS \\
&+ \frac{k_h}{a^2} \int_S \left(-\frac{\cos 2\Theta}{\sin^2 \Theta} |\mathfrak{W}|^2 + 4 \frac{\cotan \Theta}{\sin \Theta} \frac{\partial v}{\partial \lambda} u \right) dS, \quad (4.3.7)
\end{aligned}$$

where $dS = a^2 \sin \Theta \, d\Theta \, d\lambda$ is an element of the area.

Estimate now the first and the last brackets in the right-hand part of equation (4.3.7) by the Cauchy and Young inequality, and the second bracket by the Hölder inequality.

In line with the conditions of the theorem $\sin\Theta > \min(\sin\Theta_0, \sin\Theta_1) \equiv \delta > 0$. Thus, introducing the notation

$$\mathfrak{I}(t) = \tfrac{1}{2}\int_S (\mathfrak{W}^2 + gD\mathfrak{z}^2)\,dS$$

we obtain the inequality

$$\frac{d}{dt}\mathfrak{I}(t) \leqslant C\mathfrak{I}(t).$$

As $\mathfrak{I}(0) = 0$ [see equation (4.3.6)], then $\mathfrak{I}(t) = 0$ at any t and, hence, $\mathfrak{W}, \mathfrak{z} \equiv 0$ throughout the domain S.

Demonstration of the solvability of problem (4.3.1)—(4.3.6) for a finite interval of time variation $[0, T]$ is as follows.

First of all find approximate solutions for $\mathbf{w}^n$, ζ^n by the Galerkin method. Their "energy" estimations are as follows

$$\int_S \tfrac{1}{2}(\mathbf{w}^2 + gD\zeta^2)\,dS + k_h \int_0^T\!\!\int_S (\nabla\mathbf{w})^2\,dS\,dt \leqslant C_1(T);$$

$$\int_S \frac{1}{2}\left[\left(\frac{\partial\mathbf{w}}{\partial t}\right)^2 + gD\left(\frac{\partial\zeta}{\partial t}\right)^2\right]dS$$

$$+ k_h \int_0^T\!\!\int_S \left(\frac{\partial}{\partial t}\nabla\mathbf{w}\right)^2 dS\,dt \leqslant C_1(T). \qquad (4.3.8)$$

The latter are deduced from the integral law of energy conservation. The index n is omitted here, since estimations (4.3.8) are uniform with respect to n.

Then on the basis of equation (4.3.8) one can prove unique solvability of the resulting Galerkin systems of ordinary differential equations and convergence of sequences $\mathbf{w}^n$, ζ^n to certain limits which are solutions for problem (4.3.1)—(4.3.6).

The task now is to elucidate the time-varying character of the solution. We shall restrict ourselves only to the proof that at any values of the bottom friction coefficient ι problem (4.3.1)—(4.3.6) has a periodic solution.

This statement is proved in the same way as in the case discussed in Chapter 2.

Difference scheme

While numerically solving system (4.3.1)—(4.3.6) make use of the method of fractional steps (see Chapter 3). In spherical coordinates the difference scheme of this method can be written as follows:

$$\frac{v^{\kappa+1/2}-v^{\kappa}}{\tau} - \omega u^{\kappa} \cos\Theta = -\frac{gD}{2a}\frac{\partial}{\partial\Theta}(\zeta^{\kappa}-\zeta^{+\kappa+1/2})$$

$$-\frac{r_1^{\kappa}}{2}v^{\kappa+1/2}+\frac{K_h}{2a^2}\left[\frac{v_{\lambda\lambda}^{\kappa+1/2}}{\sin^2\Theta}+\frac{1}{\sin\Theta}(v_\Theta^{\kappa}\sin\Theta)_{\bar\Theta}\right.$$

$$\left.+(1-\cotan^2\Theta)v^{\kappa+1/2}-\frac{2\cos\Theta}{\sin^2\Theta}u_\lambda^{\kappa}\right];$$

$$\frac{u^{\kappa+1/2}-u^{\kappa}}{\tau} + \omega v^{\kappa+1/2}\cos\Theta = -\frac{gD}{2a}\frac{\partial}{\sin\Theta\,d\lambda}(\zeta^{\kappa}-\zeta^{+\kappa+1/2})$$

$$-\frac{r_1^{\kappa}}{2}u^{\kappa+1/2}+\frac{K_h}{2a^2}\left[\frac{u_{\lambda\lambda}^{\kappa}}{\sin^2\Theta}+\frac{1}{\sin\Theta}(u_\Theta^{\kappa+1/2}\sin\Theta)_{\bar\Theta}\right.$$

$$\left.+(1-\cotan^2\Theta)u^{\kappa+1/2}+\frac{2\cos\Theta}{\sin^2\Theta}v_\lambda^{\kappa+1/2}\right]; \qquad (4.3.10)$$

$$\frac{u^{\kappa+1}-u^{\kappa+1/2}}{\tau} + \omega v^{\kappa+1/2}\cos\Theta = -\frac{gD}{2a}\frac{\partial}{\sin\Theta\,\partial\lambda}(\zeta^{\kappa+1}-\zeta^{+\kappa+1})$$

$$-\frac{r_1^{\kappa+1/2}}{2}u^{\kappa+1}+\frac{K_h}{2a^2}\left[\frac{u_{\lambda\lambda}^{\kappa+1}}{\sin^2\Theta}+\frac{1}{\sin\Theta}(u_\Theta^{\kappa+1/2}\sin\Theta)_{\bar\Theta}\right.$$

$$\left.+(1-\cotan^2\Theta)u^{\kappa+1}+\frac{2\cos\Theta}{\sin^2\Theta}v_\lambda^{\kappa+1/2}\right]; \qquad (4.3.11)$$

$$\frac{v^{\kappa+1}-v^{\kappa+1/2}}{\tau} - \omega u^{\kappa+1}\cos\Theta = -\frac{gD}{2}\frac{\partial}{\partial\Theta}(\zeta^{\kappa+1}-\zeta^{+\kappa+1})$$

$$-\frac{r_1^{\kappa+1/2}}{2}v^{\kappa+1}+\frac{K_h}{2a^2}\left[\frac{v_{\lambda\lambda}^{\kappa+1/2}}{\sin^2\Theta}+\frac{1}{\sin\Theta}(v_\Theta^{\kappa+1}\sin\Theta)_{\bar\Theta}\right.$$

$$\left.+(1-\cotan^2\Theta)v^{\kappa+1}-\frac{2\cos\Theta}{\sin^2\Theta}u_\lambda^{\kappa+1}\right]; \qquad (4.3.12)$$

$$\frac{\partial\zeta^{\kappa+1}}{\partial\Theta} = \frac{\partial\zeta^{\kappa}}{\partial\Theta} - \frac{\tau}{a}\left\{\frac{1}{\sin\Theta}\left[\left(\frac{v^{\kappa+1}+v^{\kappa+1/2}}{2}\sin\Theta\right)_\Theta\right.\right.$$

$$\left.\left.+\left(\frac{u^{\kappa+1/2}+u^{\kappa}}{2}\right)_\lambda\right]\right\}_{\bar\Theta}. \qquad (4.3.13)$$

$$\frac{\partial\zeta^{\kappa+1}}{\partial\Theta} = \frac{\partial\zeta^{\kappa}}{\partial\lambda} + \frac{\tau}{a}\left\{\frac{1}{\sin\Theta}\left[\left(\frac{v^{\kappa+1/2}+v^{\kappa}}{2}\sin\Theta\right)_\Theta\right.\right.$$

$$\left.\left.+\left(\frac{u^{\kappa+1}+u^{\kappa+1/2}}{2}\right)_\lambda\right]\right\}_{\bar\lambda}, \qquad (4.3.14)$$

where
$$u, v|_{\Gamma_h} = 0;$$

$$u, \quad v, \quad \frac{\partial \zeta}{\partial \Theta}, \quad \frac{\partial \zeta}{\partial \lambda}\bigg|_{t=0} = 0. \qquad (4.3.15)$$

Here τ is the time step, κ is its number, $r_1 = (t/D_2)(u^2+v^2)^{1/2}$, Γ_h is the boundary of the grid domain S_h, lower indices denoting directed difference relations (those without a bar are forward, those barred are backward-directed relations).

Let us begin investigating scheme (4.3.9)—(4.3.15) with regard to its unique solvability. For this purpose consider a homogeneous algebraic system of equations (4.3.9)—(4.3.10) at the layer $\kappa+1/2$:

$$\frac{v^{\kappa+1/2}}{\tau} + \frac{r_1^\kappa}{2} v^{\kappa+1/2} = \frac{\kappa_h}{2a^2}\left(\frac{1}{\sin^2 \Theta} v_{\bar{\lambda}\lambda}^{\kappa+1/2} - \frac{\cos 2\Theta}{\sin^2 \Theta} v^{\kappa+1/2}\right); \qquad (4.3.16)$$

$$\frac{u^{\kappa+1/2}}{\tau} + \omega v^{\kappa+1/2} \cos \Theta + \frac{r_1^\kappa}{2} u^{\kappa+1/2} = \frac{\kappa_h}{2a^2}\left[\frac{1}{\sin \Theta}(u_\Theta^{\kappa+1/2} \sin \Theta)_{\bar{\Theta}}\right.$$

$$\left. - \frac{\cos 2\Theta}{\sin^2 \Theta} u^{\kappa+1/2} + \frac{2\cos \Theta}{\sin^2 \Theta} v_\lambda^{\kappa+1/2}\right]. \qquad (4.3.17)$$

Multiply equation (4.3.16) by $\tau h^2 v^{\kappa+1/2}$ (here h is the angular grid step) and sum up the resulting expression over the whole domain S_h. Employing the boundary conditions from equation (4.3.15) and known algebraic relations, we get

$$\|v^{\kappa+1/2}\|^2 + \frac{\tau h^2}{2}\sum_{S_h} r_1^\kappa (v^{\kappa+1/2})^2 = \frac{\tau \kappa_h}{2a^2}\left(\left\|\frac{v_\lambda^{\kappa+1/2}}{\sin \Theta}\right\|^2\right.$$

$$\left. - h^2 \sum_{S_h} \frac{\cos 2\Theta}{\sin^2 \Theta}(v^{\kappa+1/2})^2\right). \qquad (4.3.18)$$

At $\tau < (2a^2/\kappa_h)\delta^2$ it yields $v^{\kappa+1/2} \equiv 0$ throughout the domain S_h. It can be now easily proved [see equation (4.3.17)] that $u^{\kappa+1/2} \equiv 0$ in S_h. Solutions for the algebraic system (4.3.11), (4.3.12) at the layer $\kappa+1$ can be demonstrated to equal zero in an analogous way. Thus, unique solvability for scheme (4.3.9)—(4.3.15) can be considered proved.

Let us now analyse the stability of the difference scheme. For this purpose multiply equation (4.3.9) by $2\tau h^2 v^{\kappa+1} \sin \Theta$, equation (4.3.10) by $2\tau h^2 u^{\kappa+1/2} \sin \Theta$, equation (4.3.11) by $2\tau h^2 u^{\kappa+1} \sin \Theta$ and equation (4.3.12) by $2\tau h^2 v^{\kappa+1} \sin \Theta$. Add the resulting expressions and sum them up over the domain S_h. Summation by parts yields the following identity:

$$\|\sqrt{\sin\Theta}\,\mathbf{w}^{\kappa+1}\|^2 - \|\sqrt{\sin\Theta}\,\mathbf{w}^\kappa\|^2 + \tau^2(\|\sqrt{\sin\Theta}\,\mathbf{w}_{\bar{t}}^{\kappa+1/2}\|^2$$
$$+ \|\sqrt{\sin\Theta}\,\mathbf{w}_{\bar{t}}^{\kappa+1}\|^2) + \tau h^2 \sum_{S_h} [r_1^\kappa(\mathbf{w}^{\kappa+1/2})^2$$
$$+ r_1^{\kappa+1/2}(\mathbf{w}^{\kappa+1})^2] + \frac{\tau K_h}{2a^2}\left(\left\|\frac{\mathbf{w}_\lambda^{\kappa+1/2}}{\sqrt{\sin\Theta}}\right\|^2\right.$$
$$+ \|\sqrt{\sin\Theta}\,\mathbf{w}_\Theta^{\kappa+1/2}\|^2 + \|\sqrt{\sin\Theta}\,\mathbf{w}_\Theta^{\kappa+1}\|^2$$
$$\left.+ \left\|\frac{\mathbf{w}_\lambda^{\kappa+1}}{\sqrt{\sin\Theta}}\right\|^2\right) = -\frac{g\tau}{2a}\mathfrak{T}$$
$$+ \frac{\tau K_h}{2a^2}\left[\tau h^2 \sum_{S_h}\left(v_\Theta^\kappa v_{\Theta\bar{t}}^{\kappa+1/2}\sin\Theta + \frac{1}{\sin\Theta}u_\lambda^\kappa u_{\lambda\bar{t}}^{\kappa+1/2}\right.\right.$$
$$\left.+ u_\Theta^{\kappa+1/2}u_{\Theta\bar{t}}^{\kappa+1}\sin\Theta + \frac{1}{\sin\Theta}v_\lambda^{\kappa+1/2}v_{\lambda\bar{t}}^{\kappa+1}\right)$$
$$- \frac{\cos 2\Theta}{\sin\Theta}(\|\mathbf{w}^{\kappa+1/2}\|^2 + \|\mathbf{w}^{\kappa+1}\|^2)$$
$$+ h^2 \sum_{S_h}(v_\lambda^{\kappa+1/2}u^{\kappa+1/2} - u_\lambda^\kappa v^{\kappa+1/2} + v_\lambda^{\kappa+1/2}u^{\kappa+1}$$
$$\left.- u_\lambda^{\kappa+1}v^{\kappa+1})\cotan\Theta\right] + \frac{g\tau h^2}{a}\sum_{S_h}D\left[\left(v^{\kappa+1/2}\frac{\partial\zeta^{+\kappa+1/2}}{\partial\Theta}\right.\right.$$
$$+ v^{\kappa+1}\frac{\partial\zeta^{+\kappa+1}}{\partial\Theta}\right)\sin\Theta + \left(u^{\kappa+1/2}\frac{\partial\zeta^{+\kappa+1/2}}{\partial\lambda}\right.$$
$$\left.\left.+ u^{\kappa+1}\frac{\partial\zeta^{+\kappa+1}}{\partial\lambda}\right)\right], \qquad (3.4.19)$$

where

$$\mathfrak{T} = h^2 \sum_{S_h}D\left[\left(v^{\kappa+1/2}\frac{\partial\zeta^\kappa}{\partial\Theta} + v^{\kappa+1}\frac{\partial\zeta^{\kappa+1}}{\partial\Theta}\right)\sin\Theta\right.$$
$$\left.+ \left(u^{\kappa+1/2}\frac{\partial\zeta^\kappa}{\partial\lambda} + u^{\kappa+1}\frac{\partial\zeta^{\kappa+1}}{\partial\lambda}\right)\right].$$

Transform the expression for $\mathfrak{T}$ using relations (4.3.13), (4.3.14) as well as formulae for summation by parts. In this case

$$\mathfrak{I} = \frac{\tau}{4}\bigg[||\mathfrak{D}^{\kappa+1}||^2 - ||\mathfrak{D}^{\kappa}||^2 + \frac{1}{2}(||\mathfrak{d}_1^{\kappa+1} - \mathfrak{d}_2^{\kappa+1}||^2 - ||\mathfrak{d}_1^{\kappa+1/2} - \mathfrak{d}_2^{\kappa+1/2}||^2$$

$$- 2h^2 \sum_{S_h} (\mathfrak{d}_2^{\kappa+1}\mathfrak{d}_1^{\kappa+1/2} + \mathfrak{d}_1^{\kappa+1}\mathfrak{d}_2^{\kappa+1/2})\bigg]$$

$$+ h^2 \sum_{S_h} \mathfrak{D}^{\kappa+1}[(v^{\kappa+1} + v^{\kappa+1/2})D_\Theta \sqrt{\sin\Theta}$$

$$+ (u^{\kappa+1} + u^{\kappa+1/2})\frac{D_\lambda}{\sqrt{\sin\Theta}}\bigg] - h^2 \sum_{S_h}\bigg[(\mathfrak{d}_1^{\kappa+1} + \mathfrak{d}_2^{\kappa+1}$$

$$+ \mathfrak{d}_1^{\kappa+1/2} + \mathfrak{d}_2^{\kappa+1/2} + \mathfrak{d}_2^{\kappa})v^{\kappa+1/2}D_\Theta \sqrt{\sin\Theta}$$

$$- (\mathfrak{d}_1^{\kappa+1} + \mathfrak{d}_2^{\kappa+1} + \mathfrak{d}_1^{\kappa+1/2} + \mathfrak{d}_2^{\kappa+1/2} + \mathfrak{d}_1^{\kappa})u^{\kappa+1/2}\frac{D_\lambda}{\sqrt{\sin\Theta}}$$

$$- \bigg(\mathfrak{d}_1^{\kappa+1}v^{\kappa+1}D_\Theta \sqrt{\sin\Theta} + \mathfrak{d}_1^{\kappa+1}u^{\kappa+1}\frac{D_\lambda}{\sqrt{\sin\Theta}}\bigg)\bigg], \quad (4.3.20)$$

where

$$\mathfrak{d}_1^{\kappa} = \sqrt{\frac{D}{\sin\Theta}}(v^{\kappa}\sin\Theta)_\Theta,$$

$$\mathfrak{d}_2^{\kappa} = \sqrt{\frac{D}{\sin\Theta}}v_\lambda^{\kappa};$$

$$\mathfrak{D}^{\kappa} = \sum_{j=0}^{\kappa}(\mathfrak{d}_1^j + \mathfrak{d}_2^j) + \sum_{j=0}^{\kappa-1}(\mathfrak{d}_1^{j+1/2} + \mathfrak{d}_1^{j+1/2});$$

$$\mathfrak{D}^{\kappa+1/2} = \sum_{j=0}^{\kappa}(\mathfrak{d}_1^j + \mathfrak{d}_2^j) + \sum_{j=0}^{\kappa}(\mathfrak{d}_1^{j+1/2} + \mathfrak{d}_2^{j+1/2}).$$

Substitute now equation (4.3.20) into equation (4.3.19) and estimate the right-hand part of this relation by the Cauchy and Young inequalities, and by inequality $||u_\Theta|| \leq 2/h\,||u||$. Summation over j from 0 to κ and omitting the positive terms in the left-hand part yields the following *a priori* estimation:

$$||\mathbf{w}^{\kappa+1}||^2 + \frac{g\tau^2}{4}||\mathfrak{D}^{\kappa+1}||^2 + \frac{\tau\kappa_h}{a^2}\bigg(1 - \frac{2\tau\kappa_h}{(ah\delta)^2}\bigg)\sum_{j=0}^{\kappa}(||\mathbf{w}_\Theta^{j+1}||^2 + ||\mathbf{w}_\lambda^{j+1}||^2$$

$$+ ||\mathbf{w}_\Theta^{j+1/2}||^2 + ||\mathbf{w}_\lambda^{j+1/2}||^2) \leq C_3\tau \sum_{j=0}^{\kappa+1}\bigg(||\mathbf{w}||^2 + \frac{g\tau^2}{4}||\mathfrak{D}^j||^2$$

$$+ ||\mathbf{w}^{j-1/2}||^2 + \frac{g\tau^2}{4}||\mathfrak{D}^{j-1/2}||^2\bigg) + C_4\tau, \quad (4.3.21)$$

where

$$C_3 = \frac{2g\tau^2 \max_{S_h} \mathfrak{D}}{(ah\delta)^2} + \frac{\omega\tau}{\delta} + \frac{\tau M_h}{\delta^2} \sqrt{\frac{g}{\min_{S_h} \mathfrak{D}}} + \frac{5g\tau^2 M_h}{ah\delta^2} + \frac{2\tau\kappa_h}{a^2h^2\delta^2};$$

$M_h = 1/a \max |\nabla \mathfrak{D}|$; C_4 is a κ-independent constant.

An analogous inequality can be written for fractional time levels as well:

$$\|\mathbf{w}^{\kappa+1/2}\|^2 + \frac{g\tau^2}{4}\|\mathfrak{D}^{\kappa+1/2}\|^2 + \frac{\tau\kappa_h}{a^2}\left(1 - \frac{2\tau\kappa_h}{(ah\delta)^2}\right)\sum_{j=0}^{\kappa}(\|\mathbf{w}_\Theta^{j+1/2}\|^2$$

$$+\|\mathbf{w}_\lambda^{j+1/2}\|^2 + \|\mathbf{w}_\Theta^j\|^2 + \|\mathbf{w}_\lambda^j\|^2) \leq C_3\tau \sum_{j=0}^{\kappa} \left(\|\mathbf{w}^{j+1/2}\|^2\right.$$

$$\left.+\frac{g\tau^2}{4}\|\mathfrak{D}^{j+1/2}\|^2 + \|\mathbf{w}^j\|^2 + \frac{g\tau^2}{4}\|\mathfrak{D}^j\|^2\right) + C_4\tau. \qquad (4.3.22)$$

Let the following conditions be fulfilled:

$$\frac{2\tau\kappa_h}{(ah\delta)^2} < 1; \qquad (4.3.23)$$

$$\frac{2g\tau \max_{S_h} \mathfrak{D}}{(ah\delta)^2} + \frac{\tau M_h}{\delta^2}\sqrt{\frac{g}{\min_{S_h} \mathfrak{D}}} + \frac{5g\tau^2 M_h}{ah\delta^2} + \frac{2\tau\kappa_h}{a^2h\delta^2} + \frac{\omega\tau}{\delta} < 1. \qquad (4.3.24)$$

Note incidentally that the above conditions are naturally generalized conditions (4.3.20), (4.3.22) for the case when the Earth's sphericity is taken into account.

Then for

$$\|\mathfrak{I}^{\kappa+1}\|^2 = \frac{1}{2}\left(\|\mathbf{w}^\kappa\|^2 + \frac{g\tau^2}{4}\|\mathfrak{D}^\kappa\|^2\right)$$

equations (4.3.21), (4.3.22) yield the following estimation:

$$\|\mathfrak{I}^\kappa\| \leq C_5(T), \quad \kappa = 0, 1/2, 1, \ldots \qquad (4.3.25)$$

The estimation obtained guarantees stability of the difference scheme (4.3.9)—(4.3.15) at any finite interval of time change [0, T].

Convergence of scheme (4.3.9)—(4.3.15) to the exact solution of problem (4.3.1)—(4.3.6) is proved in the same manner as it was carried out in the preceding chapter for the difference equation of the method of fractional steps presented in Cartesian coordinates.

Results of calculation and their correlation with observational data from islands in the open sea

Numerical integration of the system of equations (4.3.1)—(4.3.6) was carried out on a regular latitude–longitudinal grid covering a part of the sphere between the 65°N and 70°S parallels. This zonal belt did not include the continental regions; the latter were combined into two groups, one of them including the mainlands of America, Africa, Eurasia, and Australia, the other comprising the Antarctic Continent. The boundary of the first continental region was drawn along the 1000-metre isobath and then approximated by segments of arcs coinciding with the meridians and parallels. The boundary of the second region was drawn along the circle of 70°S latitude. The grid nodes located on this boundary were also ascribed to the depth equal to 1000 metres. Depths at the inner nodes were determined by the chart of smoothed bottom contours presented in the paper by Pekeris and Accad (1969).

Note that matching of the ocean boundary with the 1000-metre isobath means, in fact, neglecting the continental shelf zone which cannot be accurately allowed for by a regular grid. This difficulty is usually eliminated by the use of irregular grids with node concentrations in continental shelf zones.

However, in real conditions, when the basin under investigation can have any shape whatever, and the shelf width varies from point to point, construction of such a grid appears to be much more complex. Besides, in this case we are concerned with the loss of accuracy in approximation.

In the calculations under consideration a grid step was assumed constant in space and equal to 5°, with a time step equal to 6 minutes. Values of the coefficients of the bottom and horizontal turbulent friction were set equal to 3×10^{-3} and 10^7 cm²/s, respectively. Note, that the chosen value of the coefficient of the horizontal turbulent friction was one to two orders lower than that characterizing the intensity of macroturbulent exchange in tidal motions. This deliberate lowering of the κ_h value was aimed at weakening the effect of the horizontal turbulent friction and at approximating the initial equations of tidal dynamics to their traditional form used for analytical solution.

Calculations of tidal motion from an initial state of rest were carried out until the solution became periodic. The criterion for the periodic regime to be set was the condition of coincidence of the mean-kinetic energy throughout the region in two successive tidal periods, to the accuracy of 1%. For the M_2 harmonic this condition was fulfilled as soon as twenty-two tidal periods were over.

The resulting chart of isoamplitudes and cotidal lines of the M_2 wave is given in Fig. 4.9. Before analysing this picture of tides in the World Ocean, let

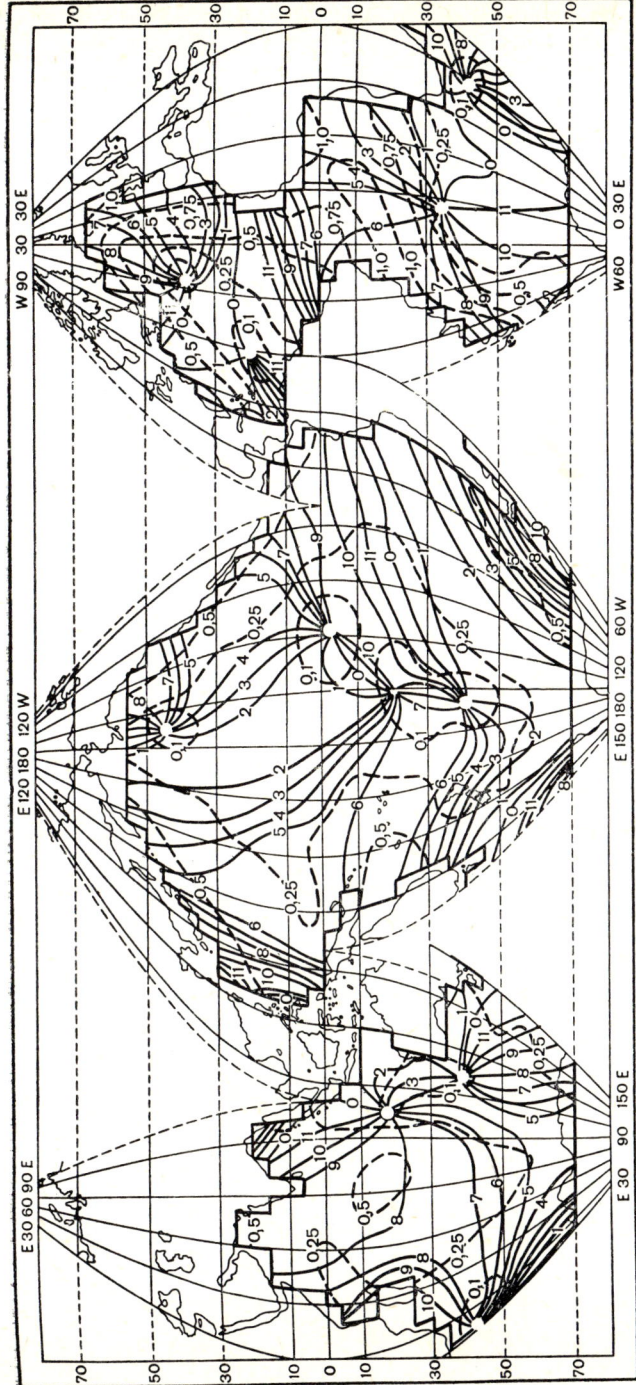

FIG. 4.9. Tidal chart of the M_2 wave in the World Ocean (according to Gordeev, Kagan and Rivkind, 1973). Solid lines denote cotidals, dashed lines denote isoamplitudes. The time of high water with respect to the moment of the Moon's transit over the Greenwich meridian is given in mean lunar hours, the tidal amplitudes in metres.

us correlate the results of calculation with the observational data in the open sea, which is no less important.

We have already mentioned that such observational data are rather scarce and their lack cannot be compensated by increasing the number of data from coastal elevations because of shelf effects unaccounted for in calculations. These effects (coastal dissipation, refraction and multiple reflection of the tidal wave from the shore and the outer shelf edge) are known to influence tidal phenomena in coastal oceanic regions appreciably. They can distort characteristics of the tidal wave on the shelf to such an extent that the data from coastal elevations will not correspond to the conditions of the open sea at all.

That is why correlation of calculated charts of tides in the World Ocean with empirical charts constructed from observational data, partly insular but mainly coastal, requires a certain degree of precaution. In all probability, one can speak only of qualitative agreement of these charts; which is, to judge by Figs. 4.5—4.7, though not faultless (which is in principle impossible), nevertheless not too bad.

Let us now try to elucidate the possibility of numerical modelling from the standpoint of reproducing regular quantitative features. The results of comparing calculated values of the harmonic constants with observational data from island stations and with deepwater measurements in the open sea are listed in Tables 4.2, 4.3, respectively. As may be seen, greatest discrepancies are found at the points located on the Mayotta and Mauritius islands (the Indian Ocean), on Ascension, Saint Helena and Trinidad islands (the Atlantic Ocean), on the Kuril, Nuku Hiva and Florida islands, on the Josie II station, on the Arno and Millie atolls (northern part of the Pacific Ocean) and on the New Zealand islands (southern part of the Pacific Ocean). These discrepancies can be accounted for by many factors, in particular, by the influence of local peculiarities of the bottom contours and by the coastline configuration.

To elucidate the above statements, let us consider the way of presenting islands in our calculations including such big ones as Madagascar and New Zealand. The first, for instance, was combined with the mainland, while the New Zealand islands were modelled by elevating the bottom to the depth of 1000 m. It is quite clear that in this case one cannot take account of all peculiarities of tide-generation in the Mozambique channel, with the Mayotte island situated at its entrance, or in the region of the New Zealand plateau.

Discrepancies of calculated and observational values of the harmonic constants from the Arno and Millie atolls seem to be accounted for by resonance phenomena on the island shelf. The latter conclusion can be regarded as a verisimilar hypothesis brought about by investigations of trapping of long waves by isolated cylindrical islands.

TIDES IN THE WORLD OCEAN 139

TABLE 4.2. *Comparison of calculated values of harmonic constants of elevation with the observational data from islands*

Observational site	Coordinates		Observations		Calculations	
	latitude	longitude	amplitude, cm	phase, degree	amplitude, cm	phase, degree
Indian Ocean						
Middle Andaman	12°19′N	92°43′E	51.8	70	46	0
South Andaman	11 41	92 46	61	95	45	356
Minicoy	8 16	73 01	26.3	183	38	260
Nicobar Islands	8 03	93 29	49	84	37	341
Ihavandu	6 57	72 55	23.5	189	38	260
Sri Lanka	6 02	80 13	16.7	257	36	270
Simalur	4 35	95 32	13.1	330	21	337
Add atoll	0 34 S	73 13	29.3	254	33	255
Mahe	4 37	55 27	40.2	13	24	232
Diego Garciaatoll	7 21	72 57	52	265	47	249
Agalega Islands	10 25	56 40	28.2	290	32	230
Christmas	10 30	105 40	36.6	7	22	34
Mayotte	12 46	45 12	70	32	20	300
Cocos Islands	12 05	96 53	27.1	312	19	345
Madagascar	12 15	49 19	42.7	352	22	290
Mauritius	20 09	57 29	13.2	268	35	233
Reunion	20 55	55 17	14.4	328	33	237
Saint Paul	38 43	77 35	37.7	232	37	225
Prince Edward islands	46 53	37 51	10	65	13	217
Kerguelen	49 08	70 11	43.6	229	28	176
Herd	53 01	73 23	23	195	25	160
Northern part of the Pacific Ocean						
Aleutian islands	52°59′N	178°27′W	20.6	67	28	52
Honshu	36 56	140 55 E	31	210	61	213
Amami islands	28 27	129 39	64	300	70	310
Midway atoll	28 12	177 22 W	11.2	82	23	62
Okinawa	26 12	127 40 E	57.5	303	71	330
Oahu	21 18	157 52 W	15.9	62	18	57
Hawaii	19 44	155 04	22.3	31	18	56
Pagan	18 08	145 46 E	17.1	294	29	174
Rota	14 08	145 08	15.8	290	29	172
Fassarai atoll	9 55	139 40	37	294	29	190
Riau	8 36	152 15	12	113	25	174
Olimarao atoll	7 45	134 38	52	293	32	341
Oroluk atoll	7 40	155 10	18	129	25	162
Poulap atoll	7 39	149 25	4	130	25	171
Lamotraque atoll	7 28	146 23	14.9	297	25	171
Voleau atoll	7 22	143 54	18.9	302	27	174
Aten	7 22	151 53	6.4	129	26	172
Arno atoll	7 08	171 48	59	128	23	154
Ponape	6 59	158 13	24.6	125	24	161
Milie atoll	6 14	171 42	57	131	23	154
Palmyra	5 52	162 06 W	30.4	105	18	53
Kusaye	5 20	163 01 E	42	130	25	159
Fanning atoll	3 51	159 22 W	18.6	113	18	52
Christmas	1 59	157 28	26.1	73	16	52

OCEAN TIDES

TABLE 4.2. (cont.)

Observational site	Coordinates		Observations		Calculations	
	latitude	longitude	amplitude, cm	phase, degree	amplitude, cm	phase, degree
Southern part of the Pacific Ocean						
Galapagos islands	0°26′S	90°17′W	72	247	43	290
Ocean	0 52	169 35 E	48.9	149	27	171
Manus	2 01	147 16	8.8	274	27	176
Phoenix islands	2 50	171 43 W	35.4	168	20	132
Ellis islands	8 31	179 12 E	58.2	152	35	178
Nuku Hiva	8 55	140 06 W	47	36	10	330
Florida	9 11	160 13 E	9	149	47	187
Cook islands (North)	9 00	157 59 W	10.4	143	16	60
Danger islands	10 52	165 50	14.6	189	18	138
Rotuma	12 30	177 05	48.2	167	30	177
Upolu	13 49	171 46	37.7	172	19	170
Ache atoll	14 32	146 21	12	87	11	14
Espiritu	15 35	166 59 E	28.7	211	54	190
Society islands	16	150 W	12.2	8	10	28
Molecula	16 25	167 47 E	39.6	184	51	189
Vit Levu	17 21	177 49	58	176	36	187
Efate	17 35	168 15	38.7	191	51	187
Hao atoll	18 15	140 55 W	6.6	80	8	296
Cook islands (South)	18 51	159 47	17	209	8	160
New Caledonia	20 29	164 11 E	45	270	55	180
Tonga islands (Friendship)	21 08	175 13 W	52	192	28	190
Rarotonga	21 12	159 46	26.2	215	10	186
Gambier islands	23 07	134 58	27	348	16	320
Easter	27 09	109 27	20.8	16	26	26
Norfolk	29 00	167 55 E	57.9	263	35	153
Lord Howe	31 32	159 04	59	304	40	120
New Zealand (North I.)	38	174	64–121	178–289	23	120
New Zealand (South I.)	45	167	73–114	11–321	26	51
Karnley	50 52	166 05	33.5	36	16	28
Campbell	52 33	169 13	42.7	65	26	30
Macquarie	54 31	158 58	27.5	55	27	5
Northern part of the Atlantic Ocean						
Iceland	64°19′N	22°06′W	125	182	92	213
Lewis	58 12	6 23	135.2	179	102	167
Flores	39 23	31 11	39.6	60	33	120
San Miguel	37 44	25 41	47.3	68	48	118
Santa Maria	36 57	25 09	51	60	48	118
Portu-Monish	32 52	17 10	73.1	50	74	80
Madeira islands	32 38	16 54	72	45	74	80
Bermuda Is	32 19	64 50	37.6	359	27	25
Salvateck	30 02	16 03	76.2	27	70	35
Eleuthera	24 56	76 09	32.1	20	51	56

TABLE 4.2. (cont.)

Observational site	Coordinates		Observations		Calculations	
	latitude	longitude	amplitude, cm	phase, degree	amplitude, cm	phase, degree
	Northern part of the Atlantic Ocean					
Ackmens	22°10′	74 18	29.6	7	37	300
Saint Thomas	18 20	64 56	3.7	338	14	60
Saint Vincent	16 53	25 00	30.5	253	63	310
Guadeloupe	16 14	61 32	8.8	233	10	340
Martinique	14 36	61 05	5.5	211	11	335
Barbados	13 06	59 37	24.1	236	12	338
Grenadines islands	12 38	61 21	12	210	12	335
	Southern part of the Atlantic Ocean					
Fernando de Noronha	3°50′S	32°24′W	79	207	57	180
Ascension	7 55	14 25	32.8	177	103	154
Saint Helena	15 55	5 42	32	81	88	80
Trinidad	20 30	29 20	33.1	210	104	183
Tristan da Cunha	37 02	12 18	23.1	11.6	13	10
Falkland islands	51 42	57 51	45.4	261	94	270
South Georgia	54 31	36 01	22.6	285	17	292
Lorie	60 44	44 39	46.4	261	33	298
Deception	64 33	61 57	38	286	48	303

TABLE 4.3. *Comparison of calculated values of harmonic constants of elevation with deep-sea measurements in the open ocean*

Observational site	Coordinates		Observations		Calculations	
	latitude	longitude	amplitude, cm	phase degree	amplitude, cm	phase degree
Filloux	24°46.9′N	129°01.1′W	18.78	107.14	32	125
Katie	27 45	124 25.8	28.64	128.03	46	136
Josie I	31 01.7	119 47.9	42.59	142.07	54	138
Flicki	32 14.4	120 51.4	42.5	149.06	55	136
Josie II	34 00.3	144 59.7	27	267	19	125

Note one more detail in the tidal chart presented in Fig. 4.9, with the centre of the amphidrome in the South Atlantic located at 33°S, 18°W. According to observational data (see Fig. 4.8) its coordinates are approximately 28°S, 18°W. Thus, the centre of the amphidrome appeared to be shifted southwards by only 5°, i.e. by a grid step length. However, even this, generally speaking, allowable displacement was enough to cause a more than three-fold change in the amplitudes of the tidal oscillations at Ascension, Saint Helena and Trinidad islands. An analogous situation is observed in the south-eastern part of the Pacific Ocean, where the northwards displacement of the amphidrome caused a 4-hour change in the time of high water.

We can with some satisfaction state that the calculation errors enumerated above are quite rare (see Fig. 4.10), errors within $\pm(5-25)$ cm interval for the amplitudes and $\pm(0-30)$ degrees for the tidal phases being typical.

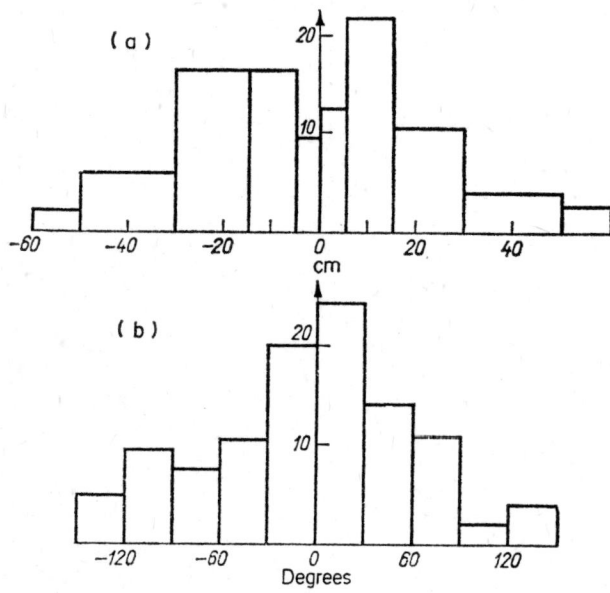

FIG. 4.10. Distribution of calculation discrepancies of amplitudes (a) and phases (b) of tidal elevations in the World Ocean. The abscissa is the difference of calculated and real values of the characteristic under discussion (amplitudes, in centimetres; phases, in degrees), the ordinate is the number of cases (in % from their total number), when this difference is observed.

To evaluate our calculations of tidal oscillations in each specific ocean and in the World Ocean as a whole, average discrepancies and means quare deviations for amplitudes and phases were calculated; these qualitative statistics are presented in Table 4.4. In scrutinizing this table, one cannot help noticing a comparatively large discrepancy between calculated and observational values of tidal amplitudes. This, however, is unavoidable on account of all the possible sources of error due to the simplified formulation of the problem; their number increased during the course of numerical solution.

TIDES IN THE WORLD OCEAN

TABLE 4.4. *Estimation of the quality of calculations of harmonic constants of elevation for the M_2 wave in some oceans and in the World Ocean as a whole*

Region of investigation	Mean calculation error		Mean-root-square deviation	
	amplitude, cm	phase, degree	amplitude, cm	phase degree
Indian Ocean	−2.8	−23	16.7	73
Atlantic Ocean	6.9	32.4	28.1	56.5
Northern part of the Pacific Ocean[†]	4.0	−17.8	17.2	61.9
Southern part of the Pacific Ocean	−10.4	−27.3	24.9	67.5
World Ocean	−0.8	−9.5	22.0	65

† The data on deep-sea level measurements in the open sea (see Table 4.3) were also used here.

Systematic errors of this kind are certain to be characteristic of all attempts at numerical modelling of tides in the World Ocean, based on "classical" equations of tidal dynamics with simplified boundary conditions.

4.4. Some Other Calculations of Tides in the World Ocean

The problem of quantitative description of tidal motions in the World Ocean has always been central in the theory of tides. In recent years it has developed especially rapidly due to the use of computers and the expanding network of recording stations. A number of publications referring to the problem may serve as indirect evidence of this. For instance, only during the last 10 years six tidal charts of the World Ocean for the semidiurnal wave M_2 have been constructed from calculations, outnumbering the charts made during the 130 years since the publication of the first cotidal chart by Whewell. The same principles were used in calculating these tidal charts, but the approaches to the problem are different. They may be arbitrarily classified in the following way.

Semi-empirical approach

This approach arose from the limited possibilities of the empirical approach and from the search for an objective means of analysing tides on vast areas of the World Ocean where no observational data are available.

The general trend of the works employing this approach (Bogdanov and Magarik, 1967; Tiron, Sergeev and Michurin, 1967; Hendershott and Munk,

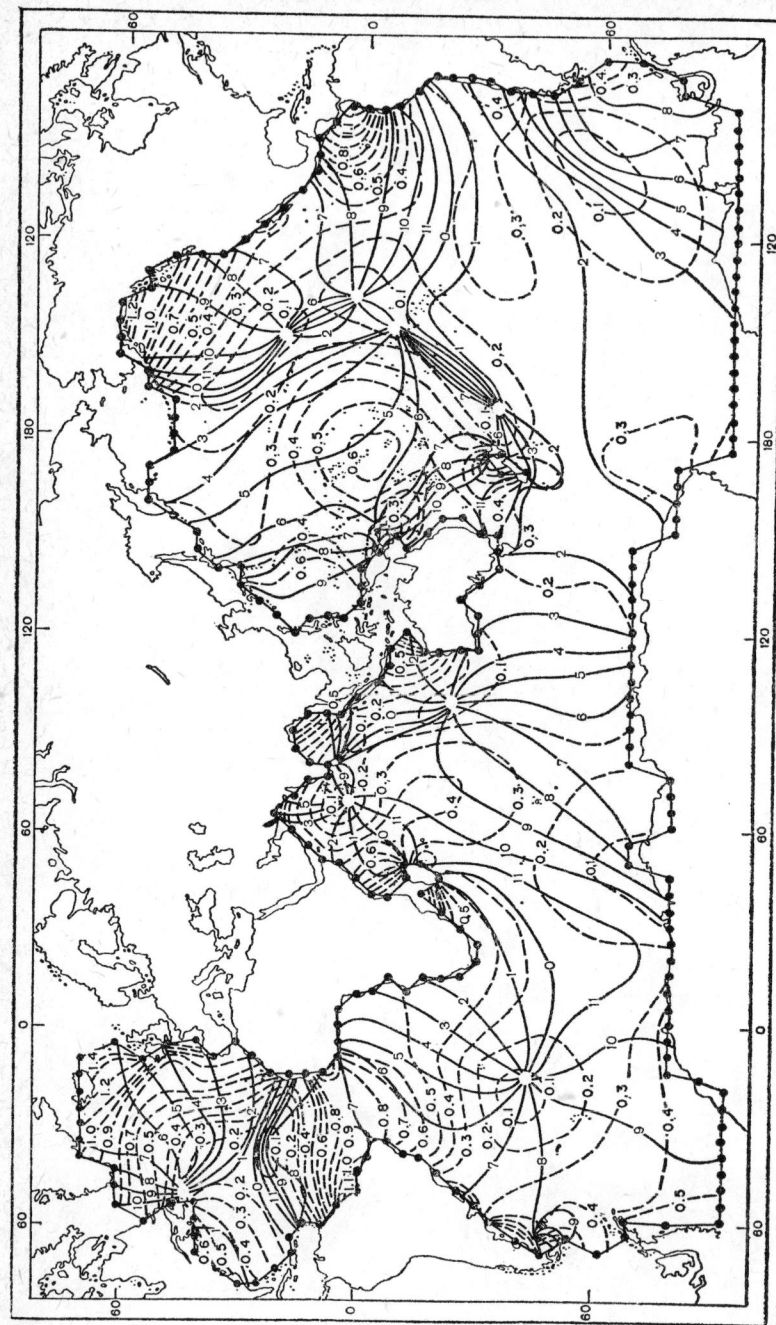

FIG. 4.11. Tidal chart of the M_2 wave in the World Ocean (according to Bogdanov and Magarik, 1967). Dots represent the points at which the values of the harmonic constants were considered given. The other notations are as in Fig. 4.9.

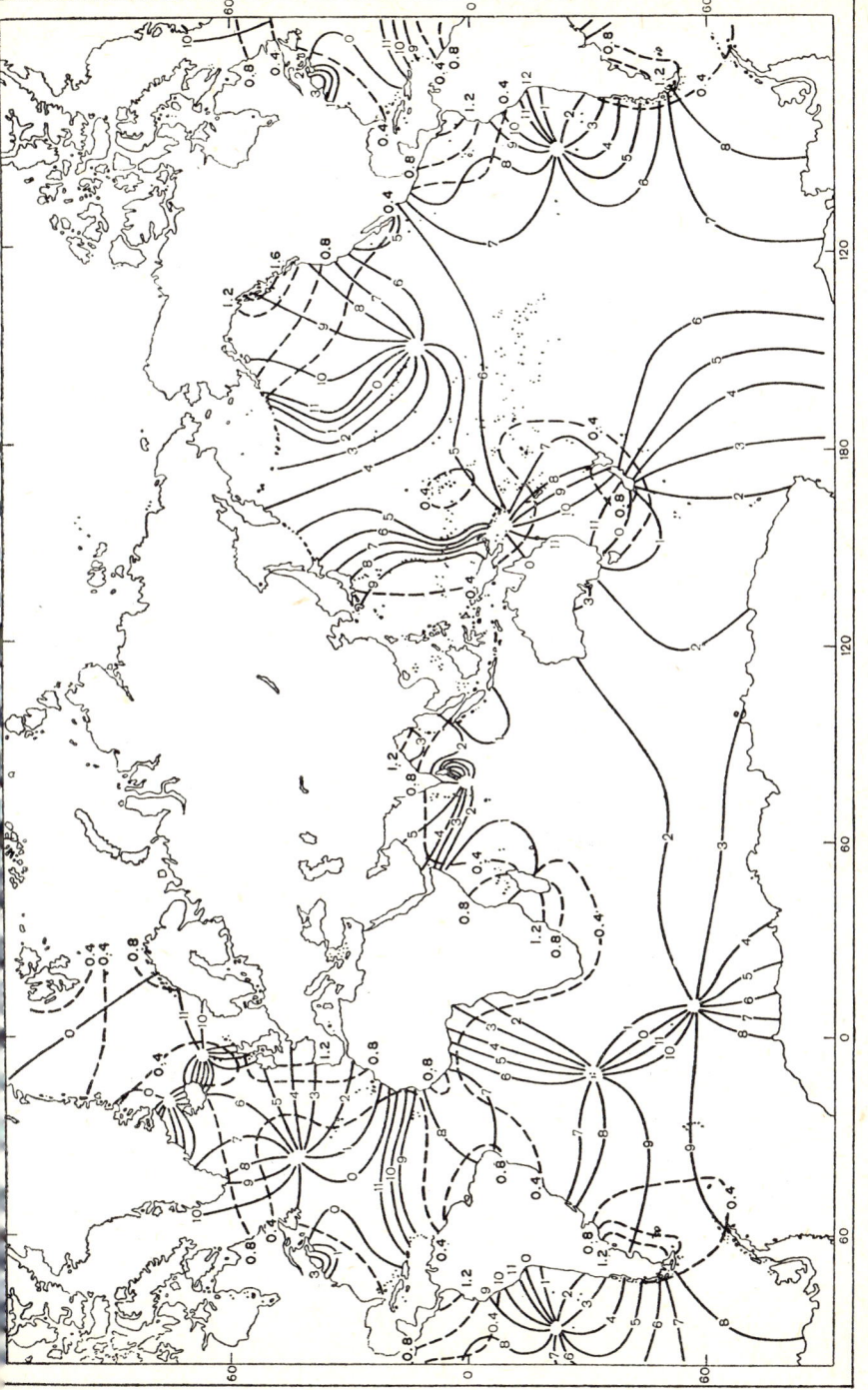

FIG. 4.12. Tidal chart of the M_2 wave in the World Ocean (according to Tiron, Sergeev and Michurin, 1967). For details see Fig. 4.9.

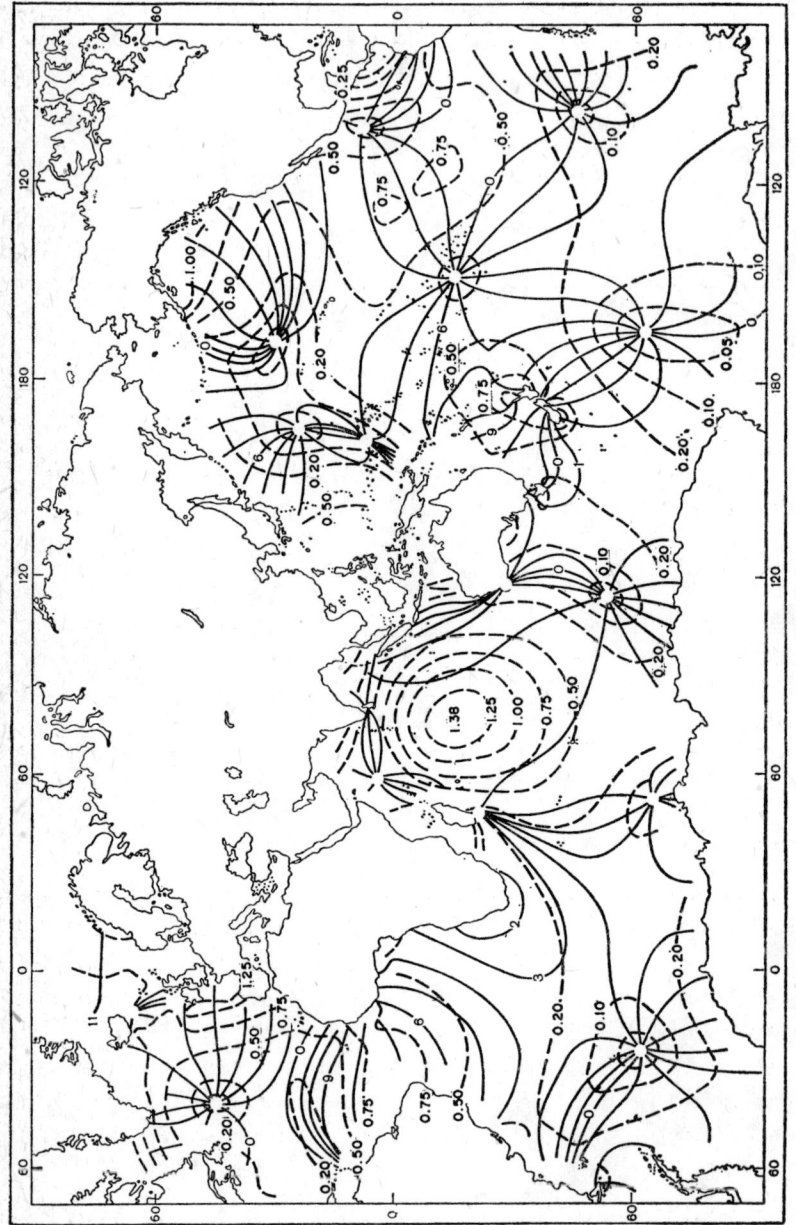

FIG. 4.13. Tidal chart of the M_2 wave in the World Ocean (according to Hendershott, 1972). For details see Fig. 4.9.

1970, see also Hendershott, 1972) is the use of the non-dissipative variation of the method of boundary values. Its essence, as has already been mentioned in Chapter 3, is reduced to numerical integration of the Laplace tidal equations under the supposition that the oscillations of tidal elements are harmonic in time, harmonic constants are considered set on the mainland coastlines and, when necessary, on islands (Tiron, Sergeev and Michurin, 1967).

Combination of the results of theoretical calculations with direct measurements should have resulted in making up a sufficiently detailed picture of tides in the World Ocean. However, this method was hampered by two difficulties which are as yet impossible to overcome: first, the fact that observational data from the coast of the Antarctic Continent[†] are scarce and often non-representative and, second, by "distortion" of harmonic constants on mainland coasts and islands due to tidal wave trapping on the shelf. The latter circumstance again raises the problem of the possibility of employing the available coastal data to describe tides in the open ocean. Hence, one cannot judge the validity of the tidal charts of the World Ocean made up by the semi-empirical approach until the above problem is solved (Figs. 4.11—4.13).

However, despite evident drawbacks, such calculations were not carried out in vain. The charts given in Figs. 4.11—4.13, due to their objectivity, depict regular qualitative features of the spatial structure of the tides better and more completely than the empirical charts. Another merit of the approach is that it provided investigators with a much better understanding of the peculiarities of the distribution of tidal amplitudes in the World Ocean. It will be recalled that the empirical charts gave information only of the times of high water.

Theoretical approach

An example of this approach to the problem of calculating tides in the World Ocean has already been considered in the preceding chapter. Here we shall demonstrate the results of other theoretical calculations in which the initial information contained data only on the potential of the tide-generating force, the coastline configuration and the bottom relief.

At present, only three calculations of the aforesaid type in addition to that discussed previously are available. One of them (Ueno, 1964) was not completed; the other two (Pekeris and Accad, 1969, Zahel, 1970) resulted in making

---

[†] Tiron, Sergeev and Michurin (1967) made an attempt to eliminate this difficulty by the assumption that the derivative of elevation normal to the coast is equal to zero. This condition results from equation (3.1.7) provided the Coriolis force is neglected.

FIG. 4.14. Tidal chart of the M_2 wave in the World Ocean (according to Pekeris and Accad, 1969). For details see Fig. 4.9.

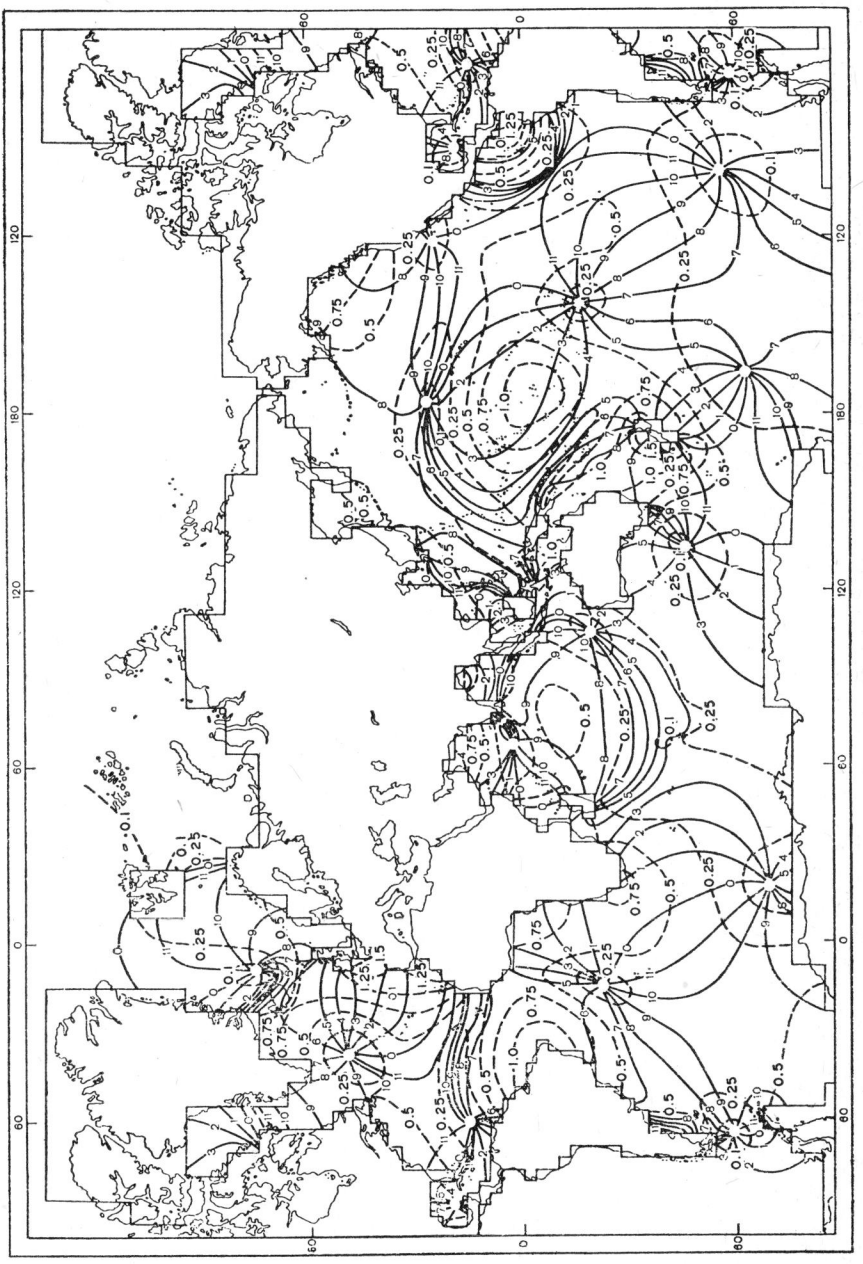

FIG. 4.15. Tidal chart of the M_2 wave in the World Ocean (according to Zahel, 1970). For details see Fig. 4.9.

up cotidal charts of the World Ocean corresponding to the M_2 harmonic of tide-generating forces (Figs. 4.14, 4.15).

These calculations were based on models which differ in fact from equations (4.3.1)—(4.3.6) only in the way of describing dissipative factors and the time variation of tidal characteristics, apart from the choice of difference schemes and methods of computer processing.

The analysis of the peculiarities of the above models is the subject of Chapter 3. Here we shall confine ourselves to mentioning the fact that in the Ueno model physical viscosity is not considered at all, while in the Pekeris and Accad model an analogous hypothesis is made only for horizontal turbulent friction. At the same time bottom friction is considered a linear function of velocity; a special expression accounting for the increase in friction in the coastal oceanic areas was introduced to determine the bottom friction coefficient. Finally, Zahel's model takes account of both mechanisms of tidal energy dissipation. Methods of presentation are traditional, but the value of the horizontal turbulent friction coefficient used in Zahel's calculations appears to be 2–3 orders greater than the conventional one.

Comparison of numerical solutions with empirical charts

We have already mentioned that in comparing calculated and empirical charts, only their qualitative correspondence can at best be derived. This is as good as one can expect, however, keeping in mind the complexity of conditions of tide generation in the World Ocean and the simplicity of the theoretical analysis tools employed.

Figures 4.5–4.7 and 4.9, 4.11–4.15 are seen to have much in common. For instance, all the charts present the North Atlantic and the Caribbean amphidromes, the zone of sharp change of phase in the tropical Atlantic and the antiamphidromic areas adjoining it to the north and south. All numerical solutions, except the one obtained by Hendershott demonstrate the existence of the amphidrome having counter-clockwise rotation in the South Atlantic. It also appears in the empirical tidal chart of the central South Atlantic (see Fig. 4.8).

Most of the numerical solutions fairly well demonstrate the antiamphidromic area in the central part of the Indian Ocean and the amphidromic system near the western coast of Australia. However, the Tiron, Sergeev and Michurin charts do not present this amphidrome at all, and Hendershott's chart depicts it only as a degenerate one.

All the charts except that by Tiron, Sergeev and Michurin distinctly depict the nodal zone bordering the central part of the Indian Ocean from the east and north-east. The north end of this nodal zone either rests against the western coast of the Indonesian subcontinent (charts by Zahel, Gordeev *et al.*) or the island of Sri Lanka (charts by Dietrich and Villain), or coincides with the centre of the amphidromic system which is again treated either as degenerate (Bogdanov and Magarik, and also Hendershott) or as real (Pekeris and Accad).

The amphidrome at the entrance to the Arabian Sea is presented only in the chart by Gordeev *et al.* Some difference in the cotidal line pattern is also observed in the vicinity of Madagascar. In Dietrich's chart an amphidrome with counter-clockwise rotation is located here; in the Hendershott chart it appears to be shifted towards the coast; in all the other charts (except that by Gordeev *et al.* which does not present it at all) this amphidrome is presented as degenerate.

It is in the Pacific Ocean, however, that the difference between the tidal charts constructed by calculated and observed data is especially striking. According to a comparatively new empirical tidal chart of this ocean (Fig. 4.16), in its area there are six amphidromic systems (including a false amphidrome in the region of New Zealand) and a nodal zone embracing the Ogasawara (Bonin) and Mariana Islands. This chart agrees best with numerical solutions obtained by Bogdanov and Magarik, also with those by Gordeev *et al.*

The other charts present a different picture. According to the tidal charts by Hendershott, and by Pekeris and Accad, the nodal zone in the western part of the Pacific Ocean develops into two and three amphidromic systems, respectively, an additional amphidrome appearing in the south-east part of the Pacific Ocean. In the chart by Pekeris and Accad there is another amphidrome off the western coast of Mexico. The Zahel chart differs from the empirical chart by the presence of two amphidromic systems, one of them located in the centre of the northern Pacific Ocean, the other approximately on the latitude of Cape Horn.

It can be easily argued that the empirical cotidal charts of the Pacific Ocean are unreliable. Hence, we have to confine ourselves to comparing numerical solutions for this area.

If the amphidrome located to the south-east of New Zealand is referred to the Pacific Ocean, we have quite an interesting situation in the Southern Ocean. Indeed, the chart by Bogdanov and Magarik rules out the possible existence of any amphidromes here. According to Tiron, Sergeev and Michurin, in the Southern Ocean there is only one, and by Gordeev *et al.* two amphidromes, located on the borderline between the Atlantic and Indian Oceans and in the Australo–Antarctic sector of the Southern Ocean, respectively. Finally, in the charts by Hendershott, Zahel, and Pekeris and Accad there is in addition the third amphidrome in the Atlantic sector of the Southern Ocean.

Thus, not only the location of amphidromes but also their number and the

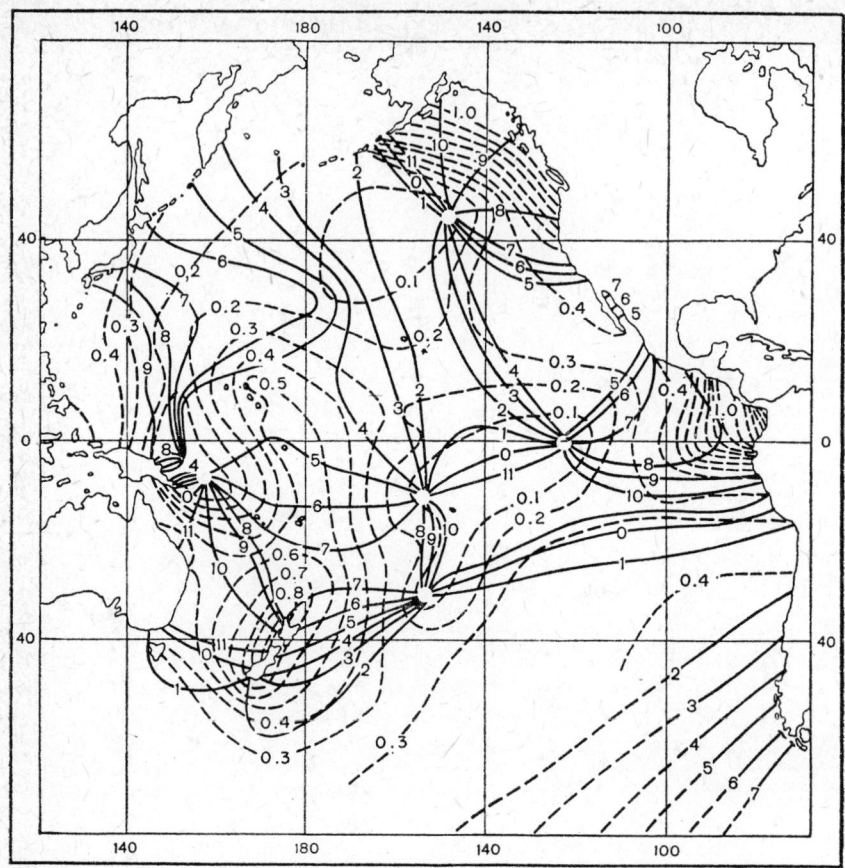

FIG. 4.16. Empirical tidal chart of the M_2 wave in the Pacific Ocean (according to Bogdanov, 1962).

direction of their rotation varies greatly according to the data reported by different investigators (see Table 4.5). It is hence self-evident that numerical solutions cannot yet claim a sufficient reliability in describing the tidal geography of the World Ocean. However, as long as we do not condemn ourselves to failure by trying to derive more than the above models can afford, the even at present, numerical solutions amplified by specially designed numerical experiments may appear to be useful in elaborating the second, perhaps the most important, problem of the theory of tides: the interpretation of the basic physical features of tide-formation in the World Ocean.

TABLE 4.5. *Location, coordinates of centres and direction of rotation of the amphidromes presented in numerical solutions of different investigators*

Ocean	Region	Coordinates of centres and direction of rotation						
		Bogdanov and Magarik (1967)	Tiron, Sergeev and Michurin (1967)	Pekeris, and Accad (1969)	Zahel (1970)	Hendershott (1972)	Gordeev, Kagan and Rivkind (1973)
Atlantic	Central region of North Atlantic	46°N, 53°W (+)	48°N, 39°W (+)	49°N, 39°W (+)	52°N, 40°W (+)	49°N, 42°W (+)	38°N, 41°W (+)
	Caribbean Sea	—	—	18°N, 64°W (+)	19°N, 61°W (+)	—	20°N, 62°W (+)
	Central region of South Atlantic	46°S, 21°W (+)	37°S, 11°W (+)	21°S, 16°W (+)	27°S, 16°W (+)	—	33°S, 18°W (+)
Indian	Arabian Sea	0°, 70°E (−)	2°N, 78°E (−)	2°N, 63°E (−)	3°N, 71°E (−)	7°N, 53°E (−)	—
	Bengal Bay	—	—	4°N, 80°E (+)	7°N, 76°E (+)	—	—
	Western region	—	—	—	—	25°S, 45°E (+)	—
	Eastern region	30°S, 99°E (−)	—	33°S, 107°E (−)	22°S, 107°E (−)	—	18°S, 98°E (−)
Pacific	North-east region	24°N, 147°W (+)	17°N, 149°W (+)	20°N, 155°W (+)	31°N, 121°W (+)	36°N, 168°W (+)	43°N, 158°W (+)
	North-west region	—	—	—	32°N, 177°W (−)	—	—

TABLE 4.5. (cont.)

Ocean	Region	Coordinates of centres and direction of rotation					
		Bogdanov and Magarik (1967)	Tiron, Sergeev and Michurin (1967)	Pekeris and Accad (1969)	Zahel (1970)	Hendershott (1972)	Gordeev, Kagan and Rivkind (1973)
Pacific	Western region	—	—	34°N, 143°E (+)	—	28°N, 163°E (+)	—
		—	—	10°N, 123°E (+)	2°S, 121°E (+)	—	—
		—	10°S, 156°E (−)	4°N, 140°E (−)	—	10°N, 159°E (−)	—
	Central region	12°S, 142°W (+)	—	24°S, 131°W (+)	20°S, 141°W (+)	20°S, 148°W (+)	20°S, 153°W (+)
	Eastern region	0°, 131°W (−)	—	16°S, 94°W (−)	—	12°N, 101°W (−)	2°N, 135°W (−)
		—	—	15°N, 101°W (+)	—	—	—
	South-east region	—	27°S, 90°W (−)	59°S, 100°W (−)	57°S, 98°W (−)	49°S, 96°W (−)	—
	South-west region	43°S, 170°W (−)	—	50°S, 160°W (−)	62°S, 167°W (−)	64°S, 164°W (−)	40°S, 155°W (−)
South	Atlantic sector	—	—	62°S, 58°W (−)	59°S, 62°W (−)	62°S, 28°W (−)	—
	Afro-Antarctic sector	—	57°S, 09°E (−)	64°S, 36°E (−)	65°S, 20°E (−)	64°S, 49°E (−)	41°S, 26°E (+)
	Australo-Antarctic sector	—	—	43°S, 141°E (−)	50°S, 134°E (−)	57°S, 109°E (−)	38°S, 110°E (+)

4.5. Numerical Experiments on Tidal Dynamics in the World Ocean

Let us begin our discussion with the numerical solution obtained in Section 4.3.

Interpretation of the results of tidal calculations in the World Ocean

In line with the chart of isoamplitudes and cotidal lines of the M_2 wave given in Fig. 4.9, semidiurnal tides in the World Ocean can be represented to first approximation as a result of superposition of nearly orthogonal standing oscillations.

In the Pacific Ocean there are two standing oscillations. One of them has three nodal zones: the southern one stretching from Australia through New Zealand up to the Peru coast, the central one crossing the Pacific Ocean from the Tuamotie islands to the coast of Mexico, and the northern one, less distinctly outlined, which stretches from Cape Mendocino (western coast of North America) to Japan. The other standing oscillation forms two nodal zones, let us call them the western zone and the eastern zone. The first crosses the Pacific Ocean in the north-west direction, the second is roughly parallel to the coast of America.

The superposition of the above nodal zones should result in the appearance of six amphidromes: two in the north part, one in the equatorial region and three in the north part of the Pacific Ocean. However, Fig. 4.9 presents only four amphidromes: one in the northern part, one in the equatorial region and two more in the southern part of the Pacific Ocean.

The disappearance of the amphidromes in the north-west and south-east parts of the Pacific Ocean can most probably be accounted for by the Coriolis effect, as is confirmed by the numerical experiments given below. These amphidromes which were to arise at the points of superposition of the northern and western as well as of the southern and eastern nodal zones should have a clockwise and a counter-clockwise rotation, respectively. The Coriolis effect, however, induces the opposite direction of the cotidal lines and must result in the weakening or in the total disappearance of amphidromes of clockwise rotation in the northern and those of counter-clockwise rotation in the southern hemispheres.

Note the progression of phases northward and southward from California, which indicate that there exist progressive waves travelling along the ocean boundaries and transferring energy to the Australasian Seas and to the region of the Drake Passage where it dissipates.

The tidal picture observed in the Indian Ocean can be also interpreted as a result of superposition of two standing oscillations, each having two nodal zones. The first is roughly parallel to the line connecting Sumatra with the southern part of Africa. The second crosses the Indian Ocean in the direction of south-west Australia and Queen Maud Land (Antarctica). One of the two nodal zones crossing them is located along the line connecting the Needle Cape and Enderbury Island (Antarctica); the other is roughly parallel to the eastern boundary of the Ocean.

The peculiar form of the Atlantic Ocean, approximately a bay of variable section elongated in the meridional direction, has a specific effect on the tidal picture of this basin. Here one can easily distinguish a standing oscillation orientated from north-west to south-east. This standing oscillation has two nodal zones located in the North and South Atlantic, respectively. The second system of standing waves includes a nearly zonal oscillation in the South Atlantic and two standing oscillations of different phases in the North Atlantic. Superposition of the nodal zones, belonging to the above system of standing waves, with those of the first standing oscillation gives rise to three amphidromes: in the central part of the South Atlantic, in the Caribbean Sea and in the central part of the North Atlantic.

By way of conclusion, note one particular feature of the picture of tides in the World Ocean, namely the distribution of the cotidal lines in the Southern Ocean. As is clearly seen from Fig. 4.9, an increase in phase from west to east is observed almost throughout the whole zone. It means that the zonal tidal motions in the Southern Ocean resemble in many ways a progressive wave transferring energy from west to east, i.e. in the direction opposite to the Sun.

Let us now dwell on the results of some numerical experiments on tidal dynamics in the World Ocean.

Analysis of the results of numerical experiments

All the experiments were carried out using the model described in Section 4.3. As before, integration of equations (4.3.1)–(4.3.6) was carried out on the latitude-longitude grid with a grid step 5°, a time step assumed to be equal to 6 minutes and values of the coefficients of bottom friction and macro-viscosity to 3×10^{-3} and $10^7 \, cm^2/s$, respectively.

(i) *Tides in the World Ocean of a constant depth.* Let us assume the depth in the World Ocean to be 4000 m throughout. In this case, as is seen from Fig. 4.17, an appreciable reconstruction of the field of tidal oscillations in the North Atlantic takes place. It is first of all characterized by a north-westerly shift of

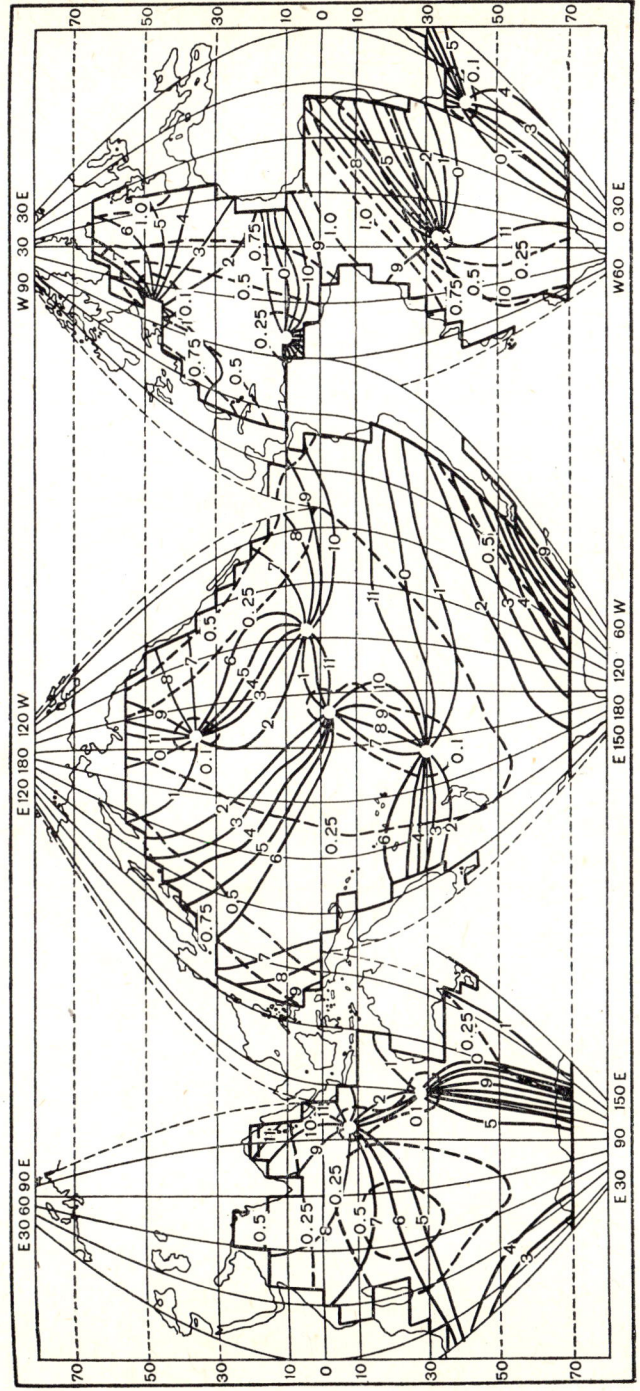

FIG. 4.17. Tidal chart of the M_2 wave in a World Ocean of constant depth. The ocean depth is 4000 m. For details see Fig. 4.9.

the amphidrome which was located in the central part of the basin in Fig. 4.9, i.e. to the region of Newfoundland, and by its transformation into a degenerate amphidrome. The Caribbean amphidrome appears to be shifted in the opposite direction, i.e. towards the north coast of South America. There is a noticeable change in the tidal phases in the tropical part of the South Atlantic (especially in the Gulf of Guinea), the basic pattern in the South Atlantic remaining unchanged.

The same is also true for the cotidal chart of the Pacific Ocean. Here, practically everywhere the time of high water changed by less than 1 hour from its initial value (Fig. 4.9), although in the region to the west of the Mariana and Ogasawara (Bonin) Islands the corresponding changes in the cotidal hours reached a quarter of the tidal period.

Similar values of tidal phase changes were also observed on the borderline of the Pacific and Indian Oceans, these changes being associated with the north-westerly shift of the system of two amphidromes located in the vicinity of the western boundary of the Indian Ocean. For the same reason there is a northward shift of the nodal zone crossing the Indian Ocean from north-east to south-west. However, generally speaking, the cotidal chart of the Indian Ocean, as well as those of the Pacific Ocean and South Atlantic, resemble in many respects the one presented in Fig. 4.9.

Note, finally, that the substitution of the real depth distribution by the mean depth of the World Ocean induces no appreciable changes in the maximum values of the tidal amplitudes.

(ii) *Tides in a World Ocean of idealized configuration.* In this experiment the World Ocean is approximated by the simplest method, i.e. as a system of adjoining spherical rectangles, limited by meridians and parallels. The distance between the boundary meridians (in angular measure) is 60° for the Atlantic Ocean and 120° and 65° for the Pacific and Indian Oceans, respectively. The northern boundaries for the Atlantic, Pacific and Indian Oceans are the 65°N, 55°N and 15°N parallels, respectively; the southern boundary is the circle of 70°S latitudes. The depth is assumed to be 4000 m throughout.

A comparison of the results of calculation, presented in Figs. 4.18 and 4.19, suggests that the existence of amphidromes having a counter-clockwise rotation in the central parts of the North and South Atlantic is practically independent of the shape of the coastline. The rectification of oceanic coastlines results in the disappearance of the Caribbean amphidrome and in the concentration of the cotidal lines in the equatorial region of the Atlantic Ocean. There is a slight shift of the North Atlantic amphidrome to the east and of the South Atlantic amphidrome to the west. There is also a noticeable phase change in the region of the North Atlantic amphidrome.

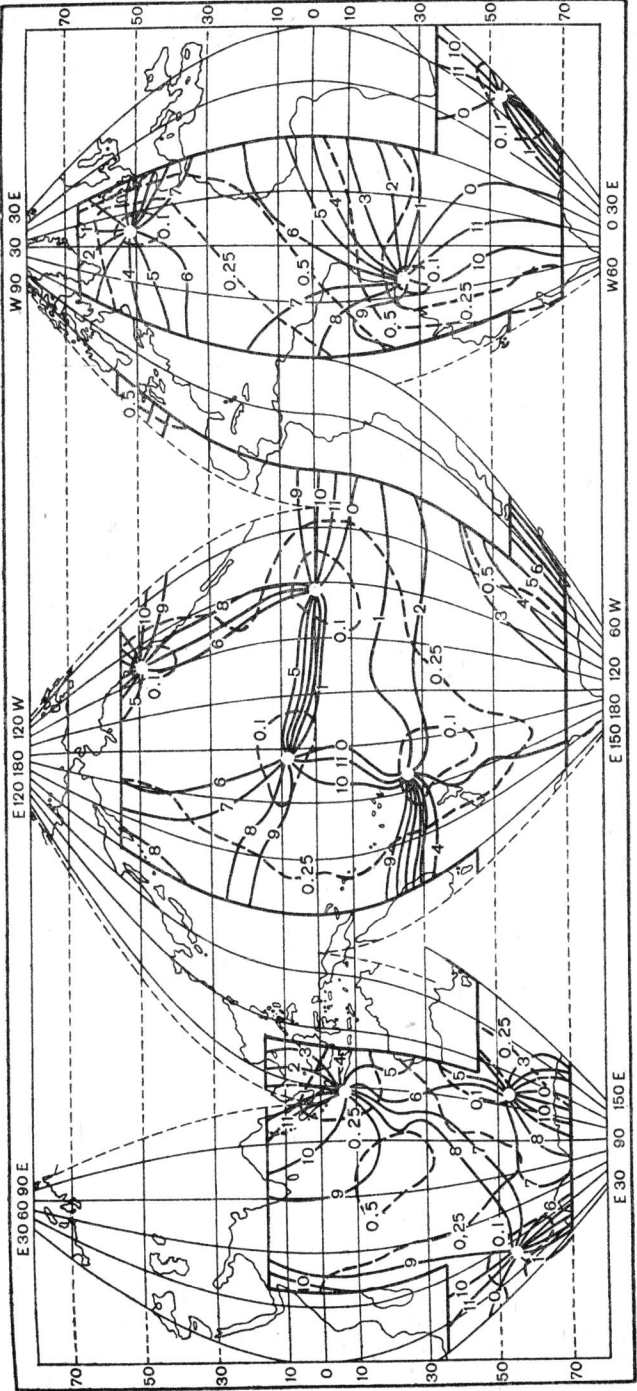

FIG. 4.18. Tidal chart of the M_2 wave in a World Ocean of an idealized configuration. The oceanic area is approximated by a system of spherical rectangles limited by meridians and parallels. For details see Fig. 4.9.

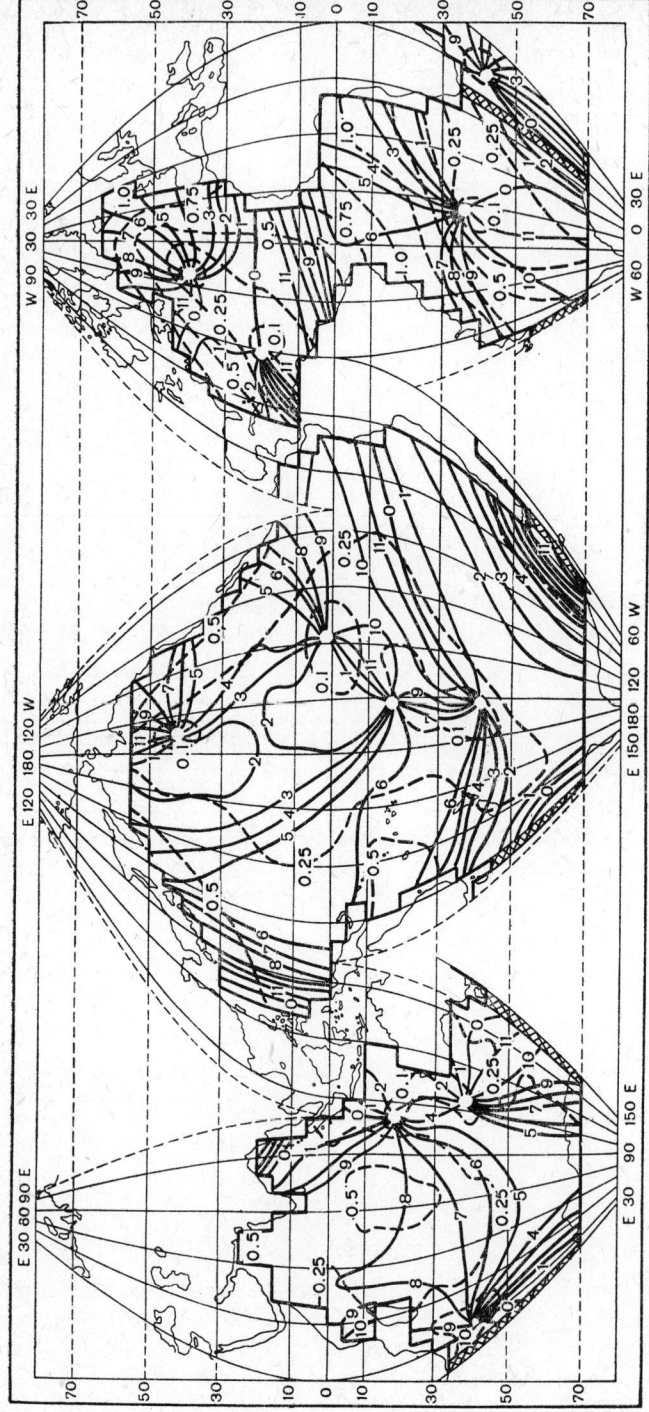

FIG. 4.19. Tidal chart of the M_2 wave in the World Ocean. Oceans are separated from one another by meridional barriers. For details see Fig. 4.9.

Marked changes in the tidal picture are observed in the Pacific Ocean. They confirm the earlier conclusion that the zonal and meridional standing oscillations here have two and three nodal zones, repectively. The nodal lines of the zonal oscillation in the ocean of an idealized configuration are located symmetrically with respect to the central 150°W meridian, while the nodal lines of the meridional oscillation are symmetrical about the equator. Allowances made for the shape of the real coastline result in the development of the system of orthogonal standing oscillations and in reconstruction of the tidal field, especially pronounced in the west of the equatorial region and in the southern part of the Pacific Ocean. The amphidromes located in this region (they are formed at the crossing points of the central and southern nodal zones with the western nodal zone) are shifted southward, this process being accompanied by a change of phase in the equatorial and south tropical regions of the Pacific Ocean.

Generally speaking, the shape of the coastline exerts a greater influence on tide generation in the Pacific Ocean than bottom morphology does, as can be easily seen by comparing Figs. 4.9, 4.17 and 4.18.

The most striking result of the numerical experiments under discussion is the fact that the coastline configuration has only an extremely weak effect upon tide-generation in the Indian Ocean, as is confirmed by the tidal chart of this ocean given in Figs. 4.9 and 4.18.

The next three numerical experiments enable one to estimate the role of tidal energy transfer from one ocean into another and that of energy exchange between the hemispheres. In these, as in the rest of the numerical experiments, the depth distribution in the World Ocean will be assumed the same as that used for modelling global tides in Section 4.3.

(iii) *Oceans arn divided from one another by meridional barriers.* Comparing Figs. 4.9 and 4.19, one sees that the presence of the meridional barriers (shaded in Fig. 4.19) is appreciable only in their vicinity, which means that in each ocean tides are generated in the ocean itself and not induced by the tidal waves from the neighbouring ocean.

(iv) *The Southern Ocean is separated from the others by a zonal barrier located at 20° latitude.* The consequences of isolating the Southern Ocean prove to be less great than might be expected. This can be demonstrated by comparing Figs. 4.9 and 4.20.

In the Atlantic Ocean one observes the shift of all three amphidromes to the north and a negligible change in phase in the region of the equator, as well as the disappearance of the region of large amplitude at points where their values exceeded 1 m. In the Pacific Ocean the equatorial amphidrome and that located in the region of the Society islands (Fig. 4.9) shift to the west, and in the

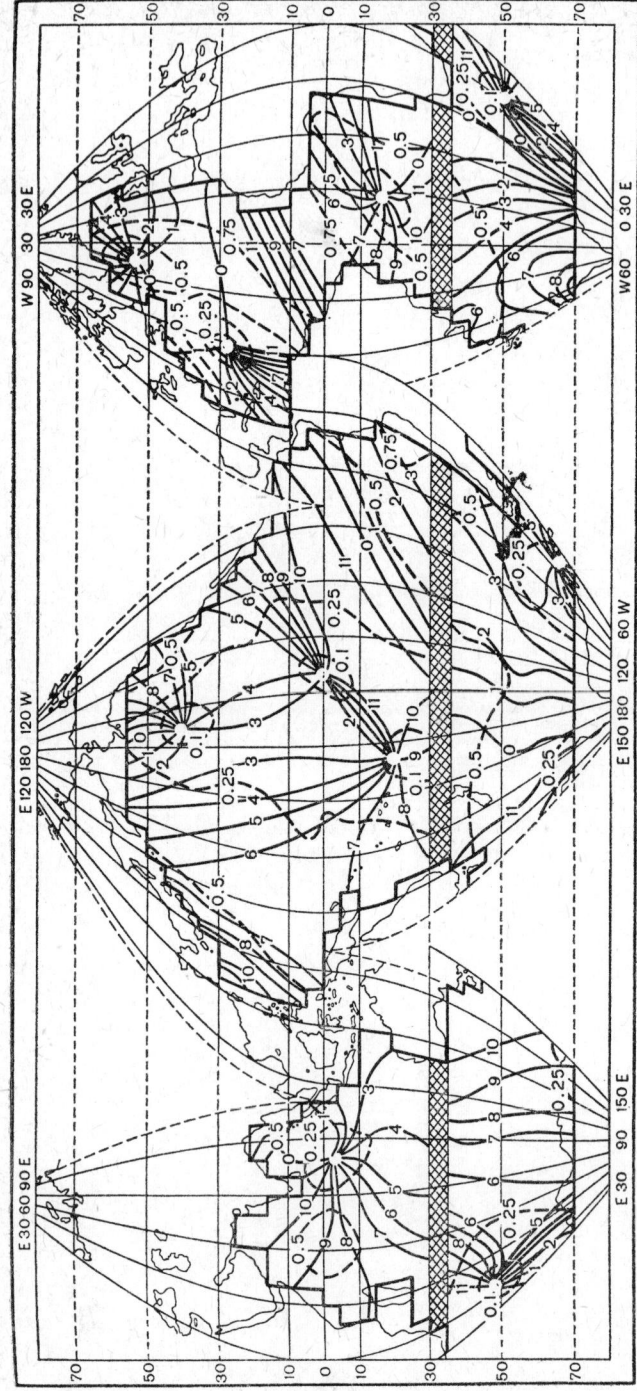

FIG. 4.20. Tidal chart of the M_2 wave in the World Ocean. The South Ocean is separated from the other oceans by a zonal barrier. For details see Fig. 4.9.

Indian Ocean the only amphidrome lying to the north of the bounding 30°S parallel shifts to the north-west.

Nevertheless, the structure of the tidal field in the Atlantic, Indian and Pacific Oceans preserves its characteristics despite the fact that there is no energy exchange with the Southern Ocean, which shows that the tides observed in these oceans are, in all probability, not externally induced.

(v) *The Atlantic and Pacific Oceans are each divided by a zonal barrier running along the epuator.* Comparing Figs. 4.9 and 4.21, one sees that the equatorial barrier, hampering the energy between the northern and southern parts of these oceans, exerts quite an appreciable influence on the tides of each ocean, particularly on the Pacific Ocean. The equatorial amphidrome disappears and the amphidromes lying in the tropical and southern parts of the Pacific Ocean shift to the east and north, respectively. Phases are changed almost throughout, and in some regions even the pattern of the cotidal lines is altered.

Isolating the Atlantic Ocean affects the pattern of the cotidal lines only in the tropical and equatorial zones of the ocean, the qualitative picture remaining, however, the same. It is interesting to note that some changes of the tidal picture can also be observed in the Indian Ocean (especially in its south-east and central parts) which indicates that damming up the Atlantic and Pacific Oceans is accompanied by an increase in the energy transfer through the channels connecting the Indian Ocean with the other oceans.

(vi) *Tides in the World Ocean in the absence of the Earth's rotation.* By means of this experiment we intended to elucidate the contribution of the Earth's rotation to the formation of the global tides.

Comparison of Figs. 4.9 and 4.22 shows that in the Atlantic Ocean there are three amphidromic systems: one (having a counter-clockwise rotation) being in the North Atlantic, the other two (having clockwise rotation), being further south. The location of these amphidromes change somewhat: the South Atlantic amphidrome shifts to the south, and the amphidrome located in Fig. 4.9 in the central part of the North Atlantic shifts to the north-east. This is accompanied by a corresponding change in the tidal amplitudes, especially noticeable in the tropical zone of the Atlantic Ocean and in the vicinity of the north-east boundary of the basin. The tidal phases change as well: everywhere (expect for the region of the equator) one can observe increased phases, and in the region of the Caribbean amphidrome even exceeding a quarter period.

However, the most striking change in the picture of tides is observed in the Pacific and Indian Oceans. In the Pacific Ocean out of four amphidromes depicted in Fig. 4.9 there remain only two: one in the north and one in the equatorial zone. Two more amphidromes in the southern Pacific Ocean, with

164 OCEAN TIDES

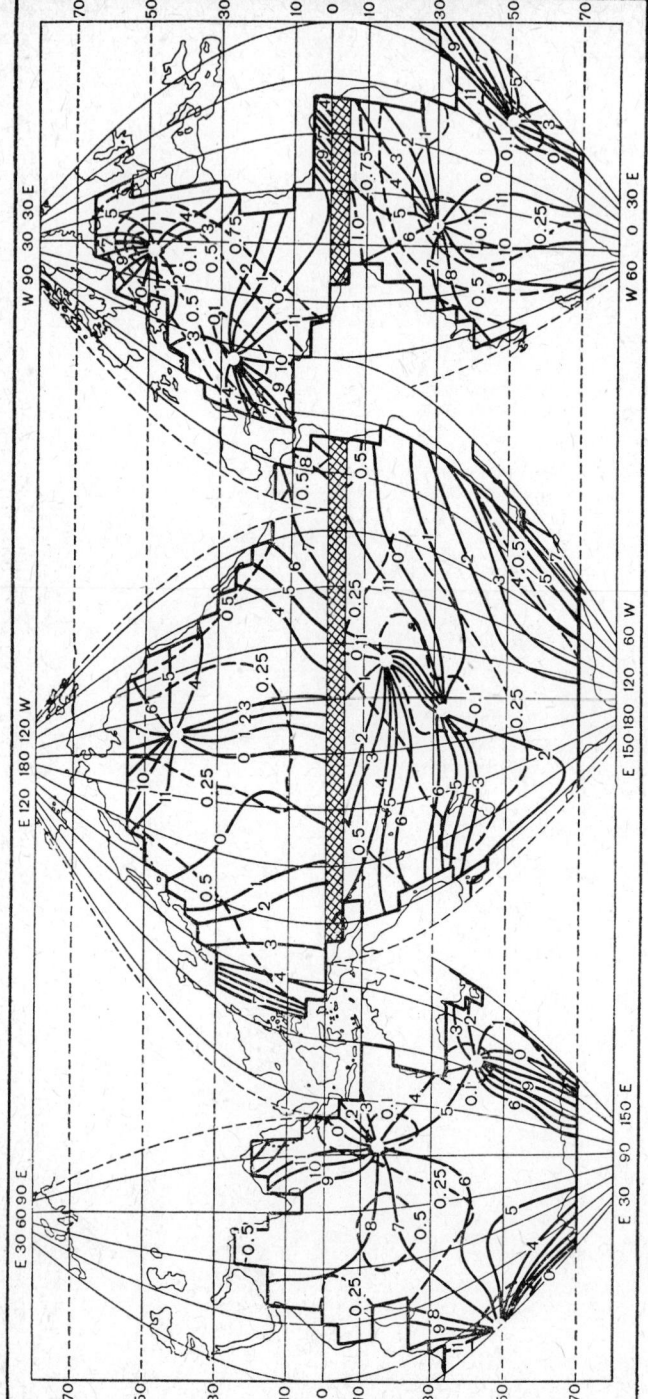

FIG. 4.21. Tidal chart of the M_2 wave in the World Ocean. The Atlantic and Pacific Oceans are separated by a zonal barrier running along the equator.

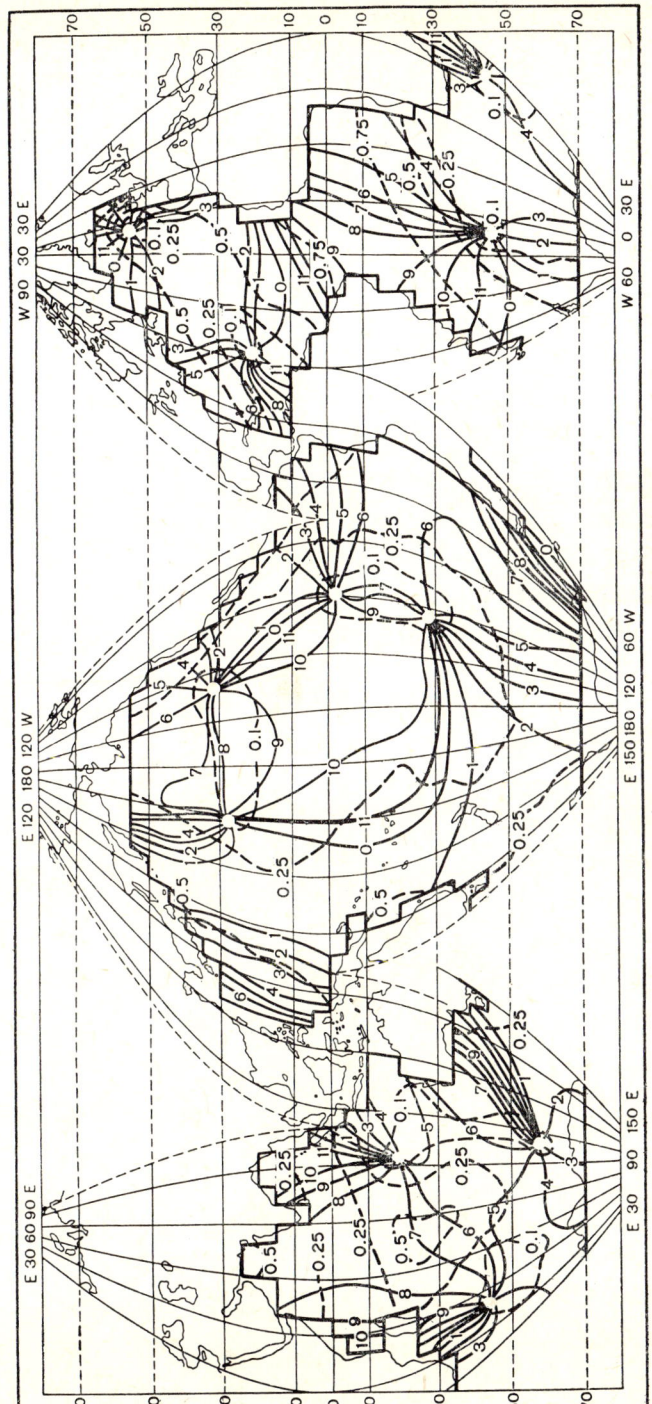

FIG. 4.22. Tidal chart of the M_2 wave in the World Ocean in the absence of the Earth's rotation. For details see Fig. 4.9.

centres located west of New Zealand and in the region of the Society islands, disappear, but two more amphidromes appear instead: one (centre 27°N, 178°E) in the north, and the other in the south, west of Easter island. Thus, the total number of amphidromes in the Pacific Ocean remains unchanged.

The Coriolis force is known to induce rotation of the cotidal lines in a counter-clockwise direction in the north and in a clockwise direction in the south hemispheres. That is why setting the Coriolis parameter equal to zero should have resulted in the weakening or even total disappearance of the clockwise amphidromes in the southern hemisphere and, vice versa, in their appearance and strengthening in the northern hemisphere. The opposite should have been observed in the case of the counter-clockwise amphidromes.

The above considerations explain all the peculiarities of the tidal picture except one: the disappearance of the counter-clockwise amphidrome in the southern part of the Pacific Ocean. One can assume that the absence of the Earth's rotation resulted in a change of the orientation of the central nodal zone and in a corresponding shift of the related amphidrome to the north where it becomes degenerate.

One cannot help noticing that even the amphidromes whose existence is not induced by the Earth's rotation (there are three, formed at the intersection of the north, central and west nodal zones), markedly shift to the east, resulting in the reconstruction of the tidal field in the central part of the Pacific Ocean.

In the Indian Ocean one can also observe some remarkable changes in the cotidal lines. The clockwise amphidrome with centre located at 57°S, 98°E appears here instead of the counter-clockwise amphidrome near south-western Australia. Thus, the absence of the Earth's rotation results not in the strengthening, as one might have expected, but in the disappearing of the counter-clockwise amphidrome and in its replacement by one with clockwise rotation. This result, which is at variance with conventional ideas on the role of the Earth's rotation, can be interpreted within the framework of the above concept of tide generation in the Indian Ocean.

As we have already mentioned, tides in the Indian Ocean can be interpreted as a result of the interference of two standing oscillations, each having two nodal zones, which form three amphidromes where they cross. Another degenerate amphidrome is found on the Antarctic coast. Under the assumption that, as in the case of the Pacific Ocean, in the absence of the Earth's rotation a change in the orientation of the standing oscillation system takes place (for instance, its counter-clockwise turn with respect to the initial position), the degenerate amphidrome is transformed into a new one of clockwise rotation, while the counter-clockwise amphidrome south-west of Australia, connected to it by general cotidal lines, will become degenerate.

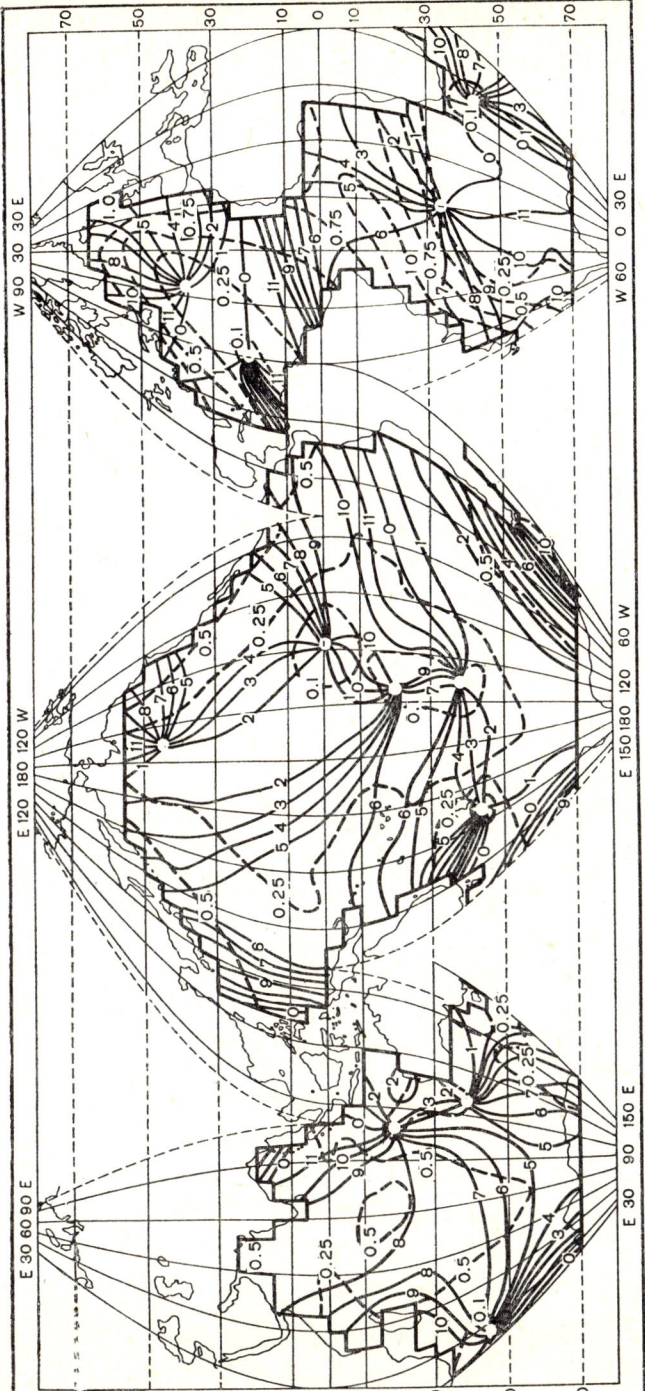

FIG. 4.23. Tidal chart of the M_2 wave in the World Ocean. New Zealand is modelled as a separate region. For details see Fig. 4.9.

Hence, the effect of the Earth's rotation results in quite appreciable (at some places major) changes in the tidal regime.

Remark. The experimental conditions were not "ideal" in a sense that we did not exclude the effect of the Earth's rotation on the gravitational force. The resulting error, however, is about an order of magnitude less than that obtained with a constant value of the acceleration due to gravity throughout.

(vii) *Tides in the World Ocean with direct account taken of the New Zealand Islands.* Until recently the World Ocean has been approximated to a binary basin. All the islands (including the largest) were either included in the system of continents, or modelled by lifting the bottom to a depth of 1000 m. In the experiment discussed in this section (results in Fig. 4.23) we made an attempt to elucidate the final picture of tides in the World Ocean by allowing for such big islands as, for instance, New Zealand.

Approximate New Zealand by a square with side equal to 5° and the centre located at 42.5°S, 172.5°E. Assume the depth at the boundary of the new region (the island) to be 1000 m, but require the condition of no-slip to be fulfilled there, as well as on the continental boundary.

Figure 4.23 shows that taking account of local peculiarities of the New Zealand basin results in the transformation of the south nodal zone (see Fig. 4.9) into a degenerate amphidromic system having counter-clockwise rotation. It is interesting that in the tidal charts constructed from observational data the pattern of the cotidal lines in this oceanic region greatly resembles that obtained from numerical calculations.

This is not the case with the pattern of isoamplitudes. According to observational data, the tidal amplitudes near the coast of New Zealand assume rather large values, increasing towards the islands. The opposite is observed in Fig. 4.23. This divergence is certainly associated with the effect of the island shelf ignored in our calculations.

In every other respect the tidal picture of the World Ocean remains in fact unchanged.

Current hypotheses on tide generation in the World Ocean and the results of numerical experiments

The Southern Ocean as a source of tide generation in the oceans. Whewell (1833) was the first to interpret tide generation in the World Ocean in this way. The most convincing reasons for the validity of this hypothesis were put forward by Goldsbrough (1927), who showed that in an ocean at high latitudes appreciable tidal oscillations can arise only with shallow depths. This showed that the Southern Ocean cannot serve as a source of the tides observed in the Atlantic, Indian and Pacific Oceans.

Whewell's hypothesis is closely associated with the interference hypothesis by Defant, Prüfer and Proudman. The most restrictive of these is that by Defant (1924, 1932), which can be applied only to the case of a basin in the form of a narrow canal; basins of the kind do occur in nature (the Arctic-Atlantic system, etc.). Defant considered the semidiurnal tides in the above oceanic system to be formed due to the interaction of two tidal waves coming from opposite directions. One of them travels from the Southern Ocean through the Atlantic Ocean into the Arctic, where it is reflected from the northern boundary of the basin and gives rise to a second wave coming towards the first wave from the Arctic to the Atlantic Ocean.

If the dissipation of the tidal energy and the effects of partial reflection throughout the whole course of these waves are not taken into account, their superposition gives rise to a pure standing oscillation with nodal lines transformed by Coriolis force into amphidromic systems with counter-clockwise rotation of the cotidal lines in the north and clockwise rotation in the southern hemispheres. With allowances made for the energy loss in the Arctic, the resulting semidiurnal tides will be characterized by a progressive standing oscillation with its progressive part moving northward.

According to the Prüfer hypothesis, tides in the Indian Ocean are induced by two systems of orthogonal oscillations, i.e. the zonal standing oscillations and the meridional cooscillational tide, originating in the Southern Ocean. Each system of oscillations has a definite number of nodal lines, dependent on the horizontal extent of the ocean and its depth. Superposition of these two oscillations forms a system of amphidromic points connected in such a way that the rotation of the cotidal lines around two neighbouring points is opposite in direction.

Another concept of tide generation in the ocean was suggested by Proudman (1944). He assumed that tides in the ocean are a combination of the tides generated by the tide-generating forces, and free waves of the Kelvin or Poincaré type. To illustrate the suggested scheme, Proudman calculated semidiurnal tides in the Atlantic Ocean, limited from the north and south by 45° and 35° parallels. He made use of the two modes from each family under the supposition that the first modes of the Kelvin and Poincaré type are formed in the Southern Ocean and travel to the north, while the second modes propagate within the basin towards the south.

The results of numerical experiment (iv) argue against the conclusion that the Southern Ocean determines tidal generation in the Atlantic, Indian and

Pacific Oceans, and thus are at variance with the Whewell hypothesis and the interference hypotheses by Defant, Prüfer and Proudman.

The resonance hypothesis by Ferrel-Harris. It is based on the supposition that there exists a coincidence of the periods of free oscillations with some components of the tide-generating force. Ferrel (1874) was apparently the first to explain the observed distribution of the semidiurnal tides in the North Atlantic by the influence of partial resonance. This idea was later used by Harris (1904) when constructing the cotidal chart of the World Ocean (see Section 4.1).

Darwin (1902) and Proudman (1944) considered this hypothesis invalid because the author took into account neither the Earth's rotation nor the effect exerted by "resonance" regions of the neighbouring oceanic parts on free oscillations.

The validity of the first supposition was discussed when analysing the results of numerical experiment (vi). The second objection to the resonance hypothesis was eliminated by subsequent investigations of "resonance" waves in complex basins. In this case the work by Lineykin (1937) is worth mentioning. He analysed the motion within a round basin of a constant depth on lifting or lowering the bottom near its centre, and also the motion in a canal with a stepped bottom, and showed that in the absence of rotation local resonances could arise in them, their effect gradually fading on moving away from the resonance frequency.

Resonance generation of tides can be judged by the reaction of the numerical solution to small perturbations of external parameters. In particular, according to Pekeris and Accad (1969) even a slight change of the coastline (induced by the reduction in the spatial grid step from 2° to 1°) results in a decrease of tidal amplitudes in the Indian Ocean by more than 3 m. According to Hendershott (1973) a change of the mean depth of the same ocean by several hundred metres is accompanied by a two- or even three-fold change of tidal amplitude.

Thus, there is every reason to believe that the semidiurnal wave of the tide in the Indian Ocean is close to resonance. The following facts, however, lead one to the conclusion that this resonance is of a "grid" rather than of a physical nature. The first is that at the same approximation of the basin but at another grid step, calculations by Pekeris and Accad give significantly different results. The second is that the numerical experiments (i) and (ii) carried out on another model do not confirm a high sensitivity of the solution to small variations of the coastline and the bottom in the Indian Ocean.

Generally speaking, resonance generation of tides under natural conditions is quite problematical, if it requires highly accurate correspondence between the periods of perturbating forces and free oscillations. The above conclusion is confirmed by the results obtained by Hough (1897, 1899) for the ocean entirely covering the Earth. According to Hough, for a 70-fold increase in the tidal amplitude, it is necessary for the difference between the period of free oscillations and that of the tide-generating forces to be less than 2–3 minutes. If this difference is 5 minutes, the tidal amplitudes show only a ten-fold increase.

Nevertheless, one should bear in mind that due to the extreme idealization of real conditions, the Hough conclusions are of a particularly approximate character and, hence, such exact "adjustment" of free oscillations to tide-generating forces is unnecessary. It was confirmed in particular by Platzman (1972) who showed that the "resonance" period in the North Atlantic can differ from the period of the M_2 harmonic of tide-generating forces by 1.6 hours.

4.6. Estimation of the Rate of Tidal Energy Dissipation in the open Ocean

It is common belief that the open area of the World Ocean is a region of quite negligible loss of tidal energy. This conclusion can be confirmed by the following considerations.

If the amplitudes of the tides observed in the open ocean are commensurable with those of the static tide and the motion here has the character of a progressive wave, then, to estimate a vertically averaged velocity of the tidal flow one can use the formula $u = (cA^+/D)\cos \sigma t$, where $c = \sqrt{gD}$ is the phase velocity of the tidal wave and A^+ is the amplitude of the static tide. In this case the rate of tidal energy dissipation per unit oceanic area ε_b due only to bottom friction (other sources of dissipation neglected) will be

$$\langle \varepsilon_b \rangle = \varrho_0 \varkappa \langle |u|^3 \rangle = \frac{4}{3\pi} \varrho_0 \varkappa \left(A^+ \sqrt{\frac{g}{D}} \right)^3, \qquad (4.6.1)$$

where ϱ_0 is the mean density of sea water; angular brackets here and elsewhere denote averaging over a tidal period.

Setting $A^+ \approx 25$ cm, $\varrho_0 \approx 1$ g/cm^3, $\varkappa = (2-3)\times 10^{-3}$, $D = 4\times 10^5$ cm, according to equation (4.6.1) we obtain $\langle \varepsilon_b \rangle \approx 2\times 10^{-3}$ erg/(cm$^2\times$s). The area of the World Ocean is approximately 3.7×10^{18} cm^2. Hence, the rate of tidal energy dissipation throughout the ocean $\langle \mathcal{E}_b \rangle$ is equal to 10^{16} erg/s in order of magnitude. Comparing this estimate with the value $(2.7\times 10^{19}$ erg/s) obtained by astronomic observational data from the Moon's secular acceleration, we see that the tidal energy dissipation in the open ocean can be in all cases neglected.

There is no other direct evidence for the validity of this conclusion except for approximate calculations of the same type.[†] We shall attempt to fill the gap by considering energy equations.

According to the equations of the model described in Section 4.3, the tidal energy equation integrated within a given domain can be represented as follows:

$$\varrho_0 \frac{d}{dt} \int_S \left(\frac{|\mathbf{v}|^2 D}{2} + \frac{g\zeta^2}{2} \right) dS = -g\varrho_0 \int_\Gamma \zeta \cdot \mathbf{v}_n D \, d\Gamma + \varrho g_0 \int_S \mathbf{v}D \cdot \nabla \zeta^+ \, dS$$

$$- \varrho_0 \iota \int_S |\mathbf{v}|^3 \, dS + \varrho_0 \frac{\kappa_h}{a} \int_\Gamma \left[\frac{\partial}{\partial \Theta} (\mathbf{v}D) \cdot \mathbf{v} \cos(n, \Theta) \right.$$

$$\left. + \frac{1}{\sin \Theta} \frac{\partial}{\partial \lambda} (\mathbf{v}D) \cdot \mathbf{v} \cos(n, \lambda) \right] d\Gamma$$

$$- \varrho_0 \frac{\kappa_h}{a^2} \int_S \left[\frac{\partial}{\partial \Theta} (\mathbf{v}D) \cdot \frac{\partial \mathbf{v}}{\partial \Theta} + \frac{1}{\sin^2 \Theta} \frac{\partial}{\partial \lambda} (\mathbf{v}D) \cdot \frac{\partial \mathbf{v}}{\partial \lambda} \right.$$

$$\left. + \frac{\cos 2\Theta}{\sin^2 \Theta} |\mathbf{v}|^2 D - \mathbf{A}_2 \frac{\partial}{\partial \lambda} (\mathbf{v}D) \cdot \mathbf{v} \right] dS, \qquad (4.6.2)$$

where $\mathbf{v}$ is the vector of the vertically averaged velocity of tidal flow, related to the vector of the total flow $\mathbf{w}$ by the expression $\mathbf{v} = \mathbf{w}/D$; $d\Gamma = a \, \partial \Gamma$ is the contour element; n is the normal to Γ; the lower index n denotes the component of the velocity vector $\mathbf{v}$ normal to the contour.

Note that for the case when the ocean has the form of a zonal belt ($d\Gamma = ad\lambda$, $\cos(n, \Theta) = 1$; $\cos(n, \lambda) = 0$), its depth remains constant throughout, and $\mathbf{v} = a \sin \Theta$; then the sum of the two last terms in the right-hand part of equation (4.6.2) characterizing the work done by forces of horizontal turbulent friction in unit time becomes zero. Hence, equation (4.6.2) meets the condition of "solid" rotation.

Rewrite now the energy equations, making allowances for the boundary condition (4.3.5). Then equation (4.6.2) loses not only the fourth term, but also

---

[†] Estimation of the rate of tidal energy dissipation by Pekeris and Accad (1969) cannot be referred to the condition of the open ocean. It was obtained by calculational data on tidal motions in the World Ocean and appeared to be more than twice the value of the lunar component of tidal dissipation. The authors attributed this result to inadequacy of their choice of resistance law, in particular, of the expression for the bottom friction coefficient, which was initially intended to ensure greater friction in coastal regions and here to reproduce shelf conditions.

the first one on the right-hand side, which can be associated with the rate of energy transfer across the contour Γ. As a result

$$\varrho_0 \frac{d}{dt} \int_S \left(\frac{|\mathbf{v}|^2 D}{2} + \frac{g\zeta^2}{2} \right) dS = g\varrho_0 \int_S \mathbf{v}D \cdot \nabla \zeta^+ \, dS - \varrho_0 \tau \int_S |\mathbf{v}|^3 \, dS$$

$$- \varrho_0 \frac{\kappa_h}{a^2} \int_S \left[\frac{\partial}{\partial \Theta}(\mathbf{v}D) \cdot \frac{\partial \mathbf{v}}{\partial \Theta} + \frac{1}{\sin^2 \Theta} \frac{\partial}{\partial \lambda}(\mathbf{v}D) \cdot \frac{\partial \mathbf{v}}{\partial \lambda} \right.$$

$$\left. + \frac{\cos 2\Theta}{\sin^2 \Theta} |\mathbf{v}|^2 D - \mathbf{A}_2 \frac{\partial}{\partial \lambda}(\mathbf{v}D) \cdot \mathbf{v} \right] dS. \tag{4.6.3}$$

Here the term on the left-hand side represents the rate of total energy change, the first term on the right-hand side describes the work performed by tide-generating forces per unit of time, and the second term represents the rate of tidal energy dissipation due to the forces of bottom friction.

To elucidate the sense of the last term, represent it as a sum of two parts

$$\varrho_0 \frac{\kappa_h}{a^2} \int_S \left[\frac{\partial}{\partial \Theta}(\mathbf{v}D) \cdot \frac{\partial \mathbf{v}}{\partial \Theta} + \frac{1}{\sin^2 \Theta} \frac{\partial}{\partial \lambda}(\mathbf{v}D) \cdot \frac{\partial \mathbf{v}}{\partial \lambda} + \frac{\cos 2\Theta}{\sin^2 \Theta} |\mathbf{v}|^2 D \right.$$

$$\left. - \mathbf{A}_2 \frac{\partial}{\partial \lambda}(\mathbf{v}D) \cdot \mathbf{v} \right] dS = \varrho_0 \kappa_h \int_S \left(D|\nabla \mathbf{v}|^2 \right.$$

$$\left. + \frac{\cos 2\Theta}{a^2 \sin^2 \Theta} |\mathbf{v}|^2 D - \frac{D}{a^2} \mathbf{A}_2 \frac{\partial \mathbf{v}}{\partial \lambda} \cdot \mathbf{v} \right) dS$$

$$- \varrho_0 \kappa_h \int_S \frac{|\mathbf{v}|^2}{2} \Delta D \, ds. \tag{4.6.4}$$

The first of these expressions ($\mathscr{E}_h$ say) characterizes the velocity of tidal energy dissipation due to macroviscosity whilst the second represents the part of the work done by horizontal friction due to irregularities in the sea bottom. Since within the domain the integrand $(|\mathbf{v}|^2/D) \Delta D$ changes sign, the second expression on the right-hand side of equation (4.6.4) generally appears to be much less in absolute value than the first, and it can be neglected when using smoothed depths.

Values of $\mathscr{E}_h$ obtained at various times in the tidal cycle as solutions of the system of equations (4.3.1)—(4.3.6), are presented in Fig. 4.2.4. The figure also shows the time variation of the rate of tidal energy dissipation due to

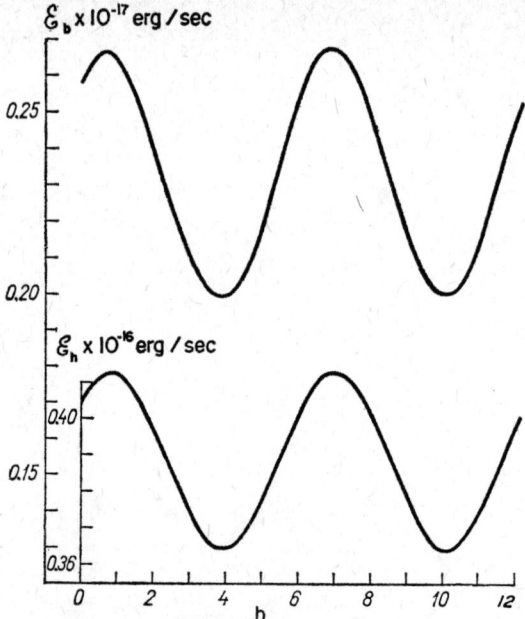

FIG. 4.24. Variation throughout the tidal cycle of the rate of tidal energy dissipation in the World Ocean, induced by the bottom and horizontal turbulent friction. For details see the text.

bottom friction

$$\mathcal{E}_b = \varrho_0 \tau \int_S |\mathbf{v}^3|\, dS.$$

Mean values of $\mathcal{E}_h$ and $\mathcal{E}_b$ with respect to the tidal period were found to be equal to 0.388×10^{16} erg/sec and 2.32×10^{16} erg/sec, respectively. Adding $\langle \mathcal{E}_h \rangle$ and $\langle \mathcal{E}_b \rangle$ we obtain the value of the total dissipation of tidal energy in the open part of the World Ocean to be $\langle \mathcal{E} \rangle = 2.71 \times 10^{16}$ erg/sec.

Thus, the above calculation confirms the validity of current estimations of $\langle \mathcal{E} \rangle$. According to our data, dissipation of the tidal energy in the open ocean is 1000 times less than the "astronomic" value.

Note that when deducing the energy equations within the framework of the theory of barotropic tides, we neglected dissipation associated with the loss of tidal energy to generate internal waves.

4.7. References

Empirical cotidal charts and the principles of their construction have been the subject of many papers [see, for instance, the reviews by Marmer (1932), Doodson (1958) and the monograph by Defant (1961)]. The derivation of the formula for the period of eigenoscillations in a closed basin can be found in the monographs by Lamb (1945) and Proudman (1953). The method of calculating tidal motions in semiclosed elongated basins, later referred to as the "narrow sea" method, was suggested by Defant (1918). The basic grounds of this method are given in the monographs by Defant (1961), Dronkers (1964) and Kagan (1968).

When discussing the basic peculiarities of the spatial distribution of the tides observed in the World Ocean, we mainly followed the paper by Villain (1952). The analysis of new data on the tides in the Southern Ocean obtained during the IGY period can be found in the papers by Titov and Shesterikov (1964) and Bogdanov (1966).

The proof of the theory of unique solvability of problem (4.3.1)—(4.3.6) and of the periodicity of its solution in time was obtained by Gordeev, Kagan and Rivkind (1973). The same work presents the conditions of stability of the difference scheme (4.3.9)—(4.3.15) represented here by equations (4.3.23), (4.3.24).

The data on deep-water tidal measurements at stations "Filloux" and "Josie II" were borrowed from the works by Filloux (1969) and Irish, Munk and Snodgrass (1971), respectively.

Information on the harmonic constants from stations "Kathy", "Josie I" and "Flikki" can be found in the papers by Munk, Snodgrass and Wimbush (1970).

Investigations on the trapping of long waves by isolated symmetrical islands of cylindrical form were carried out by Summerfield (1972).

The location of the amphidrome in the north-east part of the Pacific Ocean was derived in the paper by Irish, Munk and Snodgrass (1971) cited above.

The first attempts at a theoretical approach to the description of tidal phenomena in the World Ocean were made in classical works in which dynamic equations were solved by analytical methods. Brief reviews of the works of classical type can be found, for instance, in the papers by Doodson (1958) and Voyt (1970).

A detailed description of the results of numerical experiments on tidal dynamics in the World Ocean are also given in the paper by Gordeev, Kagan and Rivkind (1975).

Whewell's viewpoint on the cause of tide generation in the World Ocean was supported by Darwin (1898), Krümmel (1911), Warburg (1922) and See (1927) (see also Marmer, 1932). Prüfer's idea of the interference nature of tides was made use of by Dietrich (1944) while analysing the observed pattern of the cotidal lines in the World Ocean. The interpretation of the mechanism of tide generation in the ocean suggested by Proudman was successfully applied to the north-eastern part of the Pacific Ocean in the paper by Munk, Snodgrass and Wimbush (1970).

Information on the properties of the Kelvin and Poincaré waves is given in the monograph by Proudman (1953). A detailed consideration of the properties of these and other free waves is available in the review by Platzman (1971).

An approximate estimation of the tidal energy dissipation in the open ocean was borrowed from the work by Munk and McDonald (1960) (see also Jeffreys, 1970). The equation of energy (4.6.3) is presented in the work by Gordeev, Kagan and Rivkind (1974). The basic results given in the above paper serve as a basis for Section 4.6.

CHAPTER 5

The Bottom Boundary Layer in Tidal Flows

5.1. Some Definitions

Up to now we have been studying only the horizontal structure of the tidal flow, assuming the velocity of the tidal flow to be constant in depth and equal to its mean value within the oceanic depth. However, even the first measurements indicated the fact that as regards the vertical velocity distribution, the bottom oceanic layer has a number of specific peculiarities. In particular, in the immediate vicinity of the bottom the velocity of the tidal flow increases, as a rule, with height, while the vertical gradient of velocity decreases in all cases (Fig. 5.1). An analogous conclusion can also be drawn from laboratory measurements in an oscillatory flow (Fig. 5.2). The observed velocity distribution

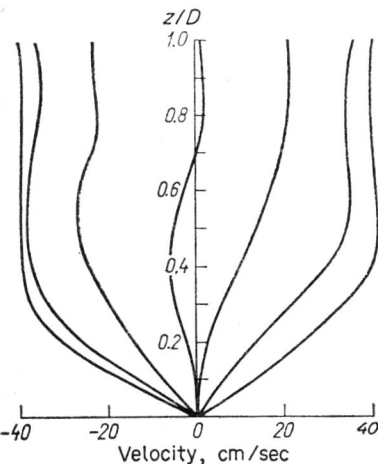

FIG. 5.1. Vertical distribution of tidal velocity at different moments of the tidal cycle (from Sverdrup, 1956).

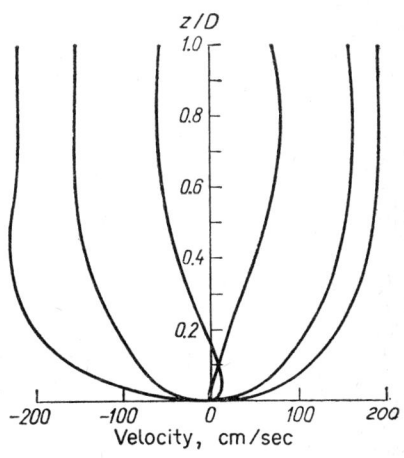

FIG. 5.2. Vertical distribution of the velocity in the boundary layer of an oscillating flow at different moments of time according to laboratory measurements (from Jonsson, 1963).

in the lower part of the flow periodically changing in time is a result of the joint action of friction and inertial forces. Let us call this portion of the flow *the bottom boundary layer*.

The regime of flows in a bottom boundary layer can be laminar and turbulent. The liquid moving under laminar regime, the scale of thickness of the bottom boundary layer is determined by the relation $h_l \approx (\nu/\frac{1}{2}\sigma)^{1/2}$. Here ν is the kinematic molecular viscosity of the fluid, σ is the oscillation frequency. A corresponding Reynolds number can be presented as $\text{Re} = Uh_l/\nu$, where U is the amplitude of the flow velocity beyond the limits of the boundary layer. In keeping with Collins (1963), the critical value of Re at which the periodic wave motion of the fluid acquires the irregular character is equal to $\text{Re}_{\text{crit}} \approx 160$. Setting $\nu = 1.8 \times 10^{-2}$ cm²/s and considering a frequency range from 0.6×10^{-7} to 3×10^{-5} c/s, corresponding to the basic harmonics of the tidal potential, we find that the critical value of the velocity of the tidal flow U_{crit} cannot exceed 0.2 cm/s.

The characteristic value of the depth-mean minimal velocity of the tidal flow in the World Ocean is seen from Fig. 5.3 to be 1–2 cm/s. The fact that in a homogeneous ocean the velocity beyond the bottom boundary layer does not greatly differ from its vertically averaged value means that in real conditions the bottom boundary layer is practically always turbulent.

Turbulence in the bottom boundary layer depends upon hydrostatic stability determined by the parameter $\mathfrak{E} = -(1/\varrho_0) d\varrho_0/dz - g/c^2$, where, together with the known notation, c is the speed of sound. Indeed, at the values of the Richardson number $\text{Ri} = g\mathfrak{E}(\partial \mathbf{u}/\partial z)^{-2}$, exceeding critical $\text{Ri}_{\text{crit}} = \frac{1}{4}$, hydrostatic forces supress vertical pulsations of the velocity; the fluctuating motions in this case become almost two-dimensional. The same tendency is observed at some other positive (but less than Ri_{crit}) values of the Richardson number, which correspond to positive $\mathfrak{E}$. Conversely, at negative $\mathfrak{E}$ and consequently negative Ri values, hydrostatic forces add energy to the vertical velocity pulsations. As a result one can observe the strengthening of all the components of velocity pulsations followed by an increase in the vertical momentum transport.

Unfortunately, there are no data on the magnitude of $\mathfrak{E}$ in the bottom oceanic layer. Moreover, one cannot be certain even of the sign of $\mathfrak{E}$. In this case one has to make the simplest assumption, that the bottom boundary layer in the ocean is neutrally stratified ($\text{Ri} = 0$).

Under the condition of neutral stratification the effect of small-scale turbulence can be evaluated by introducing an effective coefficient of turbulent viscosity κ_v which is constant in height. Under the conditions that the turbulent boundary layer remains hydrostatically stable only while its effective Reynolds number $\text{Re} = Uh_b/\kappa_v$ does not exceed Re_{crit}, and that the thickness of the

THE BOTTOM BOUNDARY LAYER 179

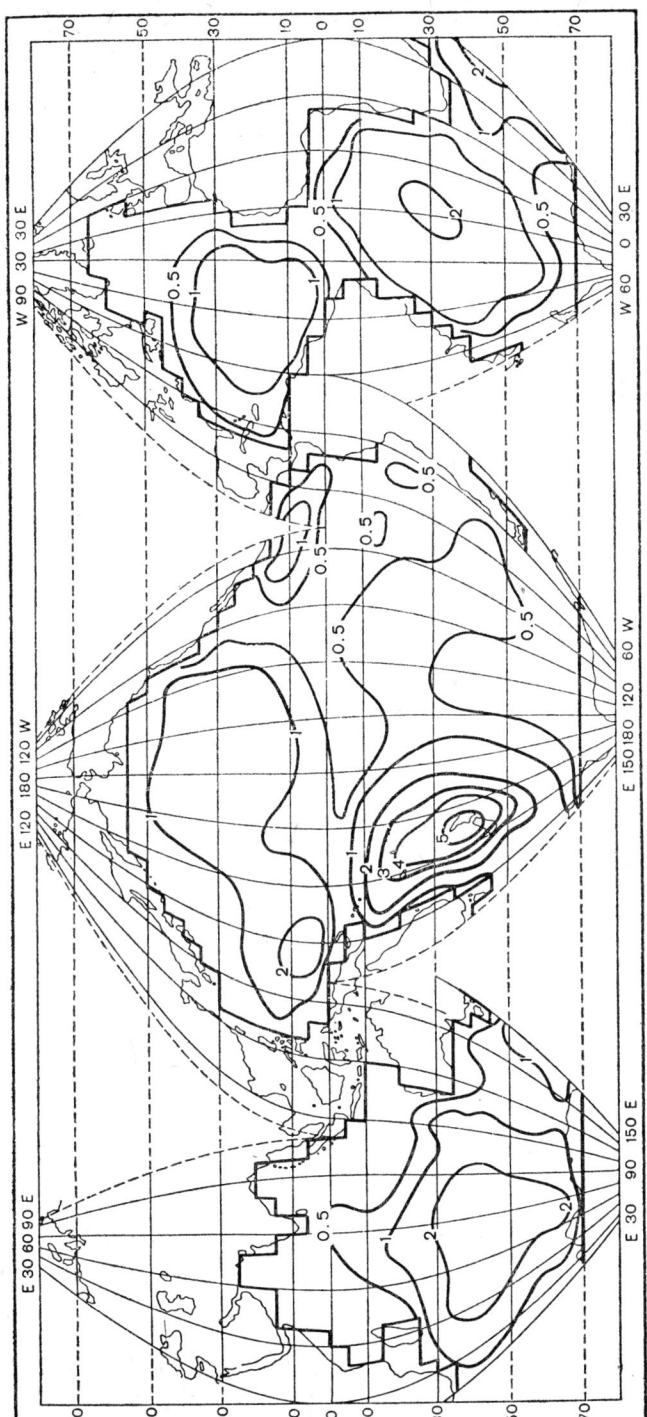

FIG. 5.3. Minimal velocity (cm/s) of the barotropic component of the tidal currents in the World Ocean. The M_2 wave. The chart was constructed from the results of calculations carried out with the theoretical model presented in Section 4.3.

turbulent boundary layer h_b can be estimated as $h_b \approx (\kappa_v/\frac{1}{2}\sigma)^{1/2}$, we obtain the following criterion:

$$h_b \geqslant \frac{2U}{\sigma \, \mathrm{Re}_{\mathrm{crit}}}.$$

Thus, the minimal thickness of the bottom boundary layer at $U = 3$ cm/s (typical of the ocean) and $0.6 \times 10^{-7} < \sigma$ (c/s) $< 3 \times 10^{-5}$ will be approximately 2 m.

In a thin fluid layer adjacent to the bottom the turbulence of any origin (including tidal) is under the direct influence of the bottom. The bottom surface being smooth, this influence is felt through viscous stresses, and in the case of a rough surface through the action of normal pressures arising when the fluid flows pass over the elements of roughness.

Determining whether surfaces are smooth or rough in the hydrodynamic sense is based on the comparison of the mean height of roughness elements with the thickness of the viscous sublayer. If the elements of roughness are so small that they are included in the viscous sublayer, then roughness does not result in an increase of resistance. In this case the surface is referred to as *hydrodynamically smooth*. In the case of a *hydrodynamically rough* surface the elements of bottom roughness are beyond the limits of the viscous sublayer and as a result exert a direct damping effect on the motion of the free-stream flow.

The thickness of the viscous sublayer h_v can be estimated only by two-dimensional parameters: the kinematic molecular viscosity v and the friction velocity (or dynamic velocity) $u_* = \sqrt{\tau_b/\varrho_0}$ (here τ_b is the tangential friction stress). From dimensional considerations

$$h_v = C_1 v/u_*; \tag{5.1.1}$$

where $C_1 \approx 12$ is a numerical constant.

Then the ratio of the height of the roughness elements h_0 and the thickness of the viscous sublayer h_v will be

$$\frac{h_0}{h_v} = \mathrm{Re}_0/12,$$

where $\mathrm{Re}_0 = h_0 u_*/v$ is the Reynolds number of the surface.

The above fact indicates that the ratio h_0/h_v and, consequently, hydrodynamic properties of the underlying surface are dependent not only on the geometrical dimensions of the elements of roughness but also on the magnitude of the mean velocity which is associated with u_*.

According to Nikuradze (1933), the surface is hydrodynamically smooth if

$$\mathrm{Re}_0 < 4,$$

and hydrodynamically rough if

$$\mathrm{Re}_0 > 60.$$

The above criteria were obtained for the flow of water in round pipes coated with homogeneous sand.

Between these two limiting Re_0 values within the range

$$4 < \mathrm{Re}_0 < 60$$

there is a transitional region in which the surface can be characterized as neither smooth nor rough. The surface resistance to the free-stream flow is in this case the sum of the viscous resistance and the resistance of the form of the roughness element, which are partially beyond the limits of the viscous sublayer.

In deducing the criteria for determining hydrodynamic properties of the underlying surface all the elements of roughness were assumed to have approximately similar form and to be closely adjacent, which is of course not the case in reality.

Photographs of the sea bottom demonstrate the existence of the whole spectrum of roughness. Sternberg (1966), for instance, divided them into four groups. To the first group he referred all the elements of roughness in the form of dunes, their height reaching approximately 30 cm, the distance between crests being up to 3 m. The elements of roughness which are ripples of a certain type are located on the dunes. The height of these waves varies from 1.5 to 2.4 cm, the mean length is approximately 16 cm. A separate group of roughness elements is coarse gravel and some elements of roughness of organic origin, their sizes varying from 5 cm to 1.4 mm. They mainly fill the gaps between the ripples. Sternberg's fourth group consisted of sand grains with the mean diameter 0.41 mm.

One should not forget that the above list of roughness elements is not complete. They are characteristic of a limited portion of the surface photographed at a small distance from the sea bottom. That is why the conditions for the Nikuradze criteria are not generally fulfilled.

The above difficulty can be overcome by introducing the roughness parameter z_0 associated with the height of the equivalent sand roughness h_S, z_0 being a certain height within the logarithmic layer (see below), at which the mean velocity of the flow becomes zero. Here h_S is the size of sand grains covering the wall such that the velocity profiles in the logarithmic layer above the wall and the real surface coincide with one another at equal u_*. Thus, if h_0 is the height of the equivalent sand roughness, then the ratio z_0/h_0 will depend not on the form, size and mutual location of the elements of roughness but will be determined only by the Reynolds number of the surface

$$\frac{z_0}{h_0} = \Phi(\mathrm{Re}_0). \qquad (5.1.2)$$

The dimensionless universal function Φ at extremely small and great values of the argument Re_0 can be determined as follows.

Let $\mathrm{Re}_0 \ll 1$. Then the underlying surface can be regarded as hydrodynamically smooth. However, if the elements of roughness are completely within the viscous sublayer, the value h_0 must be excluded from the number of determining parameters, which is possible only at $\Phi \sim \mathrm{Re}_0^{-1}$. Substituting the expression (5.1.2) for Φ we obtain

$$z_0 = C_2 \nu / u_*. \qquad (5.1.3)$$

In the other limiting case, when $\mathrm{Re}_0 \gg 1$, i.e. in the regime of full roughness, the number of determining parameters must not include the kinematic molecular viscosity ν. This means that at large Re_0 values the function Φ must tend to some constant. Denoting this constant by C_3 we obtain

$$z_0 = C_3 h_0 \qquad (5.1.4)$$

According to experimental data by Nikuradze (1933), the values of the constants C_2 and C_3 are approximately equal to 0.1 and 0.03. These data are enough to evaluate at least the limiting values of the roughness parameters of technical surfaces. As far as real surfaces (in particular, the sea bottom) are concerned, the equivalent sand roughness is unknown. Hence it has to be replaced by some other easily measurable characteristic of roughness, for instance, by the mean height of the elements of roughness. In this case, however, the coefficients of proportionality C_2 and C_3 in equations (5.1.3), (5.1.4) will not be universal constants and must depend on the form and mutual location of the roughness elements.

From the above, it is clear that the commonly used formulae for hydrodynamically smooth and rough surfaces, namely

$$z_0 \approx 0.1 \nu / u_* \quad \text{at} \quad h_0 < 0.3 h_v$$

$$z_0 \approx 0.03 h_0 \quad \text{at} \quad h_0 > 5 h_v, \qquad (5.1.5)$$

where h_0 now represents the mean height of the roughness elements, can give only approximate roughness parameters.

Table 5.1 lists the values of the roughness parameters of the sea bottom obtained by different authors. It also contains all the necessary data for checking the system of inequalities (5.1.5). Unfortunately, the data on the mean

TABLE 5.1. *Values The bottom boundery layer of the sea-bottom roughness by data of different investigators*

Authors	Sea-bottom characteristics	Height of ground accidents	z_0 cm	u_* cm/s
Revelle and Fleming (see Sverdrup, Johnson and Fleming, 1942)	—	—	2.0	—
Mosby (1949)	—	—	1.8	2 – 4
Lesser (1951)	Gravel	0.2	0.13	1.3
	Slit sand	0.15	0.16	2.3
	Slit	0.005	0.02	0.3
Bowden and Fairbairn (1952a)	Compact sand impregnated with pebbles	—	0.21	1.2 – 2.0
Bowles et al. (1958)	—	—	0.2	—
Charnock (1959)	Compact sand impregnated with pebbles	—	0.1 – 2.4	1.5 – 2.8
Bowden, Fairbairn and Hughes (1959)	The same	—	0.16	0.5 – 2.9
Sternberg (1968)	Sand unevenly covered with gravel with 6 – 10 cm rocky juts	—	$10^{-4} - 1$	—
	Gravel	2 – 3	$0.5 \times 10^{-2} - 1.5$	—
	Medium sand deformed by separate accidents on the surface	5 – 7	$10^{-3} - 0.5$	—
	Sand ripples	—	$10^{-3} - 10^{-1}$	—
	Fine sand deformed by separate surface accidents	2	$10^{-4} - 1.2$	—
Sternberg (1970)	Accidents of organic origin	0.5 – 1	$10^{-3} - 6.3$	0.066 – 0.57
	The same	0.5	$10^{-4} - 2.2$	0.073 – 0.43
	Ripples	~ 1	0.004 – 0.78	0.14 – 0.47
	Unevenly distributed sedimental elements of roughness	0.25 – 1	0.004 – 0.9	0.03 – 0.17
Wimbush and Munk (1971)	Fine sediments	< 0.1	2	0.05 – 0.35
Dyer (1972)	Dunes of gravel and sand	25 – 200	0.08 – 0.14	4.2 – 8.7
Weatherley (1972)	Fine sediments, ripples	0.5 – 2	~ 0.03	0.31 – 0.42

Remark. Dashes show that the data are not available.

height of bottom roughness were not available in all references given in the first column of Table 5.1. In order to fill the gaps, we had to estimate the roughness parameter by the formula $z_0 \approx 0.03h_0$ which holds only for the hydrodynamically rough surface. Then the criterion $h_0 \gg 5h_v$ after substituting the above ratio for z_0 and the expression for h_v can be rewritten $z_0 \gg 2v/u_*$.

The data from experimental measurements presented in Table 5.1 show that the latter inequality does not hold in all cases. One should remember, however, that these values of z_0 were determined by processing data relating to the mean velocity in the logarithmic layer and, consequently, they are characteristic of only local peculiarities of the underlying surface. In the case of the boundary layer whose thickness, as will be shown below, is several times more than that of the logarithmic layer, the sea bottom roughness parameter must be different from its local value, at least due to the fact that it will serve as a measure of non-homogeneity of the underlying surface on a larger scale.

The sea bed is known to be an inhomogeneous surface. Hence, increasing the horizontal scale of averaging, one can expect the appearance of elements of roughness whose height must be reckoned not in centimetres but in metres, especially when speaking about the scale of averaging usual for modelling global tides. Under such conditions it would not be wrong to consider the sea bed a hydrodynamically rough surface.

5.2. Experimental Data

Here we shall dwell upon the experimental data from profiles of mean velocity and the simplest statistical characteristics of turbulence in the bottom boundary layer of tidal flow. We intend also to consider here the results of measuring the spectra of velocity and temperature, and, on the basis of the analysis of their form, to draw some conclusions on the regime of turbulence in the vicinity of the sea bed.

Bearing in mind that the data presently available are extremely scarce, we will not even attempt to systematize them. We limit ourselves only to a demonstration of the most general peculiarities of the space-time distribution of those characteristics which are important for understanding the processes involved in the bottom boundary layer.

Profiles of the mean velocity of the tidal flow

Sverdrup (1926) was one of the first to study the vertical distribution of the velocity of tidal flows. While sailing on the ship *Maud*, he obtained velocity profiles at three points of the Siberian mainland shelf. The velocity was recorded by standard current meters throughout the whole tidal cycle. The observational data were then subjected to harmonic analysis, and the resulting values of harmonic constants of the velocity were used to reconstruct the vertical structure of the tidal flow.

Later, the problem of velocity profiles of tidal flow in marginal seas and straits was the subject of many papers. A review can be found in the paper by Bowden and Fairbairn (1952), which also lists values of the harmonic constants of the tidal flow velocity in Red-Wharf Bay (North Wales) which served as the basis to construct Figs. 5.4, 5.5. The fact that the dimensionless profiles of

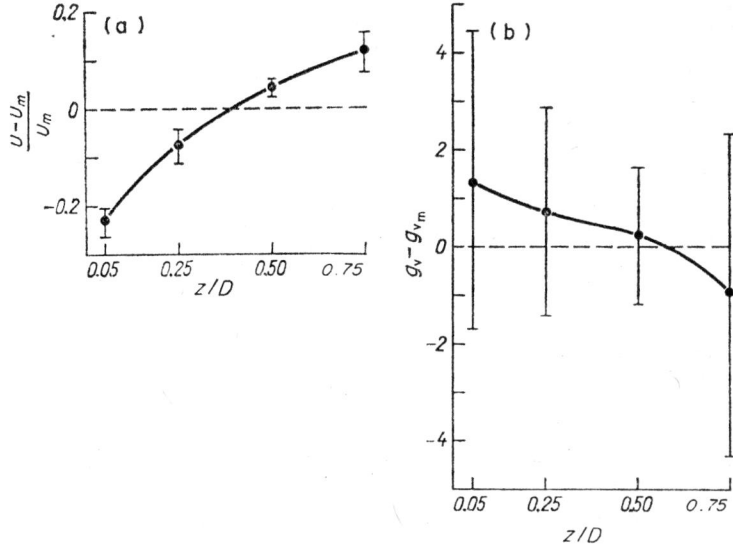

FIG. 5.4. Dimensionless profiles of the amplitudes and phases of the longitudinal component of the tidal velocity in the bottom boundary layer according to observational data by Bowden and Fairbairn (1952a). (a) the function $(U-U_m)/U_m$; (b) the function $g_u - g_{u_m}$. Index "m" indicates characteristics appropriate to the vertically averaged velocity of the tidal flow. The dots denote the average weighted values, the vertical segments denote the mean quadratic deviations due to the difference in depths.

amplitudes and phases of the horizontal velocity components of the tidal flow do not coincide in this case can be mainly attributed to the variation of the mean depth.

Special attention should be paid to the fact that these observational data (as well as the majority of the preceding ones) do not include the lowest layer immediately adjacent to the sea bottom. This lack of data was compensated for only after special constructions for measuring velocity gradients in the

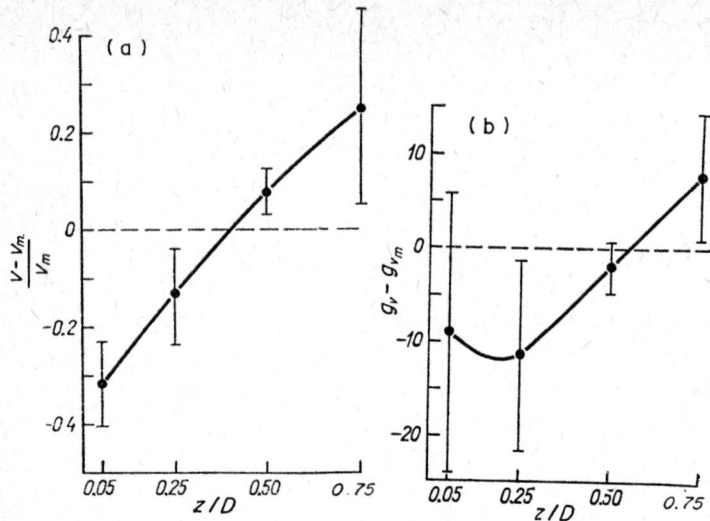

FIG. 5.5. Dimensionless profiles of the amplitudes and phases of the transverse component of the tidal velocity in the bottom boundary layer from observational data reported by Bowden and Fairbairn (1952). (a) The function $(V-V_m)/V_m$; (b) the function $g_v - g_{v_m}$. For details see Fig. 5.4.

bottom layer were devised. The first measurements of the kind were made by Mosby (1947, 1949) in the Alverströmmen flow near Bergen. Later came reports by Lesser (1951), Charnock (1959), Sternberg (1966, 1968) and finally Dyer (1972) on analogous measurements carried out in other shallow-water regions of the ocean.

The basic conclusion drawn in all the above papers was that the distribution of the mean velocity in the lowest part of the bottom boundary layer has a clearly defined logarithmic character. To illustrate the above, consider the results of the processing of the observational data obtained by Lesser (1951) for different types of sea bed (Fig. 5.6).

The above characteristic of the vertical distribution of the tidal flow in the bottom boundary layer (or as it was referred to above, the logarithmic layer) was observed not only in shallow water but evidently also in deep waters. It can be verified even by the relatively few observational data contained in papers by Sternberg (1970). Wimbush and Munk (1971) and Weatherley (1972). Figure 5.7, taken from Weatherley (1972), shows the dependence of mean velocity on height in the lowest 10 metres of the ocean (the total depth at the point of measurement was 780 m).

Unfortunately, the velocity measurements reported in the above were carried out within the logarithmic layer only. That was why they could not give a

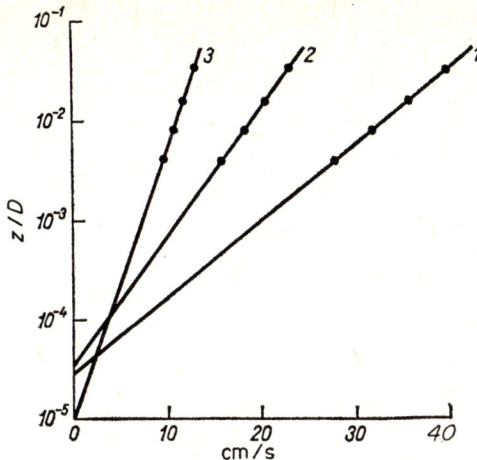

FIG. 5.6. Vertical distribution of the tidal velocity in the logarithmic layer at different soil compositions (from Lesser, 1951). (1) Sand, (2) gravel, (3) silt.

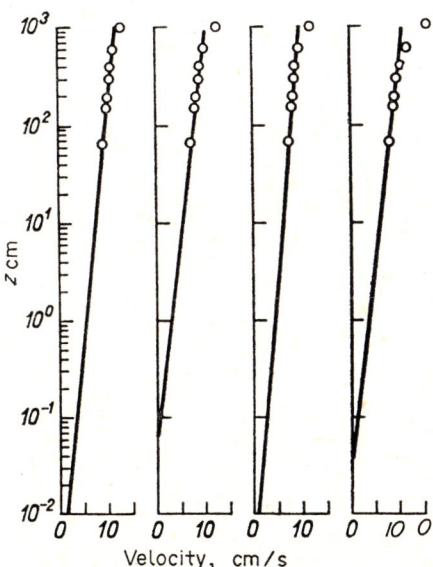

FIG. 5.7. Profiles of mean velocity in subsequent hours of the tidal period in the bottom 10-m layer of the ocean (by Weatherley, 1972), obtained by hourly averaging of measurements carried out at 25°44′N; 70°28′W at the depth of 780 m.

full representation of the behaviour of the flow velocity throughout the bottom boundary layer. To fill the gap, we made use of the results of laboratory measurements of velocity in oscillatory fluid flow published by Jonsson (1963). As is seen in Fig. 5.8, the amplitude of the longitudinal velocity component of

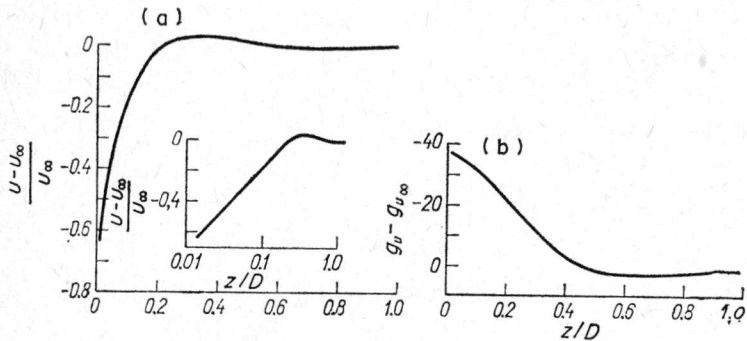

FIG. 5.8. Dimensionless profiles of the amplitudes and phases of the oscillatory flow velocity obtained in laboratory by Jonsson (1963).
(a) the function $(U-U_\infty)/U_\infty$; (b) the function $g_u - g_{u\infty}$. The inset shows the dependence of $(U-U_\infty)/U_\infty$ on z/D in the semilogarithmic scale.

the flow, which is an analogue of tidal flow, linearly increases in the logarithmic layer. Then the velocity gradient decreases and beyond the limits of the bottom boundary layer the velocity amplitude remains practically constant with height. The phase of the longitudinal velocity component changes only slightly with depth, so that the bottom flow is only a few degrees ahead of the flow beyond the limits of the boundary layer.

Before concluding the discussion on the profiles of the mean velocity of the tidal flow in the bottom boundary layer, it is necessary to decide more exactly what is meant by mean velocities. Since by their nature they continuously change in time, it would be natural to ask how the low-frequency oscillations we are interested in can be distinguished from high-frequency turbulent fluctuations.

It is clear that when determining mean values, the period of averaging must be, on the one hand, much less than the tidal period and, on the other hand, much greater than the characteristic period of the turbulence containing the principal share of energy. The case would be quite simple if in the spectrum of the flow velocity there was a minimum dividing the regions of high-frequency turbulence and comparatively low-frequency tidal oscillations. In this case [as in the surface atmospheric layer; see Van der Hoven (1957), Kolesnikova and Monin (1965)] the choice of any period of averaging within the region of the spectral minimum would guarantee the elimination of high-frequency turbulence without distorting the low-frequency velocity components.

However, the spectrum of the velocity in the bottom boundary layer of the ocean does not possess this form. According to the data given in Fig. 5.9

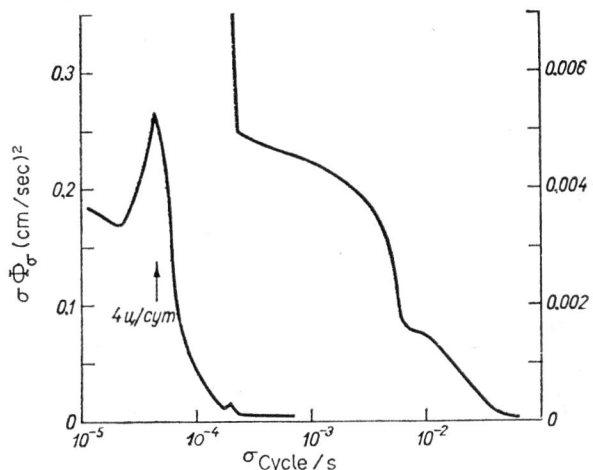

FIG. 5.9. Spectrum of the tidal velocity in the bottom boundary layer of the ocean (from Wimbush and Munk, 1971). The measurements were carried out at a depth of 3.8 km, 350 km westward of San-Diego, the data presented referring to a level 1.5 m from the bottom. Hig hfrequency portion of the spectrum has 50 times the scale of the low-frequency portion.

the peak in the spectrum of the horizontal velocity component is at the frequency of 4 c/day which corresponds to the of semidiurnal tidal oscillations. With increasing frequency the spectral density reduces monotonically to high frequencies, where its rate of decrease tails off noticeably. Nevertheless, the spectral density does not have a second maximum here.

In the case considered, to obtain definitive mean values, it would evidently be advisable to carry out some selective transformation of the observational data, damping oscillations of frequency greater than, say, 2×10^{-3} c/s, i.e. the oscillations bearing only a small proportion of the energy.

Intensity of turbulent velocity fluctuations

The first measurements of velocity fluctuations were made by Thorade in 1926. The measurements were made in the tidal estuary of the Elbe river. The Rauschelbach current meter was sensitive to changes of velocity and direction with periods of 10 s, and longer. The current meter was lowered from a pontoon and fixed 2 m above the sea bed. Data were recorded for 17 min and, when analysed, revealed oscillations with a 5-min period and 10 cm/s amplitude.

The international expedition of August 1931 made use of analogous probes while working in the Kattegat. The probes were lowered from a ship and worked continuously for a week. The velocity fluctuations induced by the ship having been eliminated, the remaining fluctuations were divided into two groups: extremely fast fluctuations increasing with the amplitude of surface waves and at the same time independent of the mean velocity of the flow, and slower oscillations. Fluctuations of the first type were attributed by Thorade to the influence of ship movement; those of the second type seemed to originate from internal waves.

In 1936 Doodson constructed a new electric current meter equipped with a photographic means of recording. Using the Doodson current meter, which allowed fluctuations with a period of 1 s and more to be recorded, measurements were made in 1938 fortnight 4 miles off the south coast of the Isle of Man. However, the recorded velocity fluctuations were again distorted by the movement of the ship.

In 1946 velocity fluctuations of the tidal flow were recorded by Bowden in Liverpool Bay. The experiment was, however, hampered again by the wind waves. Since there was little hope that these difficulties could be overcome, it was decided to look for new techniques which would make it possible to record velocity fluctuations free from distortion by ship motion and surface waves.

To measure velocity fluctuations near the sea bed Bowden and Proudman (1949) suggested the idea of inserting the Doodson current meter into a rigid frame placed firmly on the sea bottom. This construction was utilised in August 1948 to make measurements at two points located in the estuary of the Mersey. Fifty series of measurements were carried out, each lasting for about 30 min. Turbulent fluctuations of the tidal flow were recorded at various levels, the lowest being 140 cm from the bottom. Surface elevations were recorded simultaneously.

Bowden and Fairbairn (1952b) reported measurements of longitudinal turbulent fluctuations using two Doodson current meters located at different distances vertically and horizontally across the main flow. They carried out altogether 114 series, each lasting for 10 min. Of these, 36 series of measurements were made with the modernized frame, which allowed the current meters to be spaced at 110 cm vertically and 75 cm horizontally. In the remaining series the current meters were attached to a rope at various spacings, not exceeding 3 m.

Measurements of longitudinal turbulent fluctuations in tidal flows were also made by Nan'niti (1956) using a photoelectric rotor and also a light three-blade rotor constructed by the author.

The Doodson and Nan'niti current meters did not permit simultaneous measurements of horizontal and vertical turbulent fluctuation. Bowden and Fairbairn (1956) were the first to carry out measurements of this kind. To record velocity fluctuations in two directions, they employed an electromagnetic flowmeter which allowed one to detect velocity pulsations with periods above 1 s. Owing to difficulty in measuring mean flow they conducted their experiments at the entrance to Red-Wharf Bay, where the mean tidal velocity had been studied earlier (see Bowden and Fairbairn, 1952a).

Altogether they carried out 75 series of measurements, the duration of each being 5-10 min. In 51 cases longitudinal and vertical turbulent fluctuations was measured at one of two heights, i.e. either 75 or 150 cm above the sea bottom, and in 13 cases longitudinal turbulent fluctuations were recorded at two levels spaced by 25, 50, 75, 100 or 125 cm. In the remaining cases measurements of the vertical velocity pulsations at the two levels were carried out.

Observations in Red-Wharf Bay were continued for several days, yielding data for various conditions. The analysis of these data is given in paper by Bowden (1962).

These two papers, which can be regarded as an important step in studying the structure of turbulence in tidal flow, included measurements within the limits of the lowest 2 m layer. Bowden and Howe (1963) studied the turbulent structure throughout the whole depth of a shallow sea. Measurements were carried out in the estuary of the Mersey, where the velocity of spring tidal flow exceeds 200 cm/s. Two electromagnetic flowmeters were fixed on a vertical mast lowered from the pier into the water by a shot. The position of one of the flowmeters was fixed at 3 m from the surface, while the second flowmeter was movable along the mast. In the bottom layer the flowmeters were fixed but at various spacings from one another.

Now let us consider the results obtained from the experimental data contained in the above papers.

Bowden and Proudman (1949) showed that the pulsations of the longitudinal component of the tidal velocity were approximately proportional to the mean velocity. In other words, the turbulence intensity in the tidal flow is practically independent of the mean velocity. Table 5.2 shows changes in the relative intensity of longitudinal velocity pulsations $\sqrt{\overline{u'^2}}/u$ with distance from the sea bottom.

It is clear from the table that in the lowest 2 m the degree of turbulence in the direction of the mean flow decreases with height. At large heights a decrease in $\sqrt{\overline{u'^2}}/u$ becomes less noticeable, which conclusion was obtained for the case of a calm oceanic surface. Under the influence of wind waves the

TABLE 5.2. *Relative intensity of the longitudinal component of velocity pulsations as a function of distance from the bottom*
(from Bowden and Proudman, 1949)

Distance from the bottom, m	$\sqrt{\overline{u'^2}}/u$
0.6	0.144±0.022
1.1	0.113±0.009
1.4	0.109±0.012
Mean for all distances	0.117±0.007

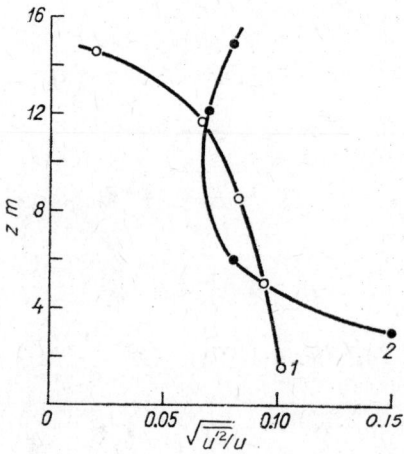

FIG. 5.10. Vertical distribution of relative intensity of longitudinal velocity fluctuations of the tidal flow in the bottom boundary layer (by Bowden and Proudman, 1949). (1) On the undisturbed surface, (2) on the disturbed surface.

distribution of $\sqrt{\overline{u'^2}}/u$ depicted by curve 1 in Fig. 5.10 will be distorted in the surface layers (see curve 2 in the same figure). Later measurements by Bowden and Fairbairn (1956) and by Bowden and Howe (1963) confirmed the above regularity.

Some idea of the order of magnitude of transverse velocity pulsations of the tidal flow can be obtained from Bowden (1962). The mean relative intensity of transverse velocity pulsations $\sqrt{\overline{v'^2}}/u$ for the lower 125 cm-thick layer was found to be equal to 0.082±0.006. No definite variation of $\sqrt{\overline{v'^2}}/u$ with height was observed.

It is interesting to compare the data by Bowden (1962) with those by Bowden and Fairbairn (1952b) obtained at the same point of Red-Wharf Bay. Processing the eight series of the flowmeter recordings obtained in 1952 showed the mean quadratic amplitude $\sqrt{\overline{u'^2}}$ of velocity pulsations at the height of 75 cm from the bottom to vary from 3.1 to 4.2 cm/s at a mean velocity of tidal flow of 28–50 cm/s. Calculating the relative intensity of the transverse velocity pulsations gives $\sqrt{\overline{u'^2}}/u = 0.102$ which is in good agreement with the data by Bowden (1962), i.e. 0.110 ± 0.008. This coincidence may serve as one more confirmation of the conclusion by Bowden and Proudman that the intensity of turbulence is independent of the mean velocity of tidal flow. In the absence of density stratification it must be primarily determined by the roughness of the sea bottom.

To verify this conclusion one can make use of the data obtained by Bowden and Howe (1963) in the Mersey estuary and by Bowden (1962) in Red-Wharf Bay. According to Bowden and Howe the magnitude of the relative intensity of longitudinal velocity pulsations $\sqrt{\overline{u'^2}}/u$ in the estuary of the River Mersey as high as 125 cm from the bottom is 0.063 which is approximately half the corresponding values of $\sqrt{\overline{u'^2}}/u$ in Red-Wharf Bay. Taking account of the fact that the bottom of the estuary is covered with silt, while Red-Wharf Bay has coarse sand and gravel, explains fairly well the difference in $\sqrt{\overline{u'^2}}/u$ value at these two places.

Note, however, that the ratio of intensities obtained for the longitudinal velocity pulsations in the estuary of the River Mersey and in Red-Wharf Bay is too great to be ascribed to the difference in sea bottom roughness only. In all probability the character of the mesoscale roughness of the sea bottom of the above two regions can also contribute to the differences in values of $\sqrt{\overline{u'^2}}/u$.

Bottom roughness (including mesoscale roughness) must affect all three components of the velocity pulsations. Unfortunately, due to the lack of data one may only judge the effect of different sea bottom roughness by making use of the longitudinal pulsations. However, there are some indications that under the same external conditions the relative intensities of longitudinal and vertical pulsations remain unchanged. Thus, according to Bowden and Proudman (1949) the $\sqrt{\overline{u'^2}}/u$ value at the depth of 4 m from the surface is 0.028. An analogous value for the relative intensity of longitudinal velocity pulsations was obtained by Bowden and Howe (1963) who made their measurements in the same region. Bowden and Fairbairn (1956) found that the relative intensity of the vertical pulsations $\sqrt{\overline{w'^2}}/u$ at 75 and 150 cm from the bottom was 0.065, while Bowden (1962) found that it was 0.066 and 0.065, respectively.

Comparing $\sqrt{\overline{u'^2}}/u$, $\sqrt{\overline{v'^2}}/u$ and $\sqrt{\overline{w'^2}}/u$ indicates that within the bottom layer the intensity of the pulsations of the longitudinal component appears to be greatest, the intensity of vertical pulsations proves to be least while $\sqrt{\overline{v'^2}}/u$ has an intermediate magnitude. From Bowden (1962), $\sqrt{\overline{u'^2}}/u \approx 0.110$, $\sqrt{\overline{v'^2}}/u \approx 0.080$ and $\sqrt{\overline{w'^2}}/u \approx 0.065$. The ratio $\sqrt{\overline{w'^2}}/\sqrt{\overline{u'^2}}$ is on average 0.51, and $\sqrt{\overline{v'^2}}/\sqrt{\overline{u'^2}}$ is equal to 0.76, these data having been obtained in Red-Wharf Bay. It is interesting that in the estuary of the River Mersey the ratio $\sqrt{\overline{w'^2}}/\sqrt{\overline{u'^2}}$ proved to be approximately the same. This conclusion can be derived from the data by Bowden and Howe (1963) who found out that as far as 125 cm from the bottom the above ratio was 0.52.

Correlation functions. Scales of turbulence

Some additional data on the structure of turbulence in tidal flows can be obtained by comparing the correlations of various velocity components. Figure 5.11 gives autocorrelation functions of the horizontal and vertical components

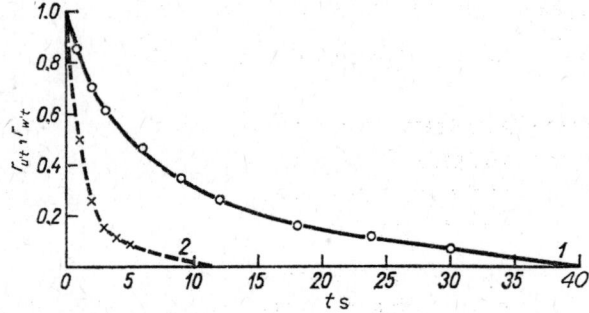

FIG. 5.11. Autocorrelation functions of the longitudinal (1) and vertical (2) components of tidal velocity fluctuations in the bottom boundary layer (by Bowden and Fairbairn, 1956).

averaged for the heights 75 and 150 cm. The figure also shows that the correlation decreases vertically faster than in the direction of the main flow. It is apparent that the vortices in the bottom layer are more stretched in the horizontal direction than vertically.

The extension of the vortices in the transverse direction is in all probability slight, as evidenced by low values of the coefficient of correlation between the longitudinal and transverse components of velocity, $r_{u'v'}$. According to Bowden (1962) the correlation coefficient $r_{u'v'}$ is 0.14 and always positive, the latter fact meaning that the contrary flow resulting from the condition of continuity can take place only in the vertical plane.

From the autocorrelation function and the relation

$$T_t = \int_0^\infty r_t \, dt,$$

where r_t is the Euler time coefficient of correlation, one can calculate the integral time scale of the velocity pulsations T_t. For the autocorrelation functions depicted in Fig. 5.11 the integral time scale of longitudinal pulsations $T_{u't}$ appeared to be 8.72 s, while the time scale of vertical pulsations was only 1.71 s.

Advantage taken of the Taylor hypothesis on "frozen turbulence" yields that for the mean flow velocity $u = 35.6$ cm/s which was observed during the experiment, the mean sizes of vortices in the longitudinal and vertical directions (or the integral longitudinal and vertical scales of velocity pulsations $L_{u'}$ and $L_{w'}$) will be 3.1 and 0.6, respectively.

With increasing distance from the bottom, the character of the autocorrelation functions of longitudinal and vertical fluctuations remains unchanged. Figure 5.12, however, shows that the autocorrelation functions referring to the

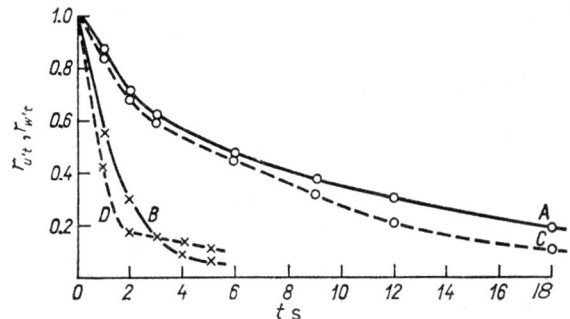

FIG. 5.12. Autocorrelation functions of the longitudinal (A, C) and vertical (B, D) components of tidal velocity fluctuations at various distances from the bottom (by Bowden and Fairbairn, 1956). (A, B) at a depth of 150 cm; (C, D) at a depth of 75 cm.

level 75 cm are located below the analogous curve for the height 150 cm, demonstrating that the role of fluctuations of longer period increases with distance from the bottom.

Fluctuations of longer period, however, may be caused only by the vortices of large size. Hence, with increased distance from the bottom the sizes of the vortices increase. This conclusion is confirmed by Bowden (1962) who calculated the integral scales of longitudinal and vertical velocity pulsations at various distances from the bottom, as shown in Table 5.1.

As to transversal velocity pulsations, at present we can evaluate only their typical scale $L_{v'}$, which is approximately 1.6 m for the height 60–125 cm used by Bowden (1962).

The values of the integral scales given in Table 5.3 were calculated according to the data obtained in Red-Wharf Bay. It is obvious that under any other external conditions the integral scales of various components of the velocity pulsations can be quite different, which is confirmed by the experimental data obtained in the Mersey estuary. According to these data the longitudinal and vertical integral scales equal 3.8 and 1.1 m at the height of 50 cm and 6.3 and 0.7 m at 125 cm (see Bowden and Howe, 1963).

TABLE 5.3. *Integral longitudinal and vertical scales of turbulent velocity fluctuations as a function of distance from the bottom*
(from Bowden, 1962)

Distance from the bottom	$L_{u'}$ m	$L_{\omega'}$ m
50, 60	3.16	0.88
75	3.59	1.25
100, 125	3.27	1.35
150, 175	4.27	1.56
Mean for all the distances	3.57	1.25

To evaluate the scales of turbulence, one can also measure the longitudinal and transversal correlation functions for various components of velocity at two distant points. Figure 5.13 (a, b) presents the transverse correlation function for longitudinal and vertical fluctuations. The upper curve in these figures corresponds to the vertical distance between the meters $\Delta z = 50$ cm, the lower curve to $\Delta z = 75$ cm. As would be expected, the curves for $\Delta z = 50$ cm are located above those for $\Delta z = 75$ cm, since with increasing distance between the meters the correlation must decrease. It is quite natural also that with increasing Δz the correlation for the vertical component decreases somewhat faster than that for the longitudinal component.

Particular emphasis should be placed on the shape of the curve of the transverse correlation function for the longitudinal component of velocity. It is generally accepted that the correlation coefficient is a symmetrical function of the distance x, its value being maximum at $x = 0$. Figure 5.13 (a), however, shows that the transverse correlation of the longitudinal velocity $r_{u'_1 u'_2}$ is not symmetrical with respect to the origin of coordinates, the maximum value of $r_{u'_1 u'_2}$ at $\Delta z = 75$ cm occurring at $x = 1.0$ m.

THE BOTTOM BOUNDARY LAYER

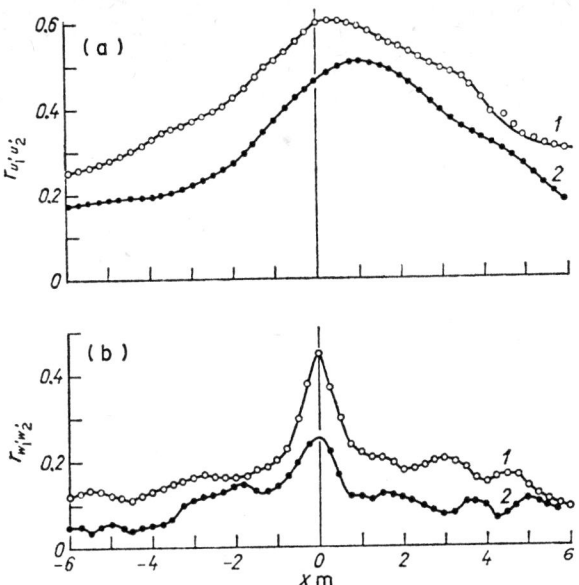

FIG. 5.13. Transverse correlation functions of the longitudinal (a) and vertical (b) components of tidal velocity fluctuations at various distances between the sensors, (1) at 50 cm, (2) at 75 cm.

The above peculiarity of the shape of $r_{u'_1 u'_2}$ is not typical of Red-Wharf Bay alone. It also appeared in results from the Mersey estuary data (see Bowden and Howe, 1963). Those results, presented in Fig. 5.14, demonstrate that the degree of asymmetry of $r_{u'_1 u'_2}$ increases with increasing distance between the devices. In other words, the greater the distance between the devices the more time is necessary for the correlation between velocity pulsations at the

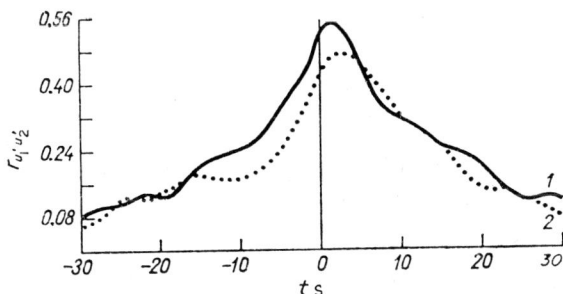

FIG. 5.14. The diagram illustrating asymmetry of the transverse correlation function of the longitudinal component of tidal velocity fluctuations with respect to the origin of coordinates (from Bowden and Howe, 1963). The distance between the sensors, (1) at 50 cm, (2) at 75 cm.

upper and lower levels to become maximum. This peculiarity can be apparently explained by the shift of the mean velocity which results in the slope of vortices with respect to the direction of the mean flow.

Let us, however, return to the determination of the scales of various components of turbulent fluctuations in tidal flow. Bowden (1962) calculated the correlation coefficients between longitudinal velocity fluctuations at two points separated by 50, 75 and 100 cm and found them equal to 0.60, 0.47 and 0.47, respectively. The correlation coefficients between the vertical velocity fluctuations for the same Δz were 0.45, 0.26 and 0.17. If we now find the distance at which the values of the coefficients of autocorrelation and cross correlation referring to the same height coincide, the value $\Delta z/x$ can be chosen as a measure of the ratio of the vertical and horizontal scales of corresponding velocity fluctuations. These correlations for the longitudinal and vertical velocity fluctuations in the bottom layer of Red-Wharf Bay are presented in Table 5.4.

TABLE 5.4. *Ratio of scales for the longitudinal and vertical tidal velocity fluctuations*
(from Bowden, 1962)

Distance between the meters, cm	Longitudinal velocity fluctuations		Distance between the meters, cm	Vertical velocity fluctuations	
	x cm at which $r_{u'}(x) = r_{u'_1 u'_2}$	$\dfrac{L_{u'y}}{L_{u'x}}$		x cm at which $r_{\omega'}(x) = r_{\omega'_1 \omega'_2}$	$\dfrac{L_{\omega'z}}{L_{\omega'x}}$
50	140	0.36	50	50	1.0
75	230	0.33	100	100	0.75
100	350	0.29	100–125	125	0.93

The table shows that the vertical scale of the longitudinal fluctuations is approximately one-third that of the horizontal fluctuations, while both longitudinal and vertical scales of the vertical fluctuations are roughly equal.

Bowden (1962) obtained two 10 min records of longitudinal velocity fluctuations with the meters orientated perpendicular to the mean flow. Their analysis showed that $L_{u'y}/L_{u'x} = 0.22$, i.e. the transverse scale of the longitudinal velocity fluctuations is nearly a quarter the corresponding longitudinal scale.

The analysis of data by Bowden and Howe (1963) in the Mersey estuary also confirms the above conclusion: strong anisotropy of the longitudinal component and approximate isotropy of the vertical component of velocity fluctuations. According to Bowden and Howe the ratio of the vertical and horizontal scales of longitudinal velocity fluctuations is 0.26, at 50 cm from the bottom, and 0.18 at 125 cm. For vertical pulsations $L_{w'z}/L_{w'x}$ is 1.0 at the height 50 cm and 0.5 at the height 125 cm.

Let us now try to imagine the shape of vortices in the bottom boundary layer. The above data allow one to conclude that in the vicinity of the bottom turbulence includes the vortices which are strongly stretched in the direction of the mean flow and flattened in the vertical and transversal directions, the main axis of the vortices forming a certain angle with the direction of the mean flow.

Spectra of velocity and temperature

The structure of turbulence in the bottom boundary layer can also be judged by the shape of one-dimensional energy spectra. The typical spectra of longitudinal, transversal and vertical components of tidal velocity are presented in Fig. 5.15.

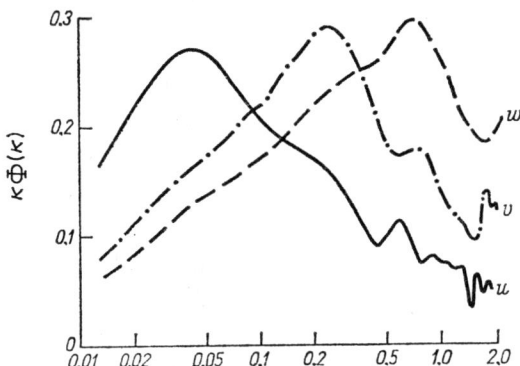

F I G. 5.15. Spectra of the longitudinal, transverse and vertical components of tidal velocity fluctuations in the bottom boundary layer (from Bowden, 1962). The wave numbers κ (c/cm) divided by $10^{-2}/2\pi$ are plotted on the x-coordinate.

First of all, one should notice that at the height of 75 cm the spectra of the separate components are completely different. Within the range of small wave numbers the energies of the transverse and the vertical components are nearly equal and approximately half that of the longitudinal components, the maximum of the energy for the longitudinal component being at wave number $\kappa \approx 6.4 \times 10^{-5}$ c/cm, and for the transversal and vertical component at wave numbers 4.0×10^{-4} and 11.2×10^{-4} c/cm, respectively.

Figure 5.15 shows the spectra within the range of large wave numbers to decrease with growing wave number. The power of this decrease, however, can hardly be evaluated, as the measurement accuracy of the electromagnetic flowmeter used falls off at large wave numbers.

200   OCEAN TIDES

The measurements by Wimbush and Munk (1971) make it possible to judge the shape of the spectra at higher wave numbers. These data were obtained by a standard Snodgrass capsule at the depth of 3.8 km in the Pacific Ocean (32°39′N, 120°33′W). Figure 5.16 gives an example of the energy spectrum for the longitudinal component of velocity fluctuations.

A remarkable peculiarity of this spectrum is the existence of a certain range (called the "inertial interval"), where the turbulence is locally isotropic. Within

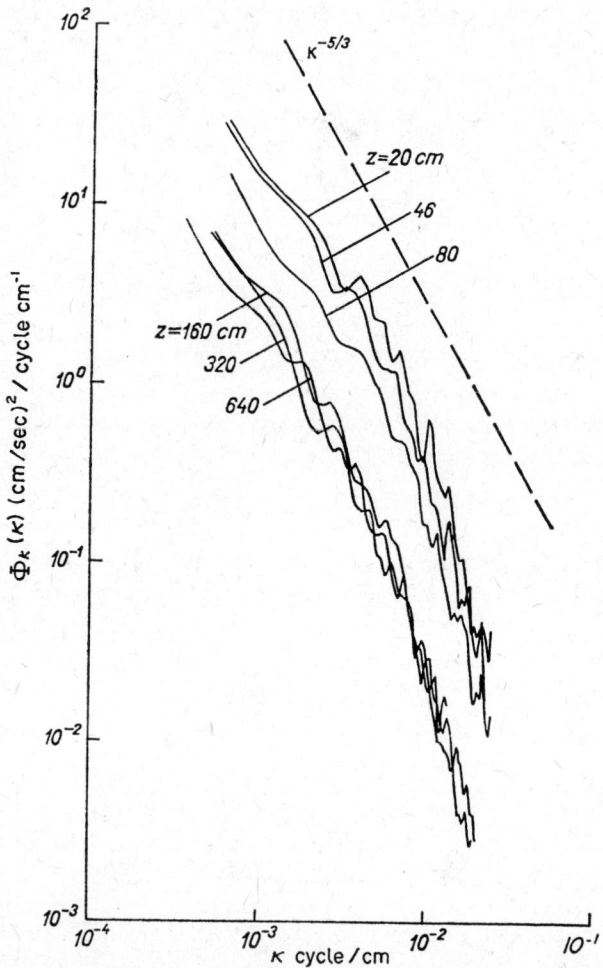

FIG. 5.16. Dimensional spectra of velocity fluctuations in the bottom boundary layer of the ocean at various distances from the bottom (from Wimbush and Munk, 1971). The measurements were carried out at 32°39′N, 120°33′W at the depth of 3.8 km. Dashed line corresponds to the $k^{-5/3}$ law.

this interval the spectrum is described by the "$k^{-5/3}$" law depicted in Fig. 5.16 by a dashed line.

A more detailed analysis of the spectrum, however, shows that the greatest wave numbers appear to be here, strictly speaking, not great enough to be associated with the inertial interval. As is well known, the inertial interval at neutral stratification is bounded at low frequencies by a wave number greater than the reciprocal distance from the underlying surface, and bounded at high frequencies by a wave number, smaller than the reciprocal of the Kolmogorov scale. In the case in question when the measurements were carried out at heights from 20 to 640 cm from the bottom, the condition $\kappa \gg z^{-1}$ is fulfilled only for the upper level. Thus, the data show that the spectrum of longitudinal velocity fluctuations obeys the "$k^{-5/3}$" law in a range of lower wave numbers than predicted by the theory of local-isotropic turbulence.

Another interesting peculiarity of the spectrum is its dependence on the height. Figure 5.16 demonstrates that at small wave numbers the energy of the longitudinal velocity fluctuations noticeably decreases with height, the width of the inertial interval increasing with height. The latter fact confirms the generally accepted rule, that the anisotropy of turbulence decreases with growing distance from the underlying surface.

The conclusion on local isotropy at large wave numbers can also be confirmed by the analysis of the cross spectrum of the longitudinal and vertical velocity fluctuations. The cross spectrum represents the contribution of the u- and w-fluctuations within the given range of frequencies to the mean product $\overline{u'w'}$ which is proportional to the stress of turbulent friction.

An example of the cross spectrum of the u- and w-components is presented in Fig. 5.17. It is clear from the figure that the greatest contribution to the cross spectrum comes from fluctuations with wave number 0.5×10^{-4} to $0.2 \times$

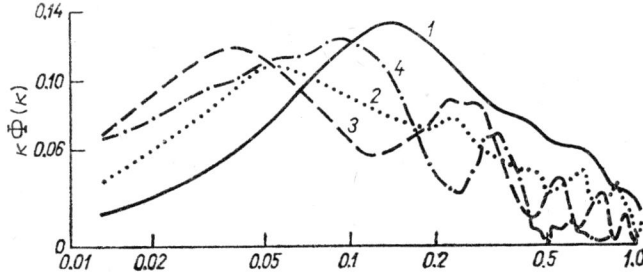

FIG. 5.17. Cross spectrum of the longitudinal and vertical components of tidal velocity fluctuations at various distances from the bottom (by Bowden, 1962). (1) 50 cm, (2) 75 cm, (3) 125 cm, (4) 150 cm.
The x-coordinate is the wave numbers κ (c/cm) divided by $10^{-2}/2\pi$.

$\times 10^{-2}$ c/cm. At lower wave numbers the values of the u-spectrum are large while the spectrum of the vertical component is small. This is due to the fact that the values of the cross spectrum in this range of wave numbers are also small.

Decrease in the cross spectrum with increasing wave number is accounted for by a sharp decrease of the u-spectrum. At wave numbers greater than $0.8 \times \times 10^{-3}$ c/cm the cross spectrum shows very small values though the vortices still possess considerable energy, as shown by the high values of the energy spectra of the longitudinal and vertical velocity pulsations in Fig. 5.16. The above peculiarity means that at large wave numbers the vortices are nearly isotropic and do not contribute sufficiently to the Reynolds stress. Thus, we have further evidence in favour of the existence of locally isotropic turbulence at large wave numbers.

Comparing Figs. 5.16, 5.17, we see that the maximum of the cross spectrum is located between the maxima of the u- and w-spectra, with the maximum of the cross spectrum shifted, however, towards the maximum values of the u-spectrum, i.e. towards lower frequencies.

At present there are available only three papers in which shear stress is determined from direct measurements of turbulence. These are the papers by Bowden and Fairbairn (1956), Bowden and Howe (1963) and Bowden (1962). The first paper says that at the distance of 75 cm from the bottom the shear stress changes from 2.1 to 4.1 dyne/cm² with mean velocity growing from 32 to 40 cm/s, while at the height 150 cm from 1.4 to 4.0 dyne/cm² with the mean velocity ranging from 32 to 50 cm/s.

In conformity with Bowden (1962), the mean magnitude of the Reynolds stress in the lowest 2 m equals 3.96 ± 0.25 dyne/cm². Table 5.5 shows variations in Reynolds stress with height. As may be seen, the tangential stress in the lower 1.5 m layer can increase by as much as twice. Incidentally, the idea of the vertical variation in the Reynolds stress within the logarithmic layer was also confirmed by the results of u_* calculations according to gradient measurements (see

TABLE 5.5. *Reynolds stress as a function of distance from the bottom* (from Bowden and Howe, 1963)

Distance from the bottom, cm	$-\varrho_0 \overline{u'w'}$ dyne/cm²	u cm/s
50	0.54	46
125	1.33	61
150	1.06	68

Wimbush and Munk, 1971). It was demonstrated in that paper that the layer of constant friction velocity was only 10 cm in height, while the height of the logarithmic layer was approximately 1 m.

This peculiarity of the behaviour of the tangential stress in the logarithmic layer is in all probability associated with the influence of the horizontal inhomogeneity of the sea bottom. Indeed, since shear stress in the lower levels corresponds to local roughness and in the upper layer to the roughness of a larger portion of the bottom, the horizontal inhomogeneity of the sea bottom results in shear stress at various levels being different, as observed.

Wimbush and Munk (1971) were the only investigators of the temperature spectra in the bottom boundary layer. Figure 5.18 presents their data, the solid curve depicting the temperature spectra at 1 m from the sea bottom. As may be

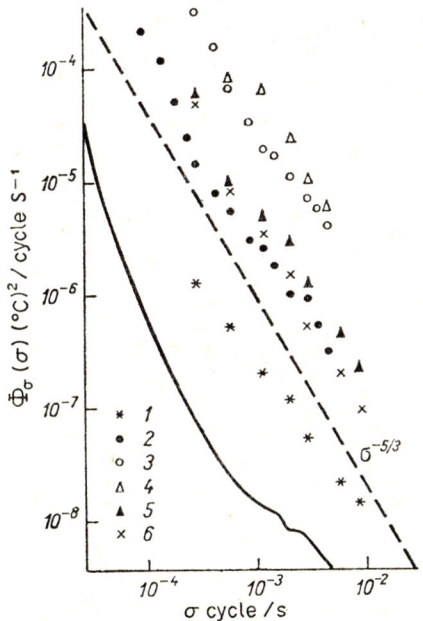

FIG. 5.18. Temperature spectra in the bottom boundary layer of the ocean (from Wimbush and Munk, 1971). The measurements were carried out: (1) at a depth of 1.7 km and at 30–145 cm from the bottom; (2) at a depth of 2 km and at 50 cm from the bottom; (3) at a depth of 3.7 km and at 120 cm from the bottom; (4) at a depth of 3.8 km and at 115 cm from the bottom; (5) at a depth of 3.8 km and at 145 cm from the bottom; (6) at a depth of 3.8 km and at 37 cm from the bottom. The data denoted by black triangles and crosses were obtained simultaneously; solid curve shows the spectrum of temperature of the ground at 1 m from the bottom; dashed line shows the spectrum corresponding to the $k^{-5/3}$ law.

seen, the spectrum of temperature pulsations in the bottom layer of the ocean is located above that of the temperature fluctuations at the sea bottom, the spectrum of the water temperature within the considered frequency range obeying the "$k^{-5/3}$" law for the inertial interval.

5.3. Theoretical Models of the Bottom Boundary Layer in Tidal Flows

Qualitative considerations on the vertical
distribution of tidal velocity
in a homogeneous ocean

The velocity distribution in a neutrally stratified bottom boundary layer depends in a complicated way on the external parameters determining the turbulent regime in this layer. In tidal flow with a vertical extension exceeding the thickness of the bottom boundary layer these external parameters are the amplitudes U, V and phases g_u, g_v of the horizontal components of the tidal flow velocity beyond the boundary layer, the Coriolis parameter $2\omega_z$, the oscillation frequency σ and the roughness parameter z_0, describing integrally the whole spectrum of the bottom roughness of the area, the linear scale of which is at least equal to the thickness of the bottom boundary layer.

By virtue of the π theorem of dimensional analysis, the above seven parameters can make up only four dimensionless combinations i.e. the ratio of the amplitudes V/U and the difference of the phases $g_u - g_v$ determining the tidal flow velocity in a frictionless oceanic layer, a modified Rossby number $Ro = U/\sigma z_0$ and the parameter $e = 2\omega_z/\sigma$ which characterizes the ratio of the forces of Coriolis and inertia. One may expect all the characteristics of the turbulent regime in the bottom boundary layer, with the mean velocity of the tidal flow included, to be associated with the external parameters only by means of these dimensionless combinations.

It has already been noted in Chapter 1 that in a homogeneous tidal flow the forces of inertia, horizontal pressure gradient, Coriolis and turbulent friction counterbalance one another. To evaluate these forces, write a simplified equation of motion (1.3.5) with allowance for the remarks given in Sections 1.1, 1.3 and 2.1, thus

$$\frac{\partial \mathbf{u}}{\partial t} + \mathbf{A}_1 \mathbf{u} = -g \cdot \nabla(\zeta - \zeta^+) + \frac{\partial}{\partial z} \kappa_v \frac{\partial \mathbf{u}}{\partial z}. \tag{5.3.1}$$

Replace in the above equation the velocity $\mathbf{u}$ by the deviation $\mathbf{W} = \mathbf{u} - \mathbf{u}_\infty$, where $\mathbf{u}_\infty$ is the vector of the tidal flow velocity in the absence of friction which can be defined in the following way

$$\frac{\partial \mathbf{u}_\infty}{\partial t} + \mathbf{A}_1 \mathbf{u}_\infty = -g \cdot \nabla(\zeta - \zeta^+). \tag{5.3.2}$$

Since $\mathbf{u}_\infty$ does not change with depth, we may subtract equation (5.3.2) from equation (5.3.1) term by term and obtain

$$\frac{\partial \mathbf{W}}{\partial t} + \mathbf{A}_1 \mathbf{W} = \frac{\partial}{\partial z} \kappa_v \frac{\partial \mathbf{W}}{\partial z}. \tag{5.3.3}$$

Introduce then the dimensionless variables

$$t_n = \sigma t;$$
$$z_n = z/H_0,$$
$$\mathbf{W}_n = \mathbf{W}/U_0,$$
$$\kappa_{vn} = \kappa_v/\kappa_0,$$

where U_0 is the characteristic amplitude of the mean velocity of the tidal flow; κ_0 and H_0 are characteristic values of the coefficient of the vertical turbulent viscosity and the vertical scale of motion, respectively (their types will be defined concretely later).

Then equation (5.3.3) can be rewritten as

$$\frac{\partial \mathbf{W}_n}{\partial t_n} + e\mathbf{A}_{1n}\mathbf{W}_n = E \frac{\partial}{\partial z_n} \kappa_{vn} \frac{\partial \mathbf{W}_n}{\partial z_n}. \tag{5.3.4}$$

Here, besides the known notation, the following symbols are introduced:

$$\mathbf{A}_{1n} = \begin{bmatrix} 0 & -1 \\ 1 & 0 \end{bmatrix},$$

and $E = \kappa_0/\sigma H_0^2$ is a modified Ekman number characterizing the ratio of the forces of turbulent friction and inertia.

At a fixed e value the whole oceanic thickness can be divided into three layers depending on the value of E. In the lowest layer adjoining the bottom $E \gg 1$. Here the velocity profile of the tidal flow is mainly formed under the influence of the force of the vertical turbulent friction. Since the sea bottom hampers free motion of the vortices in the vertical direction, the eddy coefficient in the vicinity of the bottom must be very small. With growing distance from the underlying surface the scales of the vortices increase, resulting in an increasing eddy coefficient. If we assume that the eddy coefficient increases linearly with height, then the logarithmic law of mean velocity distribution will hold within the lowest layer. The observation data presented in the preceding section confirm this conclusion.

In the second of the above three layers where $E \approx 1$, the forces of inertia, horizontal gradient of pressure, Coriolis and vertical turbulent friction are approximately equal. It seems plausible to suggest that, as confirmed by data, the decrease of production cannot be here compensated by decreasing dissipation and increasing turbulent energy due to diffusion from the lower layers. In this case the eddy coefficient decreases with height; this, together with decreasing vertical velocity gradients, results in a decrease of the turbulent frictional force, which in its turn implies an increase of tidal flow velocity with height.

If $E \ll 1$ the force of the turbulent friction becomes incommensurably less than all the other forces participating in the formation of the vertical profile of mean velocity. The layer, in which this inequality is fulfilled, is located above the bottom boundary layer. At neutral stratification the tidal flow velocity within this layer must remain constant with height and equal to $\mathbf{u}_\infty$.

These are the qualitative considerations on the vertical distribution of tidal flow velocity in a homogeneous ocean. Let us now discuss theoretical investigations of the bottom boundary layer in the tidal flow.

Models of the vertical velocity distribution at an a priori given eddy coefficient

The first models of the type were presented within a few years by Sverdrup (1926), Fjeldstad (1929) and Thorade (1931). Let us begin their description with the latest and the simplest model.

Thorade considered the case when the depth exceeds the thickness of the bottom boundary layer. Proceeding from natural considerations he assumed that the no-slip condition holds on the bottom, while at a sufficiently great distance from it the velocity of the tidal flow tends to that in the absence of friction. The coefficient of the vertical turbulent viscosity was set independent of height and given.

Let us express the above mathematically. If the complex dimensionless velocity $u_n + iv_n$ retains its former symbol $\mathbf{u}_n$ and the vertical scale of motion is equal to the thickness of the bottom boundary layer $h_b = (\kappa_v/\tfrac{1}{2}\sigma)^{1/2}$, equation (5.3.4) can be rewritten as follows:

$$\frac{\partial}{\partial t_n}(\mathbf{u}_n - \mathbf{u}_{n\infty}) + ie(\mathbf{u}_n - \bar{\mathbf{u}}_{n\infty}) = \frac{1}{2}\frac{\partial^2 \mathbf{u}_n}{\partial z_n^2}. \qquad (5.3.5)$$

THE BOTTOM BOUNDARY LAYER

Its solution obeys the boundary conditions

$$\mathbf{u}_n = 0 \quad \text{at} \quad z_n = 0, \tag{5.3.6}$$

$$\mathbf{u}_n \to \mathbf{u}_{n\infty} \quad \text{at} \quad z_n \to \infty \tag{5.3.7}$$

and the condition of clockwise rotation of the vector of the tidal velocity ($\mathbf{u}_n \sim \exp(-it_n)$), and takes the form

$$\mathbf{u}_n = \bar{\mathbf{u}}_{n\infty} \exp(-it_n)\left[1 - \exp\left(-(1+i)z_n\sqrt{e-1}\right)\right], \tag{5.3.8}$$

where $\bar{\mathbf{u}}_{n\infty}$ is the complex amplitude of the tidal velocity beyond the boundary layer.

Solution (5.3.8) is valid only for $e > 1$. We are, however, more interested in the other situation when $e < 1$. In this case

$$\mathbf{u}_n = \bar{\mathbf{u}}_{n\infty} \exp(-it_n)\left[1 - \exp\left(-(1-i)z_n\sqrt{1-e}\right)\right]. \tag{5.3.8'}$$

As is well known, any harmonic oscillation can be represented as a combination of the clockwise and counter-clockwise motions. Hence

$$\mathbf{u}_n = \mathbf{C}^- \exp(-it_n) + \mathbf{C}^+ \exp(it_n), \tag{5.3.9}$$

where $\mathbf{C}^-$ and $\mathbf{C}^+$ are some complex amplitudes, the type of which needs to be defined using equation (5.3.8) and the same relation obtained having regard to the counter-clockwise rotation of the velocity vector ($\mathbf{u}_n \sim \exp(it_n)$).

Omitting unnecessary details, write

$$\mathbf{C}^\pm = \bar{\mathbf{u}}_{n\infty}^\pm \left[1 - \exp\left(-(1\pm i)z_n\sqrt{1\pm e}\right)\right]. \tag{5.3.10}$$

Then according to equation (5.3.9) and the definition

$$u_n = \tfrac{1}{2}C_u \exp(-it_n) + \tfrac{1}{2}C_u^* \exp(it_n);$$
$$v_n = \tfrac{1}{2}C_v \exp(-it_n) + \tfrac{1}{2}C_v^* \exp(it_n), \tag{5.3.11}$$

where C_u, C_v are the complex amplitudes of the corresponding velocity components; C_u^*, C_v^* are their complex conjugates, we have

$$\mathbf{C}^- = \tfrac{1}{2}(C_u + iC_v);$$
$$\mathbf{C}^+ = \tfrac{1}{2}(C_u^* + iC_v^*).$$

Hence

$$C_u = \text{real}(\mathbf{C}^+ + \mathbf{C}^-) - i \cdot \text{imag}(\mathbf{C}^+ - \mathbf{C}^-);$$
$$C_v = \text{imag}(\mathbf{C}^+ + \mathbf{C}^-) + i \cdot \text{real}(\mathbf{C}^+ - \mathbf{C}^-). \tag{5.3.12}$$

Thus, substituting relations (5.3.10), (5.3.12) into equation (5.3.11) we obtain the required distribution of the tidal velocity.

In general, the Thorade model can be said to describe qualitatively the change of velocity of the tidal flow with height. Comparing velocity values derived from it with observational data revealed, however, quite noticeable deviations especially in the lowest oceanic layer next to the bottom. This is seen from Fig. 5.19, which depicts the ellipses of tidal flow in the bottom boundary layer calculated by the Thorade model compared with observational data from the open ocean.

The observed deviations can be probably accounted for by assuming invariability of the turbulent viscosity coefficient with height. An analogous supposition was a corner-stone of the Sverdrup model, which differs from the previous model in that the bottom boundary layer is not well defined. Instead of it, the influence of bottom friction is assumed to be exerted to a certain degree throughout the whole width of the ocean, the condition of zero shear stress being fulfilled at the free surface of the ocean.

The first model in which the eddy coefficient was considered as a function of the vertical coordinate was suggested by Fjeldstad. He generalized the Sverdrup solution for the case of the two-layer liquid, in the lower part of which the eddy coefficient was set by a power law, while in the upper layer it was set equal to zero. Fjeldstad also found solutions corresponding to exponential growth and to the decrease of the eddy coefficient with height. In all the above three cases the basic regularities of the vertical distribution of the tidal velocity remained unchanged. An increase of eddy coefficient with height was noted, however, to result in increasing vertical velocity gradients in the bottom layer and decreasing gradients in the surface layer. A certain increase in the angle of turning of the main axis of the ellipse at the bottom layer and a decrease in the shift of the velocity phases at various depths were also noted.

Bowden, Fairbairn and Hughes (1959) were on the verge of creating a model which could correctly depict all the peculiarities of the vertical distribution of the tidal flow velocity. They divided the whole thickness into two layers, i.e. the lower logarithmic layer and the above layer within which the eddy coefficient was set constant in height with the velocity profile described by the Sverdrup model. At the upper boundary of the logarithmic layer of given height, the condition of continuity of the shear stress is set. In the logarithmic layer the shear stress is assumed to be constant in height and equal to the bottom stress. The latter is determined by an empirical formula provided the value of the velocity of the tidal flow on the upper boundary of the logarithmic layer is given. To evaluate this velocity, the ratio of the velocities on a fixed level within the logarithmic layer and on the free surface are assumed known. The model by Bowden, Fairbairn and Hughes describes quite accurately the vertical distribution of the tidal flow velocity.

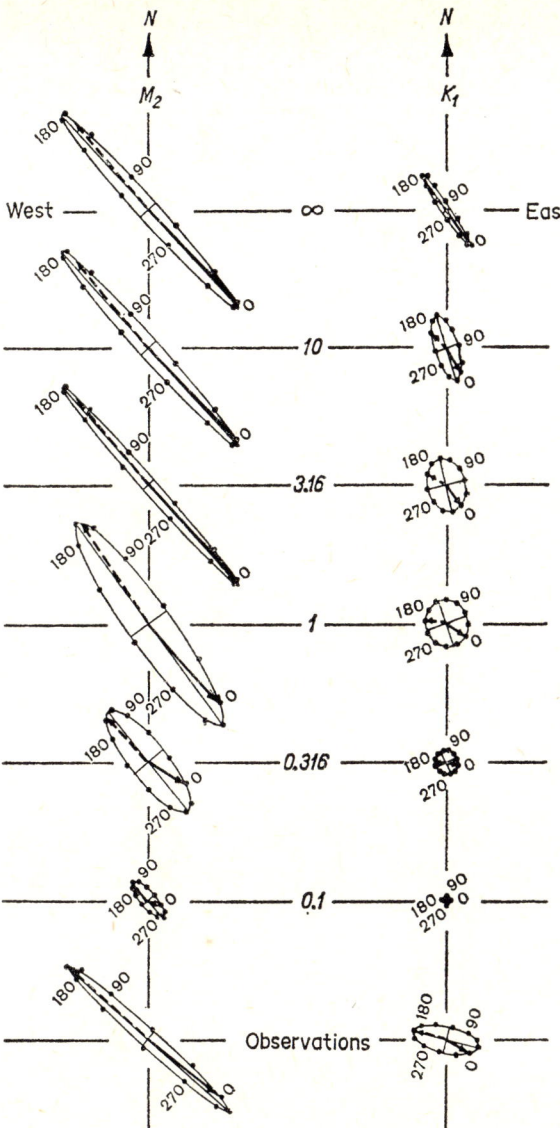

FIG. 5.19. Ellipses of the tidal currents for the M_2 and K_1 waves in the bottom boundary layer of the ocean (from Munk, Snodgrass and Wimbush, 1970). The ellipses at the upper six levels were calculated by the Thorade model, the lowest two ellipses were computed by observational data obtained at 1.7 m from the bottom. Solid arrow shows the velocity vector of the tidal currents at the moment of the Moon's transit at the Greenwich meridian, dashed arrow shows the above vector at the moment of high water (local time). Dots on the ellipses denote the positions of the vectors every other hour by Greenwich time. Numerals on the horizontal lines are dimensionless heights showing the distance from the bottom divided by the thickness of the bottom boundary layer, the latter being 3.6 m by the authors' estimation.

Its application, however, is still more limited than, for instance, the models by Thorade or Sverdrup, since besides the eddy coefficient in the upper layer, it contains two more unknown parameters, i.e. the height of the logarithmic layer and the ratio of velocities of the tidal flow in the logarithmic layer and on the surface. Arbitrary choice of these parameters is hard to justify in physical terms.

The two-layer model by Bowden, Fairbairn and Hughes was advanced by Kagan (1964, 1966), who solved the equation of motion (5.3.1) under the assumption that, averaged (over the tidal period) coefficient of turbulence changes with height by the scheme "with a knee"

$$\kappa_v = \begin{cases} \varkappa \tilde{u}_* z & \text{at} \quad z \leqslant h_{\log}; \\ \varkappa \tilde{u}_* h_{\log} & \text{at} \quad z \geqslant h_{\log}, \end{cases} \qquad (5.3.13)$$

i.e. κ_v increases linearly up to a certain level $h_{\log}$ and then remains constant up to the upper limit of the boundary layer. Here $h_{\log}$ is the height of the logarithmic layer, $\tilde{u}_*$ is the mean over a tidal period modulus of the dynamic velocity; $\varkappa$ is the Karman constant.

To evaluate these two unknown parameters $h_{\log}$ and $\tilde{u}_*$ contained in this model, we use the asymptotic equality

$$\langle |u| \rangle \approx \frac{\tilde{u}_*}{\varkappa} \ln z/z_0, \qquad (5.3.14)$$

which is valid only at small values of z (the symbol $\langle \ \rangle$ denoting an average over the tidal period), and the additional relation

$$h_{\log} \sim \tilde{u}_*/\sigma, \qquad (5.3.15)$$

which is derived from dimensional analysis. The system of equations closed in this way makes it possible to determine the velocity profiles of the tidal flow provided that only external parameters determining the vertical structure of the bottom boundary layer are given.

Calculations carried out by this model have revealed an interesting peculiarity inherent in deep-water basins with relatively weak tidal velocity. At the neutral stratification the amplitudes of both velocity components remain constant with depth beyond the limits of the boundary layer. Then they smoothly decrease in the upper portion of the boundary layer, while in the bottom layer they decrease logarithmically. The phase of the longitudinal velocity component changes with depth only slightly, but this is not the case for the phase of the transversal velocity component. The transverse component slightly

THE BOTTOM BOUNDARY LAYER 211

changes with depth beyond the boundary layer while in the boundary layer it sharply changes by up to several tens of degrees with increasing depth. In some cases even the bend of the curve is observed.

The use made of harmonic constants for constructing ellipses of the tidal flow has revealed that the sharp change in the phase of the transversal velocity component corresponds to the change in the direction of rotation of the ellipse (Fig. 5.20). Consequently, in a homogeneous ocean we can come across the

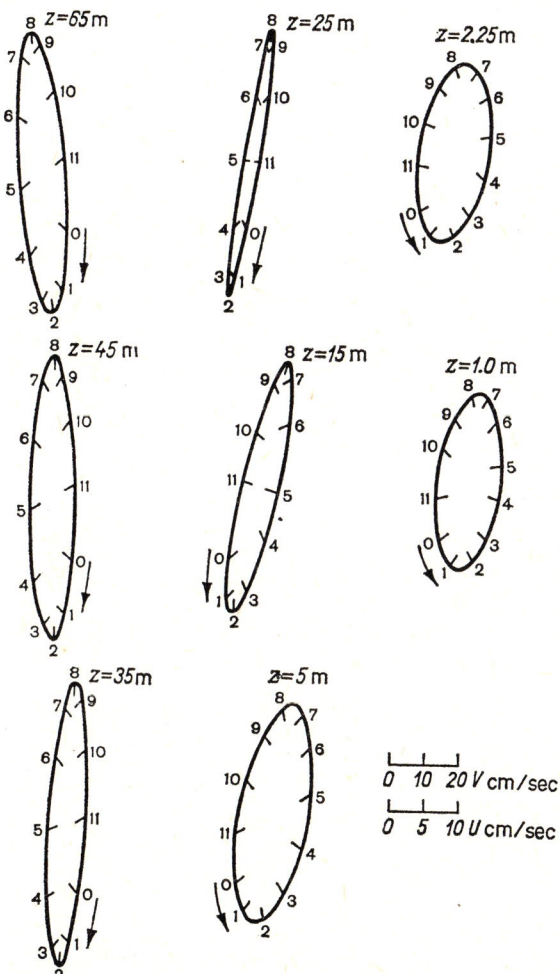

FIG. 5.20. Ellipses of the tidal currents at various distances from the bottom (by Kagan, 1968). Arrows show the direction of rotation of the ellipse, notches show the position of the edges of the velocity vector of the tidal currents at different hours of the tidal cycle.

conditions when the rotation of the ellipse in the surface and bottom layers appears to be opposite in direction.

This interesting peculiarity, which has sometimes been associated with the existence of a density interface between the layers of the fluid, can also be accounted for as follows.

At comparatively low longitudinal tidal flow, the Coriolis force induces the appearance of negligible transversal velocity components changing with depth. The opposite transversal velocity components arise from transversal pressure gradient. It might happen that at a certain depth the transverse velocity components induced by the Coriolis force and the force of the horizontal pressure gradient become equal and consequently compensated. In this case the tidal flow velocity observed at this level will be of a reversing character. Unlike the upper layer, where the force of the horizontal pressure gradient can be less than the Coriolis force, in a lower layer a reverse situation is observed, i.e. the force of the horizontal pressure gradient will exceed the Coriolis force, in which case the change in the direction of rotation the ellipse takes place, as is confirmed by calculations.

The two-layer model considered above contributes greatly to the quality of calculations from the standpoint of its correspondence with observations, though it also suffers from certain drawbacks. Let us consider only two of these, from our viewpoint rather serious. The first is associated with approximating the profile of the eddy coefficient by a known function of the vertical coordinate; the second concerns the assumption of a time-independent eddy coefficient.

In conformity with modern ideas and indirect evaluations, the eddy coefficients (see Fig. 5.21) are quite variable. A coefficient typically increases almost linearly in the lower layer, then its growth slows down and in the rest of the boundary layer it decreases monotonically with height. In all probability such

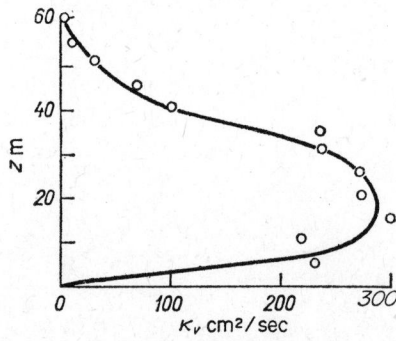

FIG. 5.21. Vertical distribution of the eddy coefficient in the bottom boundary layer of the tidal flow (by Vapniar, 1962).

a profile of an eddy coefficient may be spoken of as typical of a boundary layer in tidal flow, though the experimental data confirming this conclusion are still scarce at present.

A presentation of the character of the variability of eddy coefficient in time is given in Table 5.6 which presents the calculation data on κ_v according to the data on the vertical distribution of the mean velocity of the tidal flow. Table 5.6 clearly shows that during a tidal cycle the eddy coefficient at a fixed level can change several times, the amplitude of its oscillations increasing with height.

TABLE 5.6. *Turbulence coefficient* (cm²/s) *as function of time and height*
(from Bowden, Fairbairn and Hughes, 1959)

z/d	Hours before the moment of high water									HW
	4.5	4	3.5	3	2.5	2	1.5	1	0.5	
0.375	130	190	230	240	210	210	210	160	160	–
0.625	150	310	450	230	230	290	370	220	250	–
0.875	110	240	310	310	300	240	170	210	–	–
Mean	130	250	330	260	250	280	250	200	200	–

z/D	Hours after the moment of high water									Mean
	0.51	1.5	2	2.5	3	3.5	4	4.5	5	
0.375	–	–	–	120	160	–	80	150	120	–	169
0.625	–	–	–	150	230	130	110	130	160	150	229
0.875	–	–	50	80	120	170	130	90	100	220	178
Mean	–	–	50	120	170	150	110	120	120	180	192

Remark. Dashes indicate that data are not available, HW—high water.

Neglect of the effect of time and spatial variability of the eddy coefficient causes some important peculiarities of the turbulent regime in the boundary layer of the tidal flow to remain unclarified. In particular, still unsolved is the problem of the relationship of the time oscillation of the tidal flow velocity and the eddy coefficient. No insight has been gained into the peculiarities of the space-time distribution of the components of the turbulent energy balance. The role of density stratification in the formation of the velocity fields and the simpliest characteristics of turbulence still remains unknown.

All these problems demanded further investigation and eventually led to the construction of the models which provide a means for calculating the vertical structure of the bottom boundary layer without involving the magnitude and profile of the eddy coefficient, the latter being obtained in the process of the solution of the general problem alongside with the other turbulence characteristics and the tidal velocity.

Two models of the type will be discussed below.

Models of the bottom boundary layer based on closure of the equations through the use of semi-empirical hypotheses

Bottom boundary layer with neutral stratification. The equation of motion describing the distribution of the tidal velocity in a homogeneous ocean has the form of equation (5.3.1). The unknown variables here are the velocity vector **u** with the components u and v and the coefficient of the vertical turbulent viscosity κ_v. The horizontal gradient of the elevation $\nabla(\zeta - \zeta^+)$ is considered given.

To define κ_v, amplify equation (5.3.1) by the equation of turbulent energy,

$$\frac{\partial b^2}{\partial t} = \kappa_v \left| \frac{\partial \mathbf{u}}{\partial z} \right|^2 + \alpha_b \frac{\partial}{\partial z} \kappa_v \frac{\partial b^2}{\partial z} - \varepsilon, \qquad (5.3.16)$$

and by the relations of approximate similarity which are generally used in the semi-empirical theory of turbulence namely

$$\kappa_v = c_0 l b,$$
$$\varepsilon = c_1 b^3 / l, \qquad (5.3.17)$$

where b^2 and l are the kinetic energy and the spatial scale of velocity fluctuations; ε is the mean velocity of turbulent energy dissipation; α_b, c_0 and c_1 are numerical constants.

Close the system of equations (5.3.1), (5.3.16), (5.3.17) by means of the generalized Karman formula,

$$l = -\varkappa \frac{b}{l} \left(\frac{\partial}{\partial z} \frac{b}{l} \right)^{-1}. \qquad (5.3.18)$$

The solution of this system of equations must obey the following boundary conditions:

zero tidal velocity at the level of roughness,

$$\mathbf{u} = 0 \quad \text{at} \quad z = z_0; \qquad (5.3.19)$$

disappearance of the shear stress at the free surface

$$\kappa_v \frac{\partial \mathbf{u}}{\partial z} = 0 \quad \text{at} \quad z = D; \qquad (5.3.20)$$

absence of diffusion flows turbulent energy at the level of roughness and on the free surface

$$\alpha_b \kappa_v \frac{\partial b^2}{\partial z} = 0 \quad \text{at} \quad z = z_0, \quad z = D; \qquad (5.3.21)$$

and, finally, asymptotic transition of the expression for the turbulence scale (5.3.18) into a linear function of height at small distance from the bottom

$$l \to \varkappa z_0 \quad \text{at} \quad z \to z_0. \qquad (5.3.22)$$

We specify no initial conditions. Instead, we demand all the required functions in the tidal flow to vary periodically in time.

Thus the problem is reduced to determining for the unknown functions $\mathbf{u}$, κ_v, b^2, ε and l obeying equations (5.3.1), (5.3.16)—(5.3.18) and the boundary conditions (5.3.19)—(5.3.22).

Note that with this formulation of the problem the effect of the free surface is allowed for only by the specification of the vertical velocity distribution in the vicinity of the surface [see condition (5.3.20)]. In real conditions, however, direct influence of the free surface on the turbulent velocity pulsations, as demonstrated by suppression of the turbulent pulsations induced by the action of forces of surface tension, is observed. This fact can be taken into consideration by supposing that, within a certain layer adjoining the free surface, the scale of turbulence reduces as it approaches the surface.

The thickness of the layer which is influenced by surface tension is comparatively small, and as a rule does not exceed a few decimetres. That is why, without employing the two-layer scheme to describe the scale of turbulence, one can approximate the effect of weakening of the turbulence in the subsurface layer, setting on the free surface the condition

$$\alpha_b \kappa_v \frac{\partial b^2}{\partial z} - \mu b^2 = 0 \quad \text{at} \quad z = D, \qquad (5.3.21')$$

where μ is the coefficient of weakening of the kinetic energy of turbulence due to the damping influence of the free surface.

The coefficient μ may in all probability vary from zero ad infinitum. The upper limit of this range corresponds to the case of a solid surface damping turbulence completely. The opposite situation is observed at $\mu = 0$, in which case condition (5.3.21') reduces to condition (5.3.21), when the free surface does not directly affect the turbulent velocity pulsations. Appropriate values of μ can be determined from observational data.

No further account will be taken of the direct influence of the free surface on turbulent velocity pulsations, i.e. we shall henceforth assume $\mu = 0$.

Introduce some preliminary transformations to the system of equations (5.3.1), (5.3.16)—(5.3.22) replacing the values of velocity **u** by deviations $\mathbf{W} = \mathbf{u} - \mathbf{u}_\infty$. Remember that in tidal flow ζ and ζ^+ are periodic time functions

$$(\xi, \xi^+) = \text{real}\,(\bar{\xi}, \bar{\xi})\exp{(i\sigma t)} \tag{5.3.23}$$

and hence

$$\mathbf{u}_\infty = \text{real}\,\bar{\mathbf{u}}_\infty \exp{(i\sigma t)}. \tag{5.3.23'}$$

Substituting equations (5.3.23), (5.3.23′) into equation (5.3.2) and omitting the multiplier $\exp{(i\sigma t)}$ we obtain

$$\bar{\mathbf{u}}_\infty = -g\mathbf{B}^{-1} \cdot \nabla(\bar{\zeta} - \bar{\zeta}^+), \tag{5.3.24}$$

where $\mathbf{B} = (i\sigma\mathbf{E} + \mathbf{A}_1)$, $\mathbf{E}$ being a unit matrix.

Thus, $\mathbf{u}_\infty$ is uniquely expressed through the known values of the horizontal gradient of the elevation and consequently can replace the latter as an external parameter determining the vertical structure of the bottom boundary layer.

According to equation (5.3.23′), the components of the vector $\mathbf{u}_\infty$ can be represented as

$$u_\infty = U\cos{(\sigma t - g_u)}, \quad v_\infty = V\cos{(\sigma t - g_v)}, \tag{5.3.25}$$

where U, V and g_u, g_v are the amplitudes and phases of the corresponding velocity components, respectively. Reckon the time from the moment $t_0 = g_u/\sigma$ when u_∞ assumes its maximum value. In this case instead of equation (5.3.25) we have

$$u_\infty = U\cos{\sigma t},$$
$$v_\infty = V\cos{(\sigma t + g_u - g_v)}. \tag{5.3.26}$$

Introduce the dimensionless variables

$$t_n = \sigma t,$$
$$z_n = z/D;$$
$$\mathbf{W}_n = \mathbf{W}/U,$$
$$\kappa_{vn} = \frac{\kappa_v \,\text{Re}_{\text{crit}}}{UD};$$
$$b_n^2 = \left(\frac{c_0\,\text{Re}_{\text{crit}}}{U}\right)^2 b^2; \quad \varepsilon_n = \left(\frac{c_0^3 D\,\text{Re}_{\text{crit}}^3}{c_1 U^3}\right)\varepsilon;$$
$$l_n = \frac{l}{D}. \tag{5.3.27}$$

THE BOTTOM BOUNDARY LAYER

Now the system of equations (5.3.1), (5.3.16)—(5.3.22) can be rewritten as follows:

$$\frac{\partial W_{xn}}{\partial t_n} - eW_{yn} = E \frac{\partial}{\partial z_n} \kappa_{vn} \frac{\partial W_{xn}}{\partial z_{vn}}; \qquad (5.3.28)$$

$$\frac{\partial W_{yn}}{\partial t_n} + eW_{xn} = E \frac{\partial}{\partial z_n} \kappa_{vn} \frac{\partial W_{yn}}{\partial z_n}; \qquad (5.3.29)$$

$$\frac{c^{-1/2}}{E} \operatorname{Re}_{\mathrm{crit}}^{-2} \frac{\partial b_n^2}{\partial t_n} = \kappa_{vn} \left[\left(\frac{\partial W_{xn}}{\partial z_n}\right)^2 + \left(\frac{\partial W_{yn}}{\partial z_n}\right)^2 \right]$$

$$+ \frac{\alpha_b}{c^{1/2}} \operatorname{Re}_{\mathrm{crit}}^{-2} \frac{\partial}{\partial z_n} \kappa_{vn} \frac{\partial b_n^2}{\partial z_n} - \frac{\varepsilon_n}{\operatorname{Re}_{\mathrm{crit}}^2}; \qquad (5.3.30)$$

$$\kappa_{vn} = l_n b_n;$$

$$\varepsilon_n = \frac{b_n^3}{l_n}; \qquad (5.3.31)$$

$$l_n = -\varkappa \frac{b_n}{l_n} \left(\frac{\partial}{\partial z_n} \frac{b_n}{l_n}\right)^{-1} \qquad (5.3.32)$$

$$W_{xn} = -\cos t_n;$$
$$W_{yn} = -(V/U) \cos(t_n + g_u - g_v) \quad \text{at} \quad z_n = z_{0n}; \qquad (5.3.33)$$

$$\kappa_{vn} \frac{\partial W_{xn}}{\partial z_n} = \kappa_{vn} \frac{\partial W_{yn}}{\partial z_n} = 0 \quad \text{at} \quad z_n = 1; \qquad (5.3.34)$$

$$\alpha_b \kappa_{vn} \frac{\partial b_n^2}{\partial z_n} = 0 \quad \text{at} \quad z_n = z_{0n}, \quad z_n = 1; \qquad (5.3.35)$$

$$l_n \to \varkappa z_{0n} \quad \text{at} \quad z_n \to z_{0n}. \qquad (5.3.36)$$

Here the following notations have been used: $E = (U/\sigma D) \operatorname{Re}_{\mathrm{crit}}^{-1}$ for the Ekman number, $z_{0n} = z_0/D$ for the dimensionless roughness of the bottom and it was set that $c_1 = c_0^3 = c^{3/4}$. The sought functions W_{xn}, W_{yn}, κ_{vn}, b_n^2, ε_n, and l_n, each depending on two arguments z_n and t_n and five dimensionless combinations V/U, $g_u - g_v$, E, e and z_{0n}, provide, in principle, detailed information on the space-time distribution of all the characteristics of the turbulent regime of the bottom boundary layer.

Note, however, that the scale of length D (i.e. the depth) employed above is valid only in the case when the bottom boundary layer embraces the whole thickness of the tidal flow. If we can single out within its limits a layer, free from the direct influence of friction, then we should choose the value $U/\sigma \operatorname{Re}_{\mathrm{cr}}$

coinciding with the height of the bottom boundary layer accurate to the last multiplier as a characteristic length scale. In the latter case some of the dimensionless parameters disappear since $E = 1$, and z_{0n} assumes the form $z_{0n} = \mathrm{Re}_{\mathrm{crit}} R_0$.

To solve the non-linear system of equations (5.3.28)–(5.3.36), approximate all derivatives by directional finite differences. Assume the time step to be constant and equal to 0.1256, and discretize the variable z_n in the following way. Locate fifty nodes in the domain $z_{0n} \leqslant z_n \leqslant 0.05$ in such a way that the grid step remains constant in the logarithmic scale. Divide the range of heights $0.05 \leqslant z_n \leqslant 1$ into 200 equal intervals.

We seek the solution of the resulting system of algebraic equations at every time step by the method of successive by κ_{vn}, setting that W_{xn}, W_{yn} and b_n^2 equal zero at the initial moment of time. Then the scheme of calculating the required functions reduces to the following procedure.

Equations of motion (5.3.28), (5.3.29) with boundary conditions (5.3.33), (5.3.34) are integrated by the method of matrix factorization at an arbitrarily chosen profile κ_{vn} and zero values of W_{xn}, W_{yn} at the initial moment of time. The resulting values of the vertical gradients of the velocity deviations are substituted into the equation of the turbulent energy (5.3.30) modified by relations (5.3.31), (5.3.32). The equation of turbulent energy (5.3.30) is integrated by the factorization method, using the facts that condition (5.3.35) is fulfilled on the boundaries and $b_n^2(z_n) = 0$ at the initial moment of time.

Determine thus the new values of κ_{vn} at each node of the vertical. Use the obtained profile κ_{vn} for the solution of the equation of motion. Calculate again the profile κ_{vn} and so on until the profiles obtained in two subsequent iterations coincide to the required accuracy.

Now consider the solution for the next time step, using the profiles W_{xn}, W_{yn}, b_n^2 and κ_{vn} obtained at the preceding time step as the initial. The condition

$$\max_{\substack{z_{0n} \leqslant z_n \leqslant 1 \\ t_{1n} \leqslant t_n \leqslant t_{1n}+2\pi}} \left| \frac{\kappa_{vn}(z_n, t_n)}{\kappa_{vn}(z_n, t_n + 2\pi)} - 1 \right| < 10^{-2}$$

where t_{1n} is an arbitrary moment of time, serves as a criterion of its establishment.

In the process of calculation it was found that the time of convergence of the solution to a periodic regime is approximately 4π, which equals two tidal periods. The calculation data are given below.

Figure 5.22 presents the space-time distribution of the functions W_{xn} and W_{yn}. These functions are seen to reach their maximum values at the lower levels

THE BOTTOM BOUNDARY LAYER

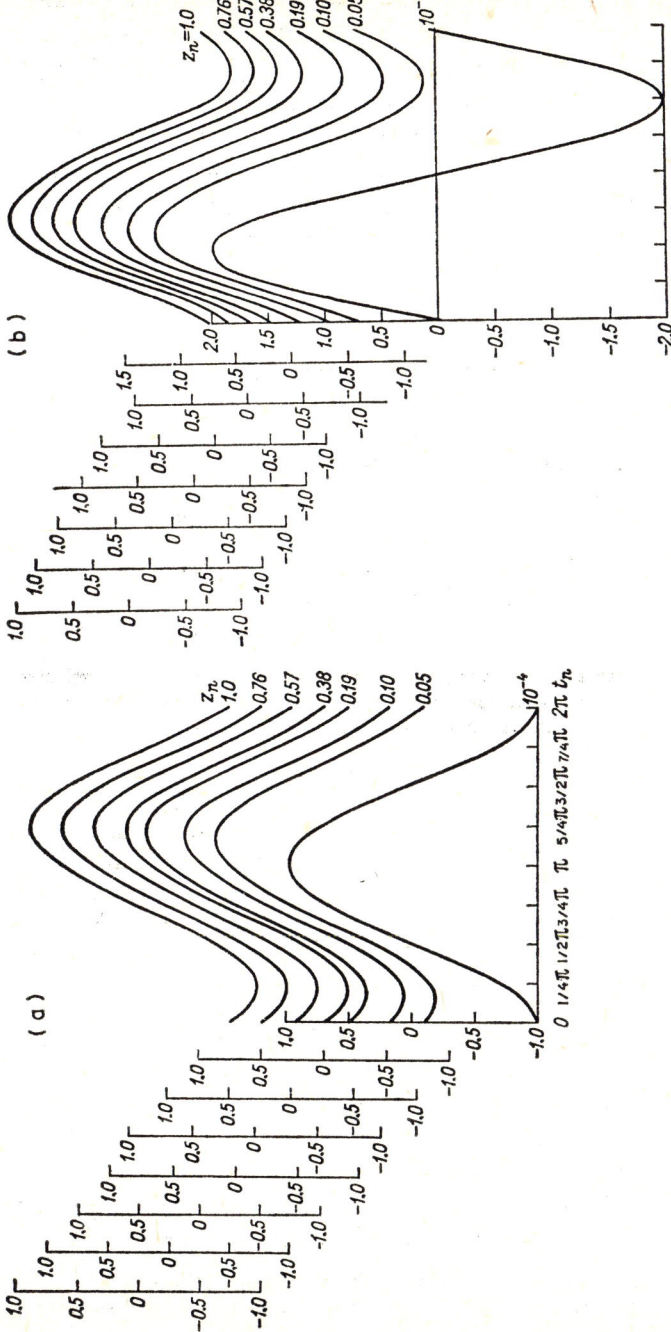

FIG. 5.22. Space-time distribution of the horizontal components of the velocity deviations in the bottom boundary layer at neutral stratification (by Vager and Kagan, 1969). (a) function W_{xn}; (b) function W_{yn}.

somewhat earlier than at the upper levels, which means that the phases of the tidal velocity ahead nearer the bottom. The latter fact, as well as the logarithmic character of the velocity distribution in the bottom layer, clearly seen in Fig. 5.23, is confirmed by observational data (see Section 5.2).

Analysing Fig. 5.24 may lead one to conclude that moduls of the tidal velocity in the bottom boundary layer increases somewhat faster than its decrease. This quite interesting peculiarity of time change of the velocity, associated with nonlinearity of the process, has been recently confirmed by experiment (Fig. 5.25).

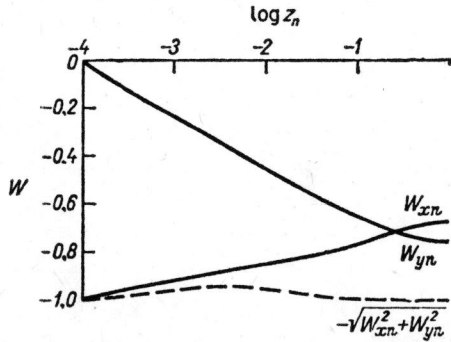

FIG. 5.23. Diagram illustrating the logarithmic character of tidal velocity distribution in the lowest portion of the bottom boundary layer at neutral stratification (by Vager and Kagan, 1969).

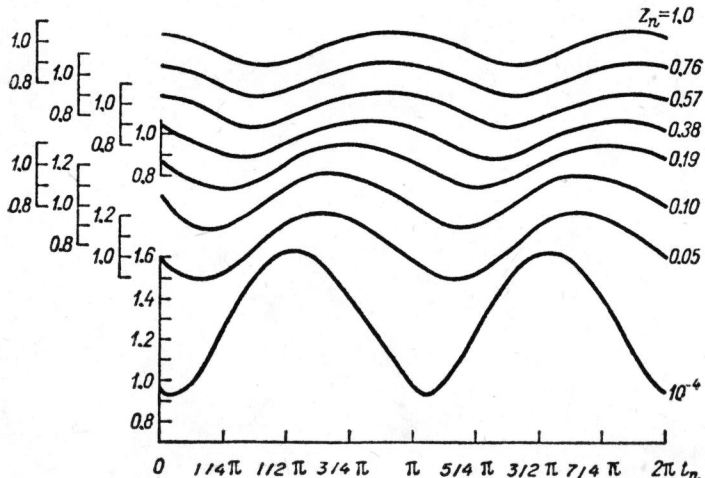

FIG. 5.24. Space-time distribution of the modulus of tidal velocity deviations in the bottom boundary layer at neutral stratification (by Vager and Kagan, 1969).

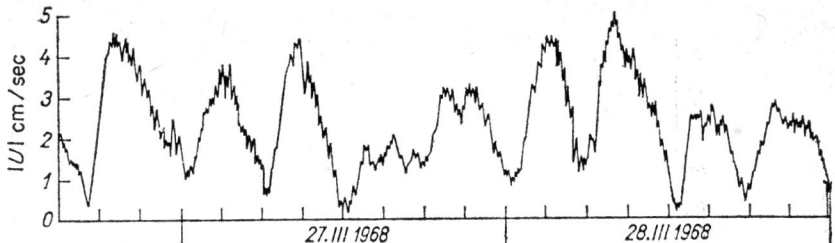

FIG. 5.25. Measurements of the modulus of flow velocity in the bottom boundary layer of the ocean (by Wimbush and Munk, 1961). The measurements were carried out at the point with the coordinates 32°13′N, 120°50′W at 1.2 m from the bottom, the depth being 3.7 km.

Figures 5.24, 5.26, 5.27 point to the asynchronism of the oscillations $|\mathbf{W}_n|$ and the values of b_n^2 and κ_{vn}. Comparing these figures shows that the kinetic energy of the velocity pulsations and the eddy coefficient change in time ahead of oscillations of the velocity deviations. Bearing in mind, however, that great values of $|\mathbf{u}_n|$ correspond to small values of $|\mathbf{W}_n|$ we have in this case, vice versa, that $|\mathbf{u}_n|$ is ahead of b_n^2 and κ_{vn}, the greatest values of the kinetic energy of the turbulent pulsations and of the eddy coefficient occurring at the moments of decrease in the modulus of the mean velocity.

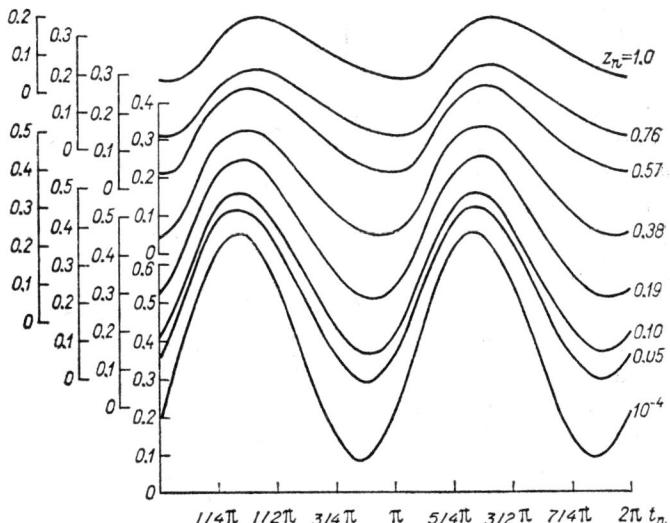

FIG. 5.26. Space–time distribution of the kinetic energy of turbulent fluctuations in the bottom boundary layer at neutral stratification (from Vager and Kagan, 1969). The y-coordinate is the $\mathrm{Re}_{\mathrm{crit}}^{-2} b_n^2$, value multiplied by 100.

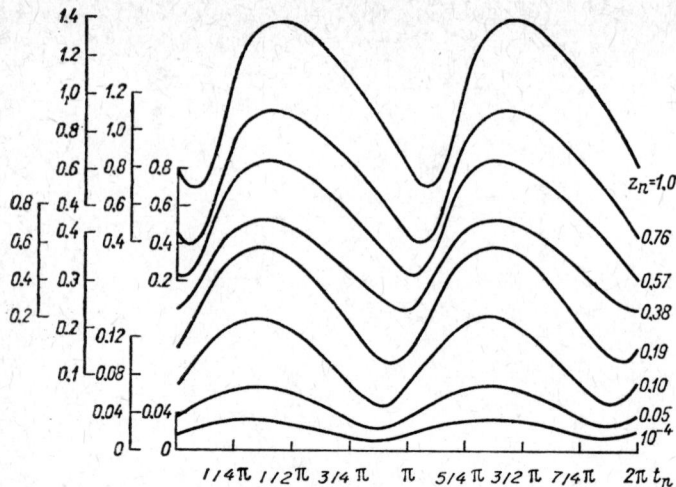

FIG. 5.27. Space–time distribution of the eddy coefficient in the bottom boundary layer at neutral stratification (from Vager and Kagan, 1969). The y-coordinate is the $\mathrm{Re}_{\mathrm{crit}}^{-1}\kappa_{vn}$ values multiplied by 100 at all levels except the lowest. The values of the function $\mathrm{Re}_{\mathrm{crit}}^{-1}\kappa_{vn}$ at the low level are multiplied by 10^4.

This conclusion, on intensification of turbulence with weakening mean velocity, is in qualitative agreement with the experimental measurements in the bottom boundary layer presented in Fig. 5.25. It can be viewed as a theoretical proof of the known empirical fact that deceleration of the flow is accompanied by decreasing of its stability.

Particular emphasis should be placed upon two peculiarities of the space–time distribution of the turbulence characteristics: first, comparatively small oscillations of the eddy coefficient in the bottom layer and, second, the asymmetry of time changes in kinetic energy and the eddy coefficient. Deviations of the eddy coefficient from simple harmonic oscillations, especially noticeable in the upper portion of the boundary layer, do not appreciably influence, however, the time dependence of W_{xn} and W_{yn}. It can be apparently attributed to the weakening effect of the friction force at a sufficiently great distance from the bottom where the formation of the mean velocity profile is chiefly determined by the action of the force of the horizontal gradient of pressure as well as the inertial and the Coriolis forces.

Figure 5.28 presents a diagram of the vertical distribution of the eddy coefficient at different moments of the tidal cycle. The eddy coefficient in the bottom layer is clearly seen to be a linear function of height. However, it increases with significant non-linearity at growing distance from the bottom.

The diagrams of the space–time distribution of the turbulent regime charac-

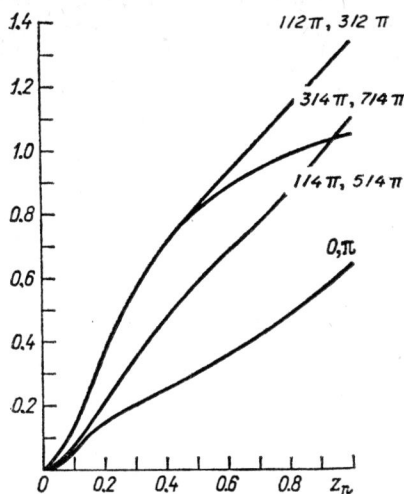

F I G. 5.28. Profiles of the eddy coefficient in the bottom boundary layer at neutral stratification at different phases of the tidal cycle (from Vager and Kagan, 1969). The y-coordinate is the $\mathrm{Re}_{\mathrm{crit}}^{-1} \kappa_{vn}$ value multiplied by 10^2.

teristics in the bottom boundary layer presented here were obtained with the following numerical constants and dimensionless parameters: $\alpha_b = 0.73$; $\varkappa = 0.4$; $c = 0.046$; $V/U = 2$; $g_u - g_v = \pi/2$; $E \cdot \mathrm{Re}_{\mathrm{crit}} = 10^2$; $z_{0n} = 10^{-4}$ and $e = 0.71$.

Stratified bottom boundary layer. Before getting down to describing the model of a stratified boundary layer in the tidal flow let us briefly review the mechanics of the formation of periodic density oscillations in the bottom boundary layer.

Unlike the boundary layer in the atmosphere, where the source generating the diurnal oscillations of the meteorological elements is the solar radiation, in the oceanic boundary layer the formation of the periodic oscillations of the hydrological elements and the turbulence characteristics follows quite a different pattern. Indeed, the vertical structure of the bottom boundary layer in a stratified liquid is known to be determined by the horizontal gradient of the elevation, the vertical gradient of mean density, hydrodynamic roughness of the sea bottom, the latitude and the oscillation frequency. Among the above external parameters only the horizontal gradient of the elevation periodically in time. Hence, considering the boundary layer in the tidal flow we must attribute the cause of the periodic oscillations of velocity, the density and the turbulence characteristics to changes in the horizontal gradient of the elevation.

The above changes induce periodic oscillations of the velocity and, eventually, of the intensity of turbulent exchange and the turbulent friction force.

Owing to friction, the vector of the tidal velocity in the boundary layer deflects from its direction in the layer undisturbed by friction. The latter fact as well as the effect of the non-stationarity of the velocity in the tidal flow result in the appearance of vertical motions which also change periodically in time. It is these vertical motions that serve as the basis for periodic deformations of the isopicnal surfaces and, consequently, of density oscillations of tidal period.

Considering the density oscillations induced by the vertical motion as a new unknown parameter complicates our problem greatly. Indeed, in this case it is necessary not only to take account of the dependence of the sought functions on the horizontal coordinates (remember, that in tidal flow the density field can be only horizontally inhomogeneous), but also to determine the velocity and the direction of the vertical motions.

To avoid these complications due to the dependence of the solutions on the horizontal coordinates, let us make the two following assumptions. First, that the influence of the horizontal variation of the density field is small compared with the influence of the other factors influencing the vertical velocity profiles of the tidal flow and density. As applied to the bottom boundary layer, such a supposition seems to be sufficiently valid. Secondly, assume that the velocity of the vertical motions is a known function of time and the vertical coordinate. This makes it possible when determining the vertical velocities to avoid the continuity equation and to consider the problem of calculating the vertical structure of a stratified bottom boundary layer as one-dimensional.

With the above simplifications, the system of equations and boundary conditions (5.3.1), (5.3.16)—(5.3.22) must be applied by the equation of density conservation

$$\frac{\partial \varrho}{\partial t} + w \left(\frac{N^2}{g/\varrho_0} + \frac{\partial \varrho}{\partial z} \right) = \alpha_\varrho \frac{\partial}{\partial z} \kappa_v \frac{\partial \varrho}{\partial z} \qquad (5.3.37)$$

and corresponding boundary conditions

$$\alpha_\varrho \kappa_v \frac{\partial \varrho}{\partial z} = 0 \quad \text{at} \quad z = z_0; \quad z = D, \qquad (5.3.38)$$

which mean the absence of mass flows through the bottom and the free oceanic surface. Here ϱ is the deviation of the density from its mean over the tidal period values w is the vertical component of the tidal velocity; $N = (g\mathfrak{S})^{1/2}$ is the Brunt–Väisälä frequency which is considered a given function of depth; α_ϱ is the ratio of the coefficients of turbulent diffusion and viscosity inversely related to the turbulent Prandtl number; g/ϱ_0 is the bouyancy parameter.

THE BOTTOM BOUNDARY LAYER

The formulation of the problem may be considered complete after certain transformations of the equation of turbulent energy (5.3.16). In the present case of a stratified boundary layer equation (5.3.16) must retain the term characterizing the inflow (or outflow) of the turbulent energy due to the bouyancy forces. Thus, rewriting equation (5.3.16) as follows:

$$\frac{\partial b^2}{\partial t} = \kappa_v \left|\frac{\partial \mathbf{u}}{\partial z}\right|^2 + \alpha_\varrho \kappa_v \frac{g}{\varrho_0}\left(-\frac{N^2}{g/\varrho_0} + \frac{\partial \varrho}{\partial z}\right)$$
$$+ \alpha_b \frac{\partial}{\partial z} \kappa_v \frac{\partial b^2}{\partial z} - \varepsilon, \qquad (5.3.16')$$

we obtain the closed system (5.3.1), (5.3.16'), (5.3.17)—(5.3.22), (5.3.37), (5.3.38) for determining the six unknown functions $\mathbf{u}, \kappa_v, b^2, \varepsilon, l$ and ϱ.

As previously, let us convert to the dimensionless variables using the depth D as a characteristic length scale, the value σ^{-1} inverse to the oscillation frequency as a time scale, the amplitude U as a velocity scale, $UD/\text{Re}_{\text{crit}}$, $(U/c_0 \text{Re}_{\text{crit}})^2$ and $c_1 U^3/c_0^3 D \, \text{Re}_{\text{crit}}^3$ as scales of the turbulent viscosity coefficient, the kinetic energy of turbulence and dissipation, respectively, and, finally, the value

$$\frac{w_0}{\sigma} \frac{N_0^2}{g/\varrho_0},$$

where w_0 and N_0 are characteristic scales of the vertical velocity and the Brunt–Väisälä frequency as a density scale.

Then the initial system of equations and boundary conditions can be rewritten in the following way:

$$\frac{\partial W_{xn}}{\partial t_n} - eW_{yn} = E \frac{\partial}{\partial z_n} \kappa_{vn} \frac{\partial W_{xn}}{\partial z_n}; \qquad (5.3.39)$$

$$\frac{\partial W_{yn}}{\partial t_n} + eW_{xn} = E \frac{\partial}{\partial z_n} \kappa_{vn} \frac{\partial W_{yn}}{\partial z_n}; \qquad (5.3.40)$$

$$\frac{\partial \varrho_n}{\partial t_n} + w_n \left(-N_n + \check{n}\frac{\partial \varrho_n}{\partial z_n}\right) = \alpha_\varrho E \frac{\partial}{\partial z_n} \kappa_{vn} \frac{\partial \varrho_n}{\partial z_n}; \qquad (5.3.41)$$

$$\frac{c^{-1/2}}{E} \text{Re}_{\text{crit}}^{-2} \frac{\partial b_n^2}{\partial t_n} = \kappa_{vn} \left|\frac{\partial u_n}{\partial z_n}\right|^2 + \alpha_\varrho \check{N}^2 \kappa_{vn}\left(-N_n^2 + \check{n}\frac{\partial \varrho}{\partial z_n}\right)$$
$$+ \frac{\alpha_b}{c^{1/2}} \text{Re}_{\text{crit}}^{-2} \frac{\partial}{\partial z_n} \kappa_{vn} \frac{\partial b_n^2}{\partial z} - \frac{\varepsilon_n}{\text{Re}_{\text{crit}}^2}; \qquad (5.3.42)$$

$$\kappa_{vn} = l_n b_n;$$
$$\varepsilon_n = b_n^3/l_n; \qquad (5.3.43)$$

$$l_n = -\varkappa \frac{b_n}{l_n} \left(\frac{\partial}{\partial z_n} \frac{b_n}{l_n} \right)^{-1}; \tag{5.3.44}$$

$$W_{yn} = -\cos t_n;$$

$$W_{yn} = -(V/U) \cos(t_n + g_u - g_v) \quad \text{at} \quad z_n = z_{0n}; \tag{5.34.5}$$

$$\kappa_{vn} \frac{\partial W_{xn}}{\partial z_n} = \kappa_{vn} \frac{\partial W_{yn}}{\partial z_n} = 0 \quad \text{at} \quad z_n = 1; \tag{5.3.46}$$

$$\alpha_b \kappa_{vn} \frac{\partial b_n^2}{\partial z_n} = 0;$$

$$\alpha_\varrho \kappa_{vn} \frac{\partial}{\partial z_n} = 0 \quad \text{at} \quad z_n = z_{0n}, \quad z_n = 1; \tag{5.3.47}$$

$$l_n \to \varkappa z_{0n} \quad \text{at} \quad z_n \to z_{0n}. \tag{5.3.48}$$

Here, as may be seen, two new dimensionless parameters, $\check{N}^2 = (DN_0/U)^2$ and $\check{n} = w_0/\sigma D$ have appeared.

The scheme of the solution of the non-linear system (5.3.9)—(5.3.48), with the equation of density conservation (5.3.4) excluded, has been described in detail above. Since, however, the method of the solution of equation (5.3.41) is as good as that used when integrating the equation of turbulent energy (in both cases the factorization method), we do not intend to dwell upon it. Note, however, that in the course of the calculations the expressions for the dimensionless velocity of the vertical motions w_n and the square of the dimensionless Brunt–Väisälä frequency N_n^2 were set equal to

$$w_n = \mathcal{W}_n \cos(t_n + g_u - g_w);$$

and

$$N_n^2 = z_n(1 - z_n),$$

where

$$\mathcal{W}_n = \frac{1 - \exp 0.1 (z_n - z_{0n})}{1 - \exp 0.1 (1 - z_{0n})}.$$

Let us now discuss the results obtained for the above values of the dimensionless parameters and the numerical constants with the four new parameters $g_u - g_w = 0$; $\check{n} = 10^{-2}$; $\check{N} = 1$; $\alpha_\varrho = 1$ added.

What is most striking even at brief scrutiny of the results and their comparison with the analogous results from the preceding sub-section is a noticeable decrease in the energy of turbulent pulsations and in the value of the eddy coefficients (Fig. 5.29 (a, b)). Especially noticeable changes induced by the effect of the buoyancy forces take place in the upper portion of the boundary layer

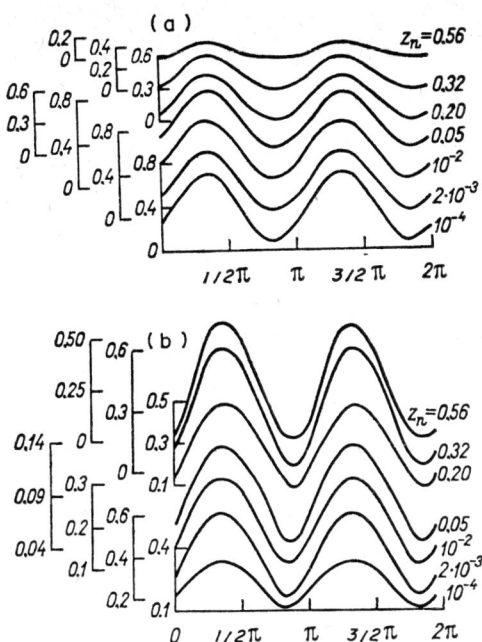

FIG. 5.29. Space–time distribution of the kinetic energy of turbulence and the eddy coefficient in a stratified bottom boundary layer (from Vager and Kagan, 1971). (a) The function $\mathrm{Re}_{\mathrm{crit}}^{-2} b_n^2$, multiplied by 10^2; (b) the function $\mathrm{Re}_{\mathrm{crit}}^{-1} \kappa_{vn}$ multiplied by 10^2 at the upper four levels, by 10^3 at the fifth level, by 10^4 and 10^5 at the sixth and seventh levels, respectively.

where the amplitudes of the oscillations of the given characteristics have decreased by a factor of about 1.5—2. And, vice versa, in the bottom layer the values of the turbulence characteristics slightly increase almost at all moments of time. It can be attributed to the shift of the domain of the maximum turbulent energy and of the viscosity coefficient from the upper to the middle parts of the boundary layer.

Figure 5.30 presents the vertical distribution of the eddy coefficient at different phases of the tidal cycle. The figure demonstrates that the maximum of the eddy coefficient is located in the lower half of the boundary layer, the location varying in the height range from 0.2 to 0.4 depending on time. The maximum value of the eddy coefficient itself does not remain constant, but changes during the tidal cycle by a factor of more than 3.

Changes in the quantity and the profile of the eddy coefficient affect the vertical distribution of mean velocity. Thus, increasing eddy coefficient results in decreasing vertical gradients of the velocity in the bottom layer. In the upper portion of the boundary layer quite a different picture is observed;

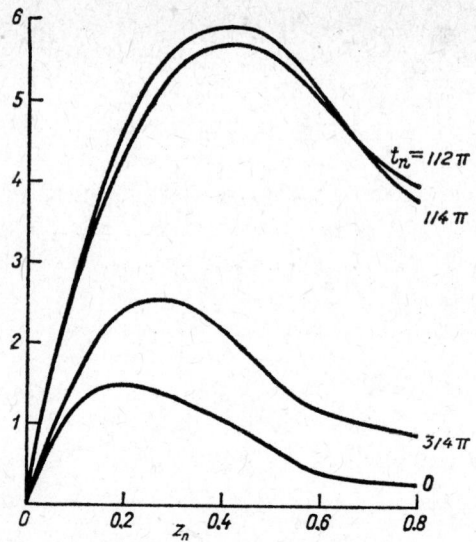

F I G. 5.30. Vertical profiles of the eddy coefficient in a stratified bottom boundary layer at different phases of the tidal cycle (from Vager and Kagan, 1971). The y-coordinate is the $\text{Re}_{\text{crit}}^{1-}\kappa_{vn}$ values multiplied by 10^3.

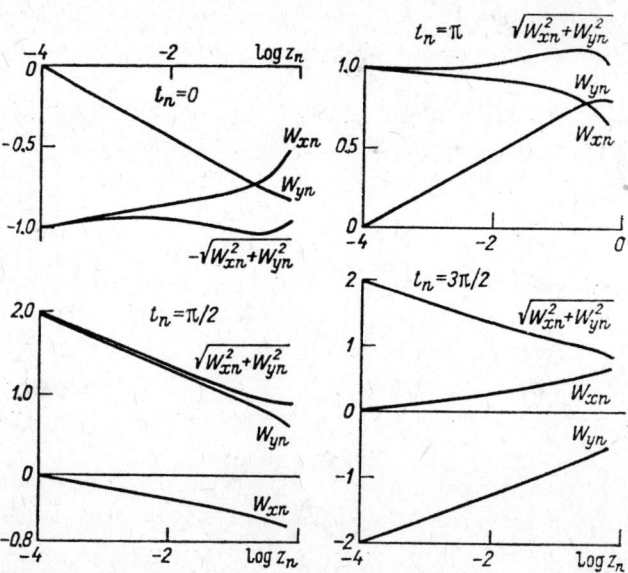

F I G. 5.31. Vertical distribution of deviations of the velocity and modulus at different moments of the tidal cycle (from Vager and Kagan, 1971).

decreasing intensity of the turbulent exchange results here in increasing vertical shift of velocity, the vertical profile of the speed transforming accordingly (Fig. 5.31).

Due to the compensation of the opposite changes in the eddy coefficient and the vertical gradient of the velocity, however, the turbulent friction force at certain levels remains practically unchanged (in comparison with the case of neutral stratification). As a result no noticeable changes in the shift of the velocity phases in the boundary layer are observed. In both cases, i.e. with and without density stratification, ahead of the tidal velocity does not exceed one-eighth of the tidal period within the whole boundary layer (Fig. 5.32).

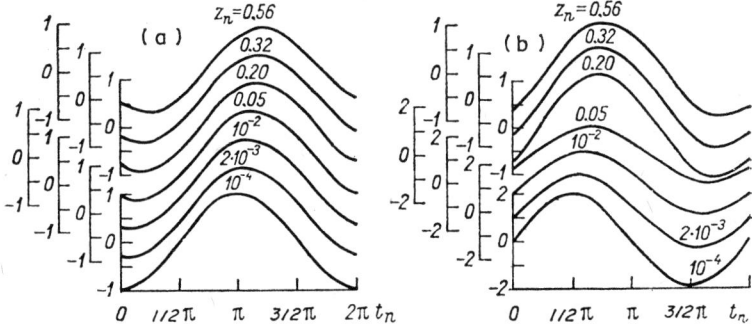

FIG. 5.32. Space–time distribution of horizontal components of the tidal velocity in a stratified bottom boundary layer (by Vager and Kagan, 1971). (a) The function u_n; (b) the function v_n.

Still smaller phase shifts are observed when analysing the space–time distribution of the tidal velocity (Fig. 5.33). The figure demonstrates that the maximum and minimum values of the modulus of velocity in the bottom layer ahead the corresponding values in the upper layers by only one-twenty-fifth of the tidal period.

The mean kinetic energy of the turbulence and the eddy coefficient change with modest delay as compared to the oscillations of the modulus of the mean velocity. It can be traced by comparing Figs. 5.29 and 5.33. The above fact demonstrates that both in stratified and in homogeneous boundary layers increase and decrease in the modulus of the mean velocity are accompanied by a certain weakening and strengthening of turbulent energy.

Let us now consider the calculated results for density oscillations. The diagram of the space–time distribution presented in Fig. 5.34 shows that the density and the vertical velocity oscillate in contraphase, the phases of density oscillations at different levels being approximately equal. The above fact indirectly shows that the contribution of the turbulent diffusion to the

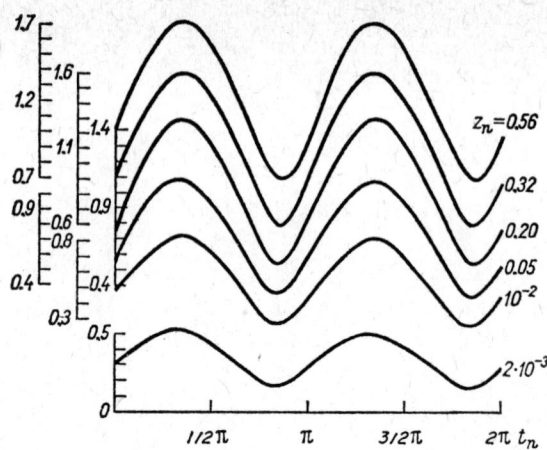

FIG. 5.33. Space–time distribution of the modulus of the tidal velocity in a stratified bottom boundary layer (from Vager and Kagan, 1971).

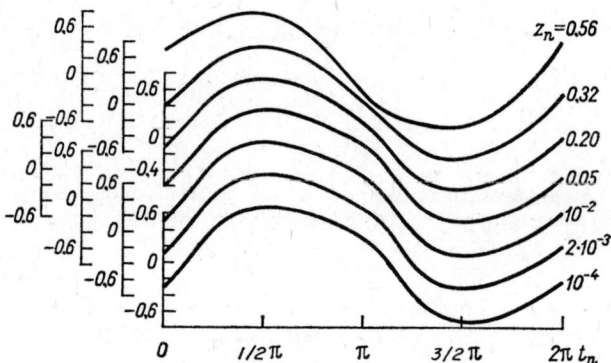

FIG. 5.34. Space–time distribution of density perturbations in a stratified bottom boundary layer (from Vager and Kagan, 1971). The y-coordinate is the ϱ_n values multiplied by 10.

formation of the density oscillations is comparatively small and, hence, the basic factor determining density changes in time is an advective mass transport.

If this is the case, then at the moments of time $t_n = \pi/2$ and $t_n = 3\pi/2$ when the velocity of the vertical motions at all the horizons become zero, the density oscillations must reach their extremal values. In line with the conditions of the problem, at the same moments of time the vertical profile of the density perturbations must nearly coincide with the neutral profile. As is seen from Fig. 5.35, the above peculiarities indeed occur.

At the same time, since at $w_n > 0$ the fluid particles are transported from the domain of greater mean density and at $w_n < 0$ from the domain of lower mean

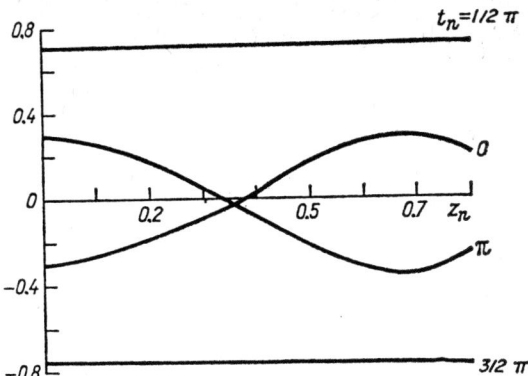

FIG. 5.35. Vertical distribution of density perturbations in a stratified bottom boundary layer at different phases of the tidal cycle (from Vager and Kagan, 1971). The y-coordinate is the ϱ_n values multiplied by 10.

density, the corresponding density perturbations must be positive in the first and negative in the second case. This is indeed observed at sufficiently great heights (Fig. 5.35; curves corresponding to the moments $t_n = 0, \pi$) where the effect of the turbulence diffusion becomes negligibly small.

Due to the fact, however, that the vertical velocity and, consequently, the vertical advective mass transport tend to zero with decreasing distance from the bottom, the effect of turbulent diffusion at the lower level must increase. It results in the appearance of the slight phase shifts in the density oscillations in the bottom layer which, nevertheless, can bring about some appreciable after-effects at the moments of maximum changes in density. Indeed, as is seen from Fig. 5.35, even slight leads of the phases of density oscillations at the lower levels result in the change of the sign of density perturbations.

Concluding the analysis of these calculated results, let us discuss the balance of the turbulent energy in a stratified boundary layer of a tidal flow.

To close the system of tidal dynamics equations, we have used the semi-empirical equation for the turbulent energy (5.3.16'). The terms of this equation characterize temporal changes in the kinetic energy of the turbulent velocity pulsations, generation of the turbulent energy, its loss in defeating buoyancy forces, the inflow of the energy due to the diffusive transport, and the viscous dissipation, respectively. Figure 5.36 presents changes of these terms in the vertical.

The figure also demonstrates that in the lower portion of the boundary layer (at $z_n \leqslant 0.15$) it is generation and dissipation that mainly contribute to the balance of the turbulent energy. A slight excess of generation of the pulsation energy over the dissipation is balanced here by the diffusive energy trans-

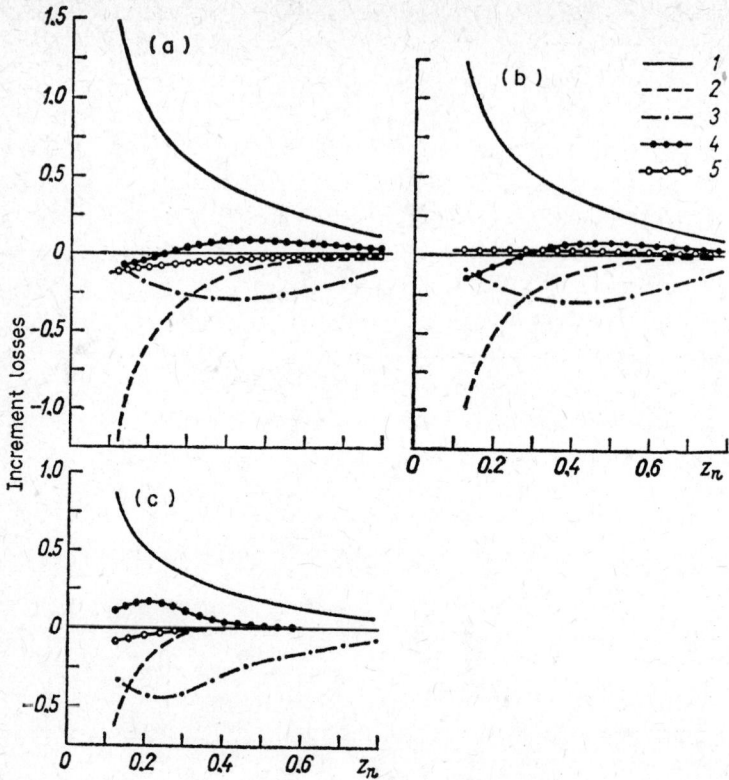

F I G. 5.36. Balance of turbulent energy in a stratified bottom boundary layer at the moment $t_n = 0.32$ (a), $t_n = \pi/2$ (b) and $t_n = 0.92\,\pi$(c). (1) Production, (2) dissipation, (3) energy loss in defeating buoyancy forces, (4) diffusion, (5) time-dependent variation in the kinetic energy of turbulent pulsations taken with the opposite sign. The values of the dimensionless components of the balance were multiplied by 10^3 (a, b) and by 10^4 (c).

port to the above layers. Energy loss in defeating buoyancy forces is generally small as compared to, say, the above-mentioned terms. An analogous conclusion may be also drawn in reference to the term describing temporal changes in the turbulent energy.

Relation of the terms in the equation of turbulent energy changes with varying height. According to the accepted expression for the Brunt–Väisälä frequency, in the lower half of the boundary layer energy loss in defeating buoyancy forces increases with height, while generation and dissipation of the turbulent energy decrease. As a result, at $z_n \approx 0.4$ generation and loss of the energy in defeating buoyancy forces become nearly equal in magnitude but opposite in sign. This is the case for two more terms characterizing dissipation and diffusion of the turbulent energy, the latter changing sign at the height $z_n = 0.26$ and remaining positive with further increase in height. From these facts it transpires that an increase in the turbulent energy takes place in the central and upper portions of the boundary layer due to turbulent diffusion.

And, finally, in the upper portion of the boundary layer the turbulent energy level is maintained mainly due to generation, diffusive influx and energy loss in defeating buoyancy forces, the balance among these three terms becoming more and more precise with growing distance from the bottom.

The above peculiarities of the vertical distribution of the components of the turbulent energy balance are typical not only of the moment $t_n = 0.32\,\pi$ when the energy of turbulent pulsations reaches its maximum value, but also of a certain portion of the period during which it decreases. It can be easily seen comparing Figs. 5.36 (a) and (b) referring to the moments of time $t_n = 0.32\,\pi$ and $t_n = \pi/2$, respectively. As far as quality is concerned, both figures are practically identical, slight differences referring to the numerical values of various balance components.

These differences become more and more striking when the energy of the turbulent pulsations reaches its minimum values (Fig. 5.36 (c)). In particular, at $t_n = 0.95\,\pi$ the numerical values of certain balanced components appear to be approximately an order less then those given in Fig. 5.36 (a) and (b). Nevertheless, as to quality, the basic peculiarities of the turbulent energy balance remain practically unchanged even at this moment of time. Indeed, energy generation at small values of z_n is mainly compensated for by viscous dissipation, while at great z_n values by the energy loss in defeating buoyancy forces. This peculiarity seems to be typical of a stratified boundary layer in tidal flow.

5.4. On the Resistance Law in Tidal Flow

In tidal dynamics the force of the bottom friction as referred to a unit mass is generally assumed to equal the square of the tidal velocity multiplied by a certain number, which is called to as the bottom friction coefficient. Many investigators (including the authors of this book, see preceding chapters) make use of another expression for the bottom stress which, instead of the velocity measured at a fixed level in the bottom layer, employs the amplitude of the vertically averaged velocity of the tidal flow. In the above cases the bottom friction coefficients differ from one another by the multiplier $8/3\,\pi$, divided by the ratio of the squared amplitudes of the averaged velocity and of the velocity at a fixed distance from the bottom.

Nowadays there exist numerical methods of evaluating the bottom friction coefficient; the most popular ones are given on p. 235.

1. Defining the bottom friction coefficient by the tidal energy equation. The equation for the tidal energy integrated with respect to time over the tidal cycle and with respect to area within a limited domain (sufficiently small to neglect the effect of the tide-generating forces) describes the balance of the energy inflow through the side boundaries of the basin and its dissipation due to bottom friction (the tidal energy dissipation due to macroviscosity is neglected). This equation, with the expression for the bottom stress substituted, is used to calculate the bottom friction coefficient by given values of the tidal elevations on a liquid contour and of the velocity throughout the considered domain.

2. Defining the bottom friction coefficient directly by the equation of motion added with the expression for the bottom stress. In this case the primary data are the vertical profiles of the mean velocity at different moments of time and the surface slope, the latter calculated either by direct measurement or by the continuity equation according to measurements of velocity.

3. Defining bottom friction coefficient as the ratio of the bottom stress to the squared velocity of the tidal flow measured at some standard level in the logarithmic layer or beyond the boundary layer. Bottom friction is evaluated by turbulence measurements or by processing the results of gradient measurements of the mean velocity in the bottom layer.

Table 5.7 presents the values of the bottom friction coefficient obtained by the above three methods. As a greater part of these values was obtained by observations in shallow bays and straits, they, naturally, characterize the bottom friction at great depths in an extremely crude way. They are quite useful, however, since they demonstrate that the bottom friction coefficient changes throughout the ocean.

Thus, the problem of establishing the universal dependence between the bottom friction coefficient and the external parameters of the bottom boundary layer gains in importance. The numerical solution of this problem (for instance, by the non-linear model of the boundary layer considered in the previous section) is difficult to obtain because of the great number of external parameters. In spite of the fact that these parameters are grouped in five dimensionless combinations, they are still too numerous for the size of calculation to remain within appropriate limits. Besides, one can always hope to obtain the solution (even an approximate one) by analytical methods; this way, however, also has its drawbacks. These drawbacks are first of all associated with the existence of periodic oscillations of the turbulence characteristics and, consequently, the data on changes in the amplitudes and phases of the tidal oscillations of the turbulent viscosity coefficients with respect to time should be available.

At present, when the lack of experimental data is appreciable, the best way

TABLE 5.7. *Values of the bottom friction coefficient by the data available in literature*

Authors	Method of estimation	Region of investigation	Bottom friction coefficient multiplied by 10^3
G. Taylor (1918)	1	Ireland sea	2.0
Grace (1936) †	2	Bristol Bay	1.4–4.1
Grace (1937) †	2	La Manche Strait	2.4–21.3
Mosby (1949) ‡ (see also Sternberg, 1968)	3	Alverstremmen Stream	7.5–9.0
Bowden and Fairbairn (1952a) †	2	Red-Wharf Bay	1.42–2.04
Bowden (1955)	3	the same	1.1–2.8
Bowden and Fairbairn (1956)	3	the same	2.0–2.5
Charnock (1959)	3	the same	3.4–3.7
Bowden, Fairbairn and Hughes (1959)	3	the same	2.5–4.4
Bowden (1962)	3	the same	2.7–3.3
Vapniar (1962) †	2	not reported	1.7–2.4 (the M_2 wave) 0.8–2.0 (the O_1 wave) 0.6–1.4 (the K_1 wave)
Sternberg (1966) ‡	3	Pickering Strait	4–9
Sternberg (1968) ‡	3	Juan de Fuca Strait Agatha Strait Kolvos Strait Pickering Strait Halle Strait	1.0–8.5 1.5–10.5 2.4–2.7 1.5–3.0 2.5–4.5
Sternberg (1970)	3	North-eastern region of the Pacific Ocean, Coordinates of the observational sites: 32°37.6′N, 118°8.6′W, 32°39′N, 120°33′W, 32°14.5′N, 120°51′W, 31°1.5′N, 119°48.5′W	1.4–16.9 1.1–10.5 1.8–6.9 1.0–4.8

† The values of the bottom friction coefficient were determined by the depth-averaged velocity of the tidal currents.
‡ The presented values of the bottom friction coefficient correspond to the conditions of hydrodynamical rough bottom.

out would be to neglect time changes in the turbulence coefficient and to limit oneself to the solution of the given task under the supposition that the turbulence regime in the bottom boundary layer remains unchanged in time, while the description of the time-averaged turbulence characteristics may be carried out in terms of the semi-empirical theory of turbulence. A similar approach was used in papers by Jonsson (1963) and Kajiura (1968).

In Jonsson (1963) the law of drag for a one-dimensional oscillating motion was obtained under the assumption of logarithmic changes in the velocity throughout the boundary layer. Values of the universal constants in the law of drag were chosen in the course of processing measurements of velocity obtained by use of a wave tank with a rough bottom. In Kajiura (1968) the law of drag was obtained by solving the system of the equations for the bottom boundary layer in a wave channel. Specific peculiarities of the turbulence structure between the roughness elements were taken into account by singling out in the bottom region a sublayer embracing all the bottom roughness. The thickness of this sublayer was considered set and equal to half the equivalent sand roughness. The turbulent transport in the sublayer near the wall was assumed to be carried out by means of the small-scale turbulence, characterized by an effective coefficient of turbulent viscosity proportional to the amplitude of friction velocity oscillations and to the thickness of the sublayer.

Let us now generalize the results obtained in the above papers for the case of a two-dimensional bottom boundary layer located above a fairly rough surface. According to the above considerations assume the turbulent viscosity coefficient in the bottom boundary layer to remain constant in time. This supposition, as will be shown below, is equivalent to neglecting all the other oscillations of the vertical momentum flow but those with frequency σ.

Assume then that the distribution of the eddy coefficient with respect to height can be approximated by a linear function of the vertical coordinate. It is apparent that in this case a distortion of the velocity profile in the middle of the boundary layer will take place; the velocity distribution in the lower layer that we are interested in remains, however, close to a real one.

With the above assumptions taken into account, the expressions for the turbulent viscosity coefficient in the bottom stress (the quantity of the latter divided by the mean density of sea water denoted as before by τ_b/ϱ_0) can be represented in the following way:

$$\kappa_v = \varkappa \tilde{u}_* z; \qquad (5.4.1)$$

$$\tau_b/\varrho_0 = \tilde{u}_* u_*. \qquad (5.4.2)$$

THE BOTTOM BOUNDARY LAYER

Let us now make use of the fact that the current velocity of the tidal flow is a periodic function of time. Since in our model the eddy coefficient is set to be time-independent, the tidal flow velocity in the boundary layer can change in time only with the frequency σ. Hence, the initial condition in this case is replaced by the condition of periodicity

$$\mathbf{u} = \text{real}\,(\bar{\mathbf{u}}\exp i\sigma t). \qquad (5.4.3)$$

Instead of u introduce a deviation of the velocity from its values beyond the boundary layer

$$\mathbf{W} = \mathbf{u} - \mathbf{u}_\infty,$$

where $\mathbf{W} = (W_x, W_y)$, and make up two new functions using W_x and W_y

$$\begin{aligned} W &= W_x + iW_y; \\ W^* &= W_x - iW_y. \end{aligned} \qquad (5.4.4)$$

Let us now pass on to the dimensionless parameters by the formulae

$$t_n = \sigma t,$$

$$z_n = \frac{\sigma}{\varkappa \tilde{u}_*} z;$$

$$(W_n, W_n^*) = \frac{\varkappa}{\tilde{u}_*}(W, W^*);$$

$$K_{vn} = \frac{\sigma}{\varkappa^2 \tilde{u}_*^2} K_v.$$

Then the initial system of equations and boundary conditions describing the distribution of complex amplitudes of velocity deviations in the bottom boundary layer takes the form

$$\frac{d}{dz_n} z_n \frac{d\overline{W}_n}{dz_n} - i(1+e)\overline{W}_n = 0;$$

$$\frac{d}{dz_n} z_n \frac{d\overline{W}_n^*}{dz_n} - i(1-e)\overline{W}_n^* = 0; \qquad (5.4.5)$$

$$\overline{W}_n = \frac{\varkappa}{\chi}\left[\exp(-ig_u) + i\frac{V}{U}\exp(-ig_v)\right];$$

$$\overline{W}_n^* = \frac{\varkappa}{\chi}\left[\exp(-ig_u) - i\frac{V}{U}\exp(-ig_v)\right]$$

$$\text{at}\quad z_n = (\varkappa\chi\,\text{Ro})^{-1}; \qquad (5.4.6)$$

$$\overline{W}_n = \overline{W}_n^* \to 0 \quad \text{at}\quad z_n \to \infty, \qquad (5.4.7)$$

where $\chi = \tilde{u}_*/U$ is the geostrophic friction coefficient,[†] the Rossby number being as determined previously.

The solution of equations (5.4.5) obeying boundary conditions (5.4.6), (5.4.7) has the form

$$\overline{W}_n = -\frac{\varkappa}{\chi} \frac{A(\eta_n)}{A_0} \left[\exp i(\alpha - g_u - \alpha_0) + i\frac{V}{U} \exp i(\alpha - g_v - \alpha_0) \right];$$

$$\overline{W}_n^* = -\frac{\varkappa}{\chi} \frac{A^*(\eta_n)}{A_0^*} \left[\exp i(\alpha^* - g_u - \alpha_0^*) - i\frac{V}{U} \exp i(\alpha^* - g_v - \alpha_0^*) \right], \quad (5.4.8)$$

where

$$A(\eta_n) = [\ker^2 \eta_n(1+e)^{1/2} + \kei^2 \eta_n(1+e)^{1/2}]^{1/2};$$

$$A^*(\eta_n) = [\kei^2 \eta_n(1-e)^{1/2} + \ker^2 \eta_n(1-e)^{1/2}]^{1/2};$$

$$\alpha = \arctan \frac{\kei \eta_n(1+e)^{1/2}}{\ker \eta_n(1+e)^{1/2}};$$

$$\alpha^* = \arctan \frac{\kei \eta_n(1-e)^{1/2}}{\ker \eta_n(1-e)^{1/2}},$$

A_0, A_0^*, α_0 and α_0^* are corresponding values of the functions $A(\eta_n)$, $A^*(\eta_n)$, α and α^* at $\eta_n = \eta_{n0}(= 2(\varkappa\chi \text{ Ro})^{-1/2})$; $\ker \eta_n(1+e)^{1/2}$, $\ker \eta_n(1-e)^{1/2}$ and $\kei \eta_n(1+e)^{1/2}$, $\kei \eta_n(1-e)^{1/2}$ are real and imaginary parts of the McDonald functions $K_0(\eta_n(1+e)^{1/2})$, $K_0(\eta_n(1-e)^{1/2})$, $\eta_n = 2z_n^{1/2}$.

Introduce the following notation:

$$\mathfrak{S} = \left[1 + \left(\frac{V}{U}\right)^2 - 2\frac{V}{U} \sin(g_u - g_v)\right]^{1/2};$$

$$\mathfrak{S}^* = \left[1 + \left(\frac{V}{U}\right)^2 + 2\frac{V}{U} \sin(g_u - g_v)\right]^{1/2};$$

$$\tan(-\beta) = \frac{\sin(\alpha - g_u - \alpha_0) + V/U \cos(\alpha - g_v - \alpha_0)}{\cos(\alpha - g_u - \alpha_0) - V/U \sin(\alpha - g_v - \alpha_0)};$$

$$\tan(-\beta^*) = \frac{\sin(\alpha^* - g_u - \alpha_0^*) - V/U \cos(\alpha^* - g_v - \alpha_0^*)}{\cos(\alpha^* - g_u - \alpha_0^*) + V/U \sin(\alpha^* - g_v - \alpha_0^*)}$$

[†] The term "geostrophic" is in this case not quite valid, as in the tidal flow the geostrophic correlation is never fulfilled. Nevertheless, bearing this in mind, we will keep to the traditional terminology.

THE BOTTOM BOUNDARY LAYER

and then pass on from $\overline{W}_n$, $\overline{W}_n^*$ to W_n, W_n^*. As a result we obtain

$$W_n = -\frac{\varkappa}{\chi} \frac{A^*(\eta_n)}{A_0} \mathfrak{S} \exp i(t_n - \beta);$$

$$W_n = -\frac{\varkappa}{\chi} \frac{A^*(\eta_n)}{A_0^*} \mathfrak{S}^* \exp i(t_n - \beta^*). \qquad (5.4.9)$$

Remember now that due to the accepted assumptions the bottom stress is determined by formula (5.4.2). On the other hand,

$$\frac{\tau_b}{\varrho_0} = K_v \left[\left(\frac{\partial W_x}{\partial z}\right)^2 + \left(\frac{\partial W_y}{\partial z}\right)^2 \right]_{z=z_0}^{1/2}. \qquad (5.4.10)$$

Equating expressions (5.4.2) and (5.4.10) and passing on to the dimensionless variables, we obtain

$$u_{*n} = \frac{\eta_{0n}}{2} \left[\left(\frac{\partial W_{xn}}{\partial \eta_n}\right)^2 + \left(\frac{\partial W_{yn}}{\partial \eta_n}\right)^2 \right]_{\eta_n = \eta_{0n}}^{1/2}, \qquad (5.4.11)$$

where $u_{*n} = u_*/\tilde{u}_*$ is the dimensionless friction velocity.

According to equation (5.4.4) we have

$$W_{xn} = \text{real } \tfrac{1}{2}(W_n + W_n^*);$$
$$W_{yn} = \text{real } \tfrac{1}{2} i(W_n - W_n^*).$$

Hence, on the basis of equations (5.4.9) we obtain

$$u_{*n} = \frac{\varkappa \eta_{0n}}{4\chi} \left(\mathfrak{S} \frac{M_0}{A_0} + \mathfrak{S}^* \frac{M_0^*}{A_0^*} \right) \left[1 - s^2 \sin^2 \left(t_n - \frac{\beta_0 + \beta_0^*}{2} \right. \right.$$
$$\left. \left. + \frac{\gamma_0 + \gamma_0^*}{2} \right) \right]^{1/2}, \qquad (5.4.12)$$

where

$$s^2 = \frac{4\mathfrak{S}\mathfrak{S}^* M_0 M_0^*/A_0 A_0^*}{(\mathfrak{S} M_0/A_0 + \mathfrak{S}^* M_0^*/A_0^*)^2};$$

$$\beta_0 = \beta|_{\eta_n = \eta_{n0}},$$
$$\beta_0^* = \beta^*|_{\eta_n = \eta_{n0}};$$

$$M_0 = (1+e)^{1/2} [\text{ker}'^2 \eta_{n0}(1+e)^{1/2} + \text{kei}'^2 \eta_{n0}(1+e)^{1/2}]^{1/2};$$

$$\gamma_0 = \frac{\text{kei}' \eta_{n0}(1+e)^{1/2} \text{ker}\, \eta_{n0}(1+e)^{1/2} - \text{kei}\, \eta_{n0}(1+e)^{1/2} \text{ker}'\, \eta_{n0}(1+e)^{1/2}}{\text{ker}\, \eta_{n0}(1+e)^{1/2} \text{ker}'\, \eta_{n0}(1+e)^{1/2} + \text{kei}\, i\eta_{n0}(1+e)^{1/2} \text{kei}'\, \eta_{n0}(1+e)^{1/2}},$$

where the prime denotes differentiation with respect to η_n, while W_0^* and γ_0^* are obtained from the expressions for W_0 and γ_0 with the substitution of $-e$ for e.

Now we can easily obtain the expression for the geostrophic friction coefficient χ. For this purpose we use the definition u_{*n} and assume that the mean value of $|u_{*n}|$ over the tidal cycle equals unity. Hence, equation (5.4.12) yields

$$\chi = \frac{\varkappa \eta_{n0}}{2\pi}\left(\mathfrak{S}\,\frac{M_0}{A_0} + \mathfrak{S}^*\,\frac{M_0^*}{A_0^*}\right) E(s, \pi/2),$$

where $E(s, \pi/2)$ is the full elliptical integral of the second type.

Now transform the above relation, taking into account the smallness of the values of the arguments $\eta_{n0}(1+e)^{1/2}$ and $\eta_{n0}(1-e)^{1/2}$. Apply the functions ker and kei and their derivatives at small values of the argument and obtain

$$\frac{M_0}{A_0} = \frac{2}{\eta_{n0}}\left[\left(\ln\frac{\chi\,\mathrm{Ro}}{1+e} - B\right)^2 + A^2\right]^{-1/2};$$

$$\frac{M_0^*}{A_0^*} = \frac{2}{\eta_{n0}}\left[\left(\ln\frac{\chi\,\mathrm{Ro}}{1-e} - B\right)^2 + A^2\right]^{-1/2} \qquad (5.4.13)$$

and hence

$$\chi = \frac{\varkappa}{\pi}\left\{\mathfrak{S}\left[\left(\ln\frac{\chi\,\mathrm{Ro}}{1+e} - B\right)^2 + A^2\right]^{-1/2}\right.$$
$$\left. + \mathfrak{S}^*\left[\left(\ln\frac{\chi\,\mathrm{Ro}}{1-e} - B\right)^2 + A^2\right]^{-1/2}\right\} E(s, \pi/2), \qquad (5.4.14)$$

where

$$s = \frac{2(\mathfrak{S}\mathfrak{S}^*)^{1/2}\left[\left(\ln\frac{\chi\,\mathrm{Ro}}{1+e} - B\right)^2 + A^2\right]^{-1/4}\left[\left(\ln\frac{\chi\,\mathrm{Ro}}{1-e} - B\right)^2 + A^2\right]^{-1/4}}{\left\{\mathfrak{S}\left[\left(\ln\frac{\chi\,\mathrm{Ro}}{1+e} - B\right)^2 + A^2\right]^{-1/2} + \mathfrak{S}^*\left[\left(\ln\frac{\chi\,\mathrm{Ro}}{1-e} - B\right)^2 + A^2\right]^{-1/2}\right\}},$$

$A = \pi/2$ and $B = (2c - \ln \varkappa)$ are the universal constants equal to 1.57 and 2.06, respectively (here c is Euler's constant).

Expression (5.4.14) relates the geostrophic friction coefficient χ to the Rossby number $\mathrm{Ro} = U/\sigma z_0$ and the other dimensionless combinations of the external parameters $e = 2\omega_z/\sigma$, $g_u - g_v$, V/U [the two latter parameters are denoted in equations (5.3.14) through $\mathfrak{S}$ and $\mathfrak{S}^*$] and is in a sense a law of drag for two-dimensional tidal flow.

As a certain limiting case, equation (5.4.14) can yield the law of drag in a deep-water tidal channel. Indeed, setting V/U, g_v and e equal to zero, we obtain $\mathfrak{S} = \mathfrak{S}^* = 1$ and $\beta_0 = \beta_0^* = g_v$. Then instead of equation (5.4.14) we have

$$\ln \text{Ro} = B + \sqrt{\left(\frac{1}{A}\frac{\varkappa}{\chi}\right)^2 - A^2}. \tag{5.4.15}$$

Also,

$$\gamma_0(=\gamma_0^*) = \arctan \frac{A}{\ln \chi \, \text{Ro} - B}. \tag{5.4.16}$$

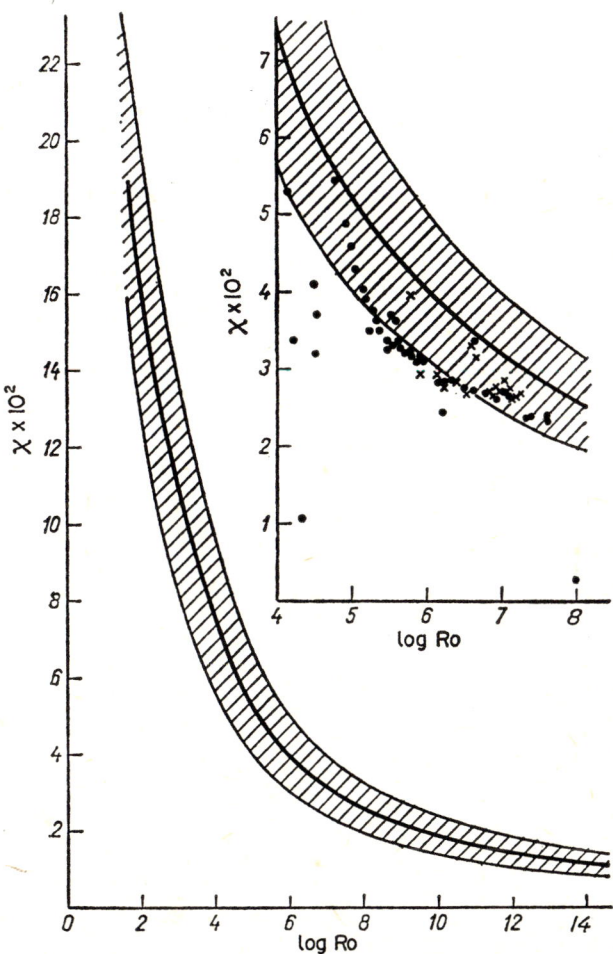

FIG. 5.37. Dependence of the geostrophic friction coefficient on the Rossby number Ro at different values of the parameters V/U and $g_u - g_v$. Shaded areas show changes in the theoretical values of χ, solid circles and crosses are experimental data by Sternberg (1970) and Weatherley (1972), respectively.

It is interesting to note that the law of drag obtained for the tidal channel (5.4.15), (5.4.16) and for the planetary boundary layer of the atmosphere (see Kazansky and Monin, 1961) are completely identical in form, the role of the angle of the total wind rotation in the atmospheric boundary layer, however, being played by the shift of phases $\gamma_0 (= \gamma^*)$ between the bottom stress and the velocity of the tidal flow beyond the boundary layer.

Ratios (5.4.15), (5.4.16) are convenient for analysis. The geostrophic friction coefficient and the phase shift appear to decrease monotonically with growing Rossby number, leading one to the conclusion that the geostrophic friction coefficient decreases with growing amplitude of the velocity as well as with decreasing bottom roughness and oscillation frequency. It accounts, in particular, for the well-known empirical fact that semidiurnal waves are more highly damped than diurnal ones due to the bottom friction. It is also interesting that with the Rossby number changing from 10^2 to 10^{14} the phase shift in the turbulent boundary layer does not exceed 45°, i.e. the value of the phase shift in a periodic laminar boundary layer.

Formula (5.4.14) presents the required dependence $\chi = \chi(V/U, g_u - g_v, \text{Ro}, e)$ in an implicit way, whereas Fig. 5.37 gives an explicit presentation of the character of dependence of the geostrophic friction coefficient on its determining parameters. The diagram, presented in this Figure and plotted at $0 \leq V/U \leq 1$, $0 \leq g_u - g_v \leq 7\pi/4$, $10^2 \leq R_0 \leq 10^{14}$ and $e = 0.7$, shows that at various fixed Rossby numbers the range of changes in the geostrophic friction coefficient affected by variations V/U and $g_u - g_v$ is not as great as might be expected. Hence, while carrying out approximate calculations, when it is sufficient to determine the geostrophic friction coefficient to an accuracy of 25%, χ can be considered a function of only one argument (the Rossby number). In other words, we can neglect dependence on V/U and $g_u - g_v$, replacing the domain of χ changes (shaded in Fig. 5.37) by the curve connecting the mean values of the geostrophic friction coefficient for each Ro.

As soon as the geostrophic friction coefficient is obtained, the components of the bottom friction stress can be found by the following relation:

$$\tau_{bx}/\varrho_0 = \frac{1}{4} \varkappa \chi U^2 \eta_{n0} \left[\mathfrak{S} \frac{M_0}{A_0} \cos(t_n - \beta_0 + \gamma_0) \right.$$
$$\left. + \mathfrak{S}^* \frac{M_0^*}{A_0^*} \cos(t_n - \beta_0^* + \gamma_0^*) \right];$$
$$\tau_{by}/\varrho_0 = \frac{1}{4} \varkappa \chi U^2 \eta_{n0} \left[\mathfrak{S} \frac{M_0}{A_0} \sin(t_n - \beta_0 + \gamma_0) \right.$$
$$\left. - \mathfrak{S}^* \frac{M_0^*}{A_0^*} \sin(t_n - \beta_0^* + \gamma_0^*) \right]. \quad (5.4.17)$$

THE BOTTOM BOUNDARY LAYER

Bearing in mind that ad $\eta_n = \eta_{n0}$ the functions M_0/A_0 and M_0^*/A_0^* have the form (5.4.13) and

$$\beta_0 = \arctan \frac{\sin g_u - V/U \cos g_v}{\cos g_u + V/U \sin g_v};$$

$$\beta_0^* = \arctan \frac{\sin g_u + V/U \cos g_v}{\cos g_u - V/U \sin g_v};$$

$$\gamma_0 = \arctan \frac{A}{\ln \dfrac{\chi \, \mathrm{Ro}}{1+e} - B};$$

$$\gamma_0^* = \arctan \frac{A}{\ln \dfrac{\chi \, \mathrm{Ro}}{1-e} - B},$$

we then obtain

$$\frac{\tau_{bx}}{\varrho_0} = \frac{1}{2} \varkappa \chi U^2 \left\{ \frac{\mathfrak{S} \cos(t_n - \beta_0 + \gamma_0)}{\left[\left(\ln \dfrac{\chi \, \mathrm{Ro}}{1+e} - B\right)^2 + A^2\right]^{1/2}} \right.$$

$$\left. + \frac{\mathfrak{S}^* \cos(t_n - \beta_0^* + \gamma_0^*)}{\left[\left(\ln \dfrac{\chi \, \mathrm{Ro}}{1-e} - B\right)^2 + A^2\right]^{1/2}} \right\};$$

$$\frac{\tau_{by}}{\varrho_0} = \frac{1}{2} \varkappa \chi U^2 \left\{ \frac{\mathfrak{S} \sin(t_n - \beta_0 + \gamma_0)}{\left[\left(\ln \dfrac{\chi \, \mathrm{Ro}}{1+e} - B\right)^2 + A^2\right]^{1/2}} \right.$$

$$\left. - \frac{\mathfrak{S}^* \sin(t_n - \beta_0^* + \gamma_0^*)}{\left[\left(\ln \dfrac{\chi \, \mathrm{Ro}}{1-e} - B\right)^2 + A^2\right]^{1/2}} \right\}. \quad (5.4.18)$$

In the particular case of a tidal channel, when V/U, g_v and e equal zero, we have

$$\frac{\tau_{bx}}{\varrho_0} = \frac{\varkappa \chi U^2}{[(\ln \chi \, \mathrm{Ro} - B)^2 + A^2]^{1/2}} \cos(t_n - g_u + \gamma_0);$$

$$\frac{\tau_{by}}{\varrho_0} = 0. \quad (5.4.19)$$

In this case the expression for the friction velocity (5.4.12) is also simplified, to take the form

$$u_{*n} = \frac{\varkappa \cos(t_n - g_u + \gamma_0)}{\chi[(\ln \chi \, \mathrm{Ro} - B)^2 + A^2]^{1/2}}. \quad (5.4.20)$$

To test the derived law of drag, it was necessary to have experimental data not only for friction velocity but also for external parameters of the bottom boundary layer. Unfortunately, as is always the case, something was lacking: either estimations of the friction velocity were not accompanied by the recording of the external parameters or, vice versa, the results of the analysis of field measurements did not include estimation of the friction velocity. Hence we had to assume that all the experimental data used to test the law of drag are adequate to the condition of a one-dimensional flow. In this case, as we have seen, to calculate the frictional stress it is sufficient to know the amplitude of the velocity, the roughness of the underlying surface and the oscillation frequency.

In four series of measurements out of five at our disposal the above assumption is more or less justified, namely the laboratory measurements in a wave tank published by Jonsson (1963) and Carlsen (1967) as well as the results of field measurements in the Kolvos and Florida straits presented by Sternberg (1968) and Weatherley (1972). One more series of data useful for our analysis was obtained by Sternberg (1970) beyond the zone of the continental slope in the open ocean where the condition of one-dimensionality of the tidal flow can hardly be valid.

In addition to the above, note that in all the papers cited above the bottom stress was defined not by equation (5.4.19) but by the following formula:

$$\frac{\tau_{bx}}{\varrho_0} = \iota |u_\infty| u_\infty \qquad (5.4.21)$$

or by its equivalent

$$\frac{\tau_{bx}}{\varrho_0} = \frac{8}{3\pi} \iota U^2 \cos(t_n - g_u), \qquad (5.4.21')$$

where ι as before is the bottom friction coefficient.

Comparing equations (5.4.19) and (5.4.21') we obtain

$$\iota = \frac{3\pi}{8} \frac{\varkappa \chi}{[(\ln \chi \, Ro - B)^2 + A^2]^{1/2}}. \qquad (5.4.22)$$

If, however, the amplitude U and the phase g_u of the tidal velocity in equation (5.4.21') refer to a certain fixed level within the boundary layer, say, to the 1-m layer, the dependence between the bottom friction coefficient and the geostrophic friction coefficient can be represented in the following way:

$$\iota_1 = \frac{3\pi}{8} \frac{\varkappa \chi U^2/U_1^2}{[(\ln \chi \, Ro - B)^2 + A^2]^{1/2}}. \qquad (5.4.23)$$

Here the index 1 denotes the values of the corresponding characteristics at the distance of 1 m from the bottom.

Equation (5.4.23) includes the unknown quantity U^2/U_1^2 defining the ratio of the squared velocity amplitudes beyond the boundary layer and at the distance of 1 m from the bottom. To estimate this unknown quantity, make use of the fact that the tidal velocity in the bottom 1 m obeys the logarithmic law. Hence,

$$\frac{U_1}{U} = \frac{\chi}{\varkappa} \ln z_1/z_0.$$

Substitute this relation into equation (5.4.23) and into the expression for the Rossby number and obtain

$$t_1 = \frac{3}{8} \frac{\varkappa^3 \chi^{-1}(\ln z_1/z_0)^{-2}}{[(\ln \chi \, \text{Ro} - B)^2 + A^2]^{1/2}}; \qquad (5.4.24)$$

$$\chi \, \text{Ro} = \varkappa \frac{U_1}{\sigma z_0}(\ln z_1/z_0)^{-1}. \qquad (5.4.25)$$

On the other hand, the dependence of $\chi(\text{Ro})$ on Ro is known from equation (5.4.15). Thus, if U, σ and z_0 are set, the Ro quantity may be considered known, allowing one to define the bottom friction coefficient as the function of the Rossby number, employing formulae (5.4.22) or (5.4.24). These are the facts that are necessary to test the law of drag according to the experimental data in a one-dimensional tidal flow.

TABLE 5.8. *Comparison of computed values of the bottom friction coefficient r and the phase shift γ with laboratory data* (by Kagan, 1971)

Ro	$\varepsilon \times 10^2$		γ degree	
	theory	experiment	theory	experiment
3.66×10^3	0.91	1.00	27	25
3.52×10^2	1.59	1.97	35	30

Table 5.8 as well as Figs. 5.37 and 5.38 present the comparison of the results of calculations with the field measurements mentioned above. Generally speaking, the theory is seen to accord roughly with the experimental data, well enough in view of the assumptions made in the above method of calculating bottom stress. As to its advantages, compared to the methods employed at present (the bottom frictional stress in them being described by a linear or a quadratic laws of drag), it is evident that (1), the bottom friction coeffi-

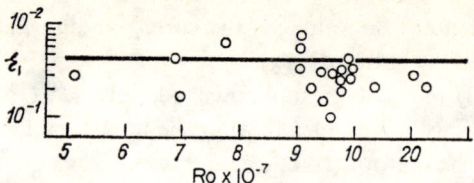

FIG. 5.38. Comparison of the computed values of the bottom friction coefficient ζ_1 with observational data (from Kagan, 1971). Solid line shows a segment of the theoretical curve, the circles are the experimental data by Sternberg (1968).

cient is not considered given *a priori* but is obtained, and (2) only real shifts of phase between the tangential stress on the bottom and the tidal velocity is taken into account.

Needless to say, we are still far from the final solution of all the problems arising in the process of investigation and parametrization of the bottom boundary layer. Certain progress, however, has been made and that is why we still have some home for a favourable outcome.

5.5 References

Experimental determination of the critical Reynolds number Re_{crit} in periodic wave motion was the subject of studies not only by Collins (1963) but also by Li (1954) and Vincent (1957). Li, while investigating the case of an oscillating flat wall, found that for the smooth underlying surface $Re_{crit} = 800$, which corresponds to $Re_{crit} = 566$ in the case of long waves. Vincent discovered that Li had overstated the upper limit of the values of the Reynolds number at which the existence of the laminar boundary layer is possible. This conclusion was confirmed by laboratory data carried out for the case of a stationary laminar boundary layer. According to Tatro and Mollo-Christiansen (1967) and Green (1968) the stationary laminar boundary layer becomes unstable when the Reynolds number exceeds its critical value, which is approximately 100.

The value of the numerical constant in the formula for the thickness of the viscosity sublayer was borrowed from the paper by Wimbush and Munk (1971). Derivation of the asymptotic formulae for the roughness parameter under different conditions of drag can be found, for instance, in the books by Zilitinkevich (1970) and Kitaigorodsky (1970).

The analysis of measurements of the turbulent velocity fluctuations carried out in early 1960s is presented in the review by Bowden (1965). While discussing the data on statistical characteristics of turbulent velocity pulsations and the temperature in tidal flow we mainly followed the monograph by Kagan (1968) and the above-mentioned paper by Wimbush and Munk (1971).

The asymmetry of the transverse correlation functions seems to have first been noticed while measuring velocity pulsations in an aerodynamic tube (see Favre, Gaviglio and Dumas, 1958) and temperature fluctuations in the atmosphere (Taylor, 1958). It was later observed while analysing the transverse correlation functions of velocity pulsations of the wind in the atmospheric surface layer (Davenport, 1961).

A modern presentation of the Torade model was borrowed from the paper by Munk, Snodgrass and Wimbush (1970).

The results referring to numerical modelling to homogeneous and stratified bottom boundary layers were obtained by Vager and Kagan (1969, 1971). The generalized Karman formula for the turbulence scale was suggested by Zilitinkevich, Laikhtman and Monin (1967), the same article presenting the values of the numerical constants α_b, c_0 and c_1 and the ratio $c_1 = c_0^3 = c^{3/4}$.

The method of matrix factorization was described in detail in the monograph by Marchuk (1961). The factorization method can be found in the book by Godunov and Ryaben'ky (1962).

Intensification of turbulence with decreasing velocity of the tidal flow was the subject of the paper by Wimbush and Munk (1971), a similar phenomenon having been observed by Bradshaw (1967) in the laboratory.

The textbook by Proudman (1953) presents arguments of inequality of the bottom friction coefficients while determining the tangential stress on the bottom through the velocity and the amplitude of the tidal velocity.

The method of evaluation of the bottom friction coefficient by the equation for tidal energy was suggested by Taylor (1918). The determination of the bottom friction coefficient directly from the equations of motion by use of the expression for the bottom stress was reported by Proudman (1923) and Grace (1929). The method was later improved by Ichije (1955), Bowden, Fairbairn and Hughes (1959) and Vapniar (1962).

The derivation of the resistance law in a two-dimensional tidal flow described in this chapter was carried out by Kagan (1971, 1972).

CHAPTER 6

Vertical Structure of Internal Tidal Waves

THEORETICAL investigations of internal waves (tidal waves included) have a long history, like studies of surface tidal waves. As early as 1847 Stokes obtained the simplest solutions for hydrodynamic equations describing the waves on an interface of fluids of different densities. Since that time the problem of calculating internal waves has attracted the constant attention of investigators. The quantity of publications devoted to the problem is amazing even from a modern standpoint. The overwhelming majority of these papers, however, reduce the above problem to simpler ones of finding eigenfunctions for quotient vertical distributions of mean density. Therefore it is not surprising that O. M. Phillips defined attempts to describe the internal waves theoretically, as "...sometimes quite speculative... not in the sense of their validity but probably in the sense of their direct relation to what is taking place in the ocean".

We do not intend to give here a detailed review of mathematical models from the viewpoint of studying the nature of the phenomenon or elaborating practical methods of calculation; we shall limit our analysis of internal tidal waves to consideration of their features and one of the effective methods of calculating their vertical structure in an arbitrarily stratified ocean.

6.1. Generation of Internal Tidal Waves

The existence of internal waves in the ocean was discovered long ago. In all probability, the first reports of their being recorded were the subject of the papers by Nansen (1902) and Helland-Hansen (1909). It was Petterson (1907), however, who was the first to point out the existence of waves of tidal nature among internal waves. He came to this conclusion on the basis of field measurements carried out by him in the Kattegat Strait. Serious experimental studies of internal tidal waves started with the oceanographic expedition on the ship

Maud in the Atlantic Ocean, followed by a number of expeditions in various parts of the World Ocean on the ships *Meteor, Snellius, Dana* and *Atlantic*. Of particular interest were the observational data on temperature oscillations and salinity accompanied by simultaneous measurements of flow velocity.

Fresh interest in internal waves was awakened in the postwar years when it became clear that all the processes taking place in the ocean, whether it is the formation of density and velocity fields or sound distribution, depend, among other things, on internal waves as well. The analysis of the data themselves is not the subject of this paper, but let us note, however, that the available data indisputably attested that internal waves with tidal period do exist, not only in coastal waters where the possibility of their existence has never been disputed, but in the open ocean as well, as is clearly seen in Figs. 6.1 and 6.2. The figures unambiguously demonstrate oscillations with the period of approximately 12 hours, which is roughly the semidiurnal tidal period. The greatest oscillations are observed in the layer of the thermocline where the internal wave-height (double-amplitude) reaches in some cases 30 metres. Above and below this layer the oscillations of the hydrological characteristics gradually die away, vanishing completely in the surface quasi-homogeneous layer.

The above facts show the inconsistency of the so-called local hypotheses of internal wave generation, which attribute the existence of the internal waves in the open ocean to their propagation from regions of continental slope or from those of "the critical" latitudes. In fact, one could hardly imagine that these waves, whose spatial scale is much less than the characteristic horizontal dimensions of the ocean (see below), could preserve their peculiarities at a sufficiently great distance from the regions of their generation, taking into account the action of the stationary currents and mechanisms of dissipation.

Quite invalid seems also the hypothesis stating that the generation of internal tidal waves is brought about by the direct action of the tide-generating forces. To prove the inconsistency of the hypothesis, assume the effect of the tide-generating forces to be negligibly small. Investigate now the properties of the internal tidal waves and, using *a posteriori* information, test the validity of the initial assumption.

Let us consider a very simplified case such that the stratified fluid fills a basin of constant depth rotating about the vertical axis with constant angular velocity; the fluid consists of two layers with different but constant density; the interface between the layers in an undisturbed state is horizontal; the effect of the forces of turbulent friction is absent.

Let the fluid density in the lower and upper layers be ϱ_1 and ϱ_2, the thickness of these layers D_1 and D_2, and the disturbances of the interface surface and of the free surface ζ_1 and ζ_2. Then the system of the equations of motion and

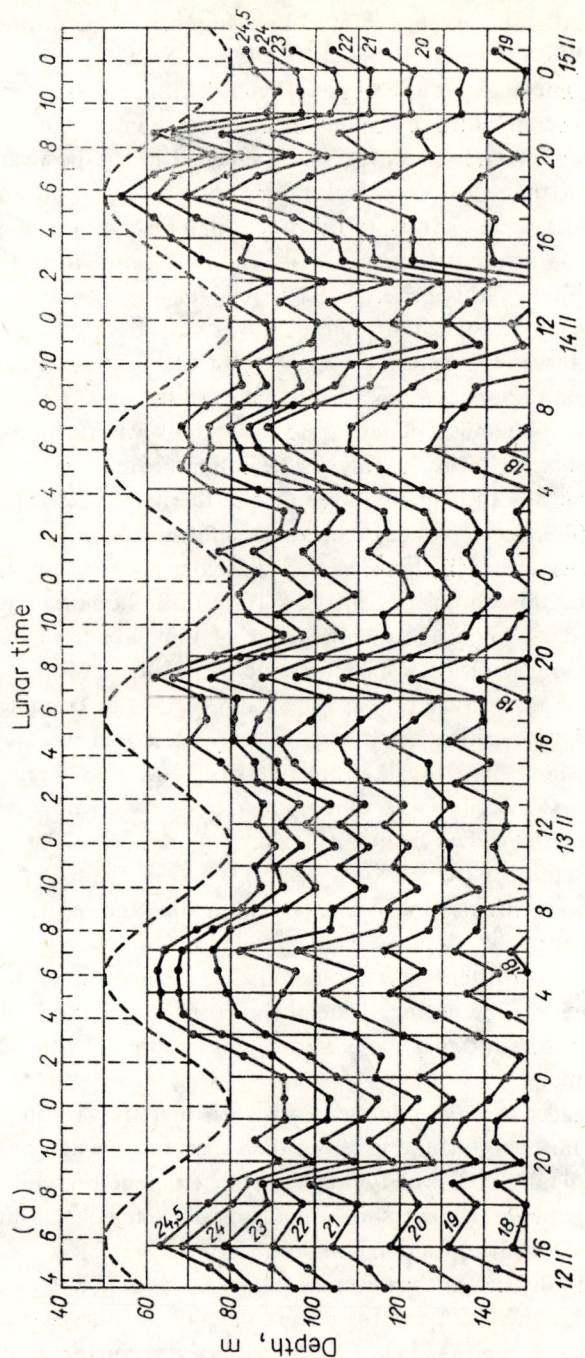

VERTICAL STRUCTURE OF INTERNAL TIDAL WAVES

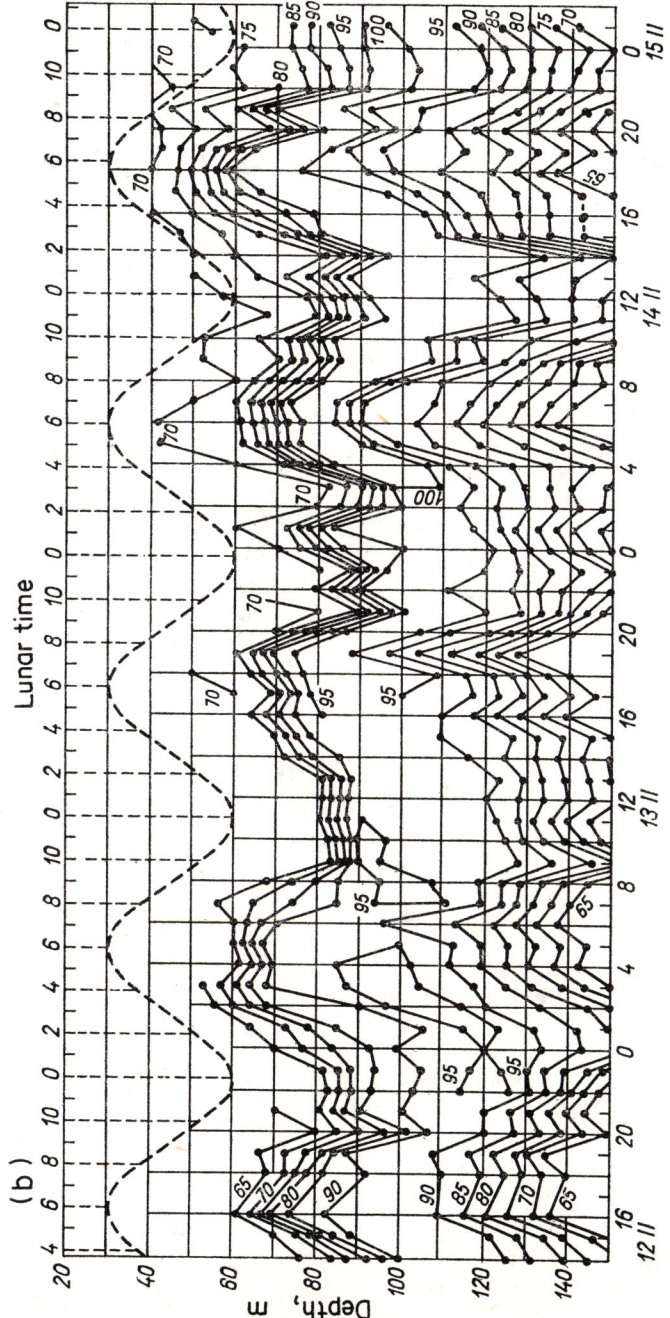

FIG. 6.1. Changes in the depth of the isotherm 18–24.5°C (a) and isohalines 36.65–37.00‰ (b) covering the period 12 to 15 February 1938. The data of the hourly measurements on the anchored station No. 385 of the research vessel *Meteor* (from Defant, 1961). Dashed line denotes a semidiurnal lunar cycle.

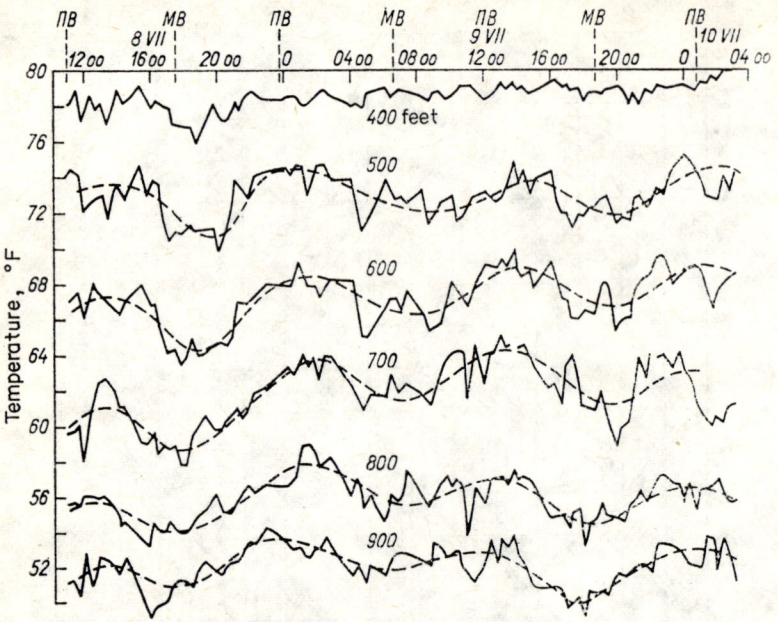

F I G. 6.2. Fluctuations of the water temperature within the layer 400–900 feet in the vicinity of the Bikini Atoll (by La Fond, 1949).

continuity (2.1.1), (2.1.2) as applied to the lower layer, will be written as

$$\frac{\partial \mathbf{v}_1}{dt} + \mathbf{A}_1 \mathbf{v}_1 = -\frac{1}{\varrho_1} \cdot \nabla p_1; \qquad (6.1.1)$$

$$p_1 = g(\varrho_1 - \varrho_2)\zeta_1 + p_2; \qquad (6.1.2)$$

$$\frac{1}{D_1} \frac{\partial \zeta_1}{\partial t} = -\operatorname{div} \mathbf{v}_1. \qquad (6.1.3)$$

In an analogous way, for the upper layer,

$$\frac{\partial \mathbf{v}_2}{\partial t} + \mathbf{A}_1 \mathbf{v}_1 = -\frac{1}{\varrho_2} \cdot \nabla p_2; \qquad (6.1.4)$$

$$p_2 = g\varrho_2 \zeta_2; \qquad (6.1.5)$$

$$\frac{1}{D_2} \frac{\partial}{\partial t}(\zeta_2 - \zeta_1) = -\operatorname{div} \mathbf{v}_2. \qquad (6.1.6)$$

Applying the operators of horizontal divergence and vorticity to equations (6.1.1), (6.1.4), exclude the expression for the vorticity from the resulting rela-

VERTICAL STRUCTURE OF INTERNAL TIDAL WAVES

tions and, substituting equations (6.1.3), (6.1.6) and integrating once with respect to time, obtain

$$\left(\frac{\partial^2}{\partial t^2} + 4\omega_z^2\right)\zeta_1 = \frac{D_1}{\varrho_1} \cdot \Delta p_1; \tag{6.1.7}$$

$$\left(\frac{\partial^2}{\partial t^2} + 4\omega_z^2\right)(\zeta_2 - \zeta_1) = \frac{D_2}{\varrho_2} \cdot \Delta p_2. \tag{6.1.8}$$

To determine the pressure disturbances in the lower and upper layers (p_1 and p_2) let us make use of relations (6.1.2), (6.1.5). Substituting them into equations (6.1.7), (6.1.8) yields

$$\left(\frac{\partial^2}{\partial t^2} + 4\omega_z^2\right)\zeta_1 = gD_1 \cdot \Delta\left(\frac{\delta\varrho}{\varrho_1}\zeta_1 + \frac{\varrho_2}{\varrho_1}\zeta_2\right); \tag{6.1.9}$$

$$\left(\frac{\partial^2}{\partial t^2} + 4\omega_z^2\right)(\zeta_2 - \zeta_1) = gD_2 \cdot \Delta\zeta_2, \tag{6.1.10}$$

where $\delta\varrho = \varrho_1 - \varrho_2$.

The above equations have elementary wave solutions of the type

$$(\zeta_1, \zeta_2) = (Z_1, Z_2) \exp i(\boldsymbol{\kappa} \cdot \mathbf{x} + \sigma t). \tag{6.1.11}$$

Here $\boldsymbol{\kappa} = (\kappa_x, \kappa_y)$ is the horizontal wave vector; Z_1, Z_2 are the complex amplitudes of vertical oscillations of the corresponding boundary surfaces; σ is the oscillation frequency.

With equation (6.1.11) taken into account, equations (6.1.9), (6.1.10) take the form

$$\left[c^2(1-e^2) - gD_1\frac{\delta\varrho}{\varrho_1}\right]Z_1 - gD_1\frac{\varrho_2}{\varrho_1}Z_2 = 0;$$
$$[c^2(1-e^2) - gD_2]Z_2 - c^2(1-e^2)Z_1 = 0, \tag{6.1.12}$$

where $c = \sigma/|\boldsymbol{\kappa}|$ is the phase velocity; $e = 2\omega_z/\sigma$.

System (6.1.12) has a solution other than zero if its determinant equals zero, i.e.

$$\begin{vmatrix} c^2(1-e^2) - gD_1\dfrac{\delta\varrho}{\varrho_1} & -gD_1\dfrac{\varrho_2}{\varrho_1} \\ -c^2(1-e^2) & c^2(1-e^2) - gD_2 \end{vmatrix} = 0.$$

Developing the above determinant we have

$$\left[c^2(1-e^2) - gD_1\frac{\delta\varrho}{\varrho_1}\right][c^2(1-e^2) - gD_2] - c^2(1-e^2)gD_1\frac{\varrho_2}{\varrho_1} = 0. \tag{6.1.13}$$

The equation obtained is a dispersion ratio. It determines the tolerance values of the phase velocity (or the wave number) corresponding to the given oscillation frequency.

Two approximate roots of equation (6.1.13), corresponding to one of the wave types, can be obtained if we set

$$c^2(1-e^2) \gg gD_1 \frac{\delta\varrho}{\varrho_1}. \tag{6.1.14}$$

Omitting the second term in the first square bracket in equation (6.1.13) and, reducing both parts of the resulting equation by $c^2(1-e^2)^\dagger$, we have

$$c^2 \approx \frac{g(D_1+D_2)}{1-e^2}. \tag{6.1.15}$$

The first type of wave is seen from equation (6.1.15) to correspond to the case of the wave motion in a homogeneous fluid with depth D_1+D_2. For any arbitrary $\delta\varrho$ the square of the phase velocity of the gravitation waves, determined by equation (6.1.15), is much greater than $gD_1(\delta\varrho/\varrho_1)$ and, hence, the assumption (6.1.14) is valid.

The values of the other roots of equation (6.1.13) can be easily found assuming

$$c^2(1-e^2) \ll gD_2. \tag{6.1.16}$$

Omit in this case the first term in the second square bracket, and regroup the terms, giving

$$c^2 \approx \frac{g}{\varrho_1} \delta\varrho \cdot (1-e^2)^{-1} \frac{D_1 D_2}{D_1+D_2}. \tag{6.1.17}$$

This solution corresponding to the second type of wave describes the internal waves propagating on the interface of the two layers of the liquid. At real values of the parameters of equation (6.1.17), for instance, $O(D_1) = 10^3$ m, $O(D_2) = 10^2$ m, $O(\delta\varrho/\varrho_1) = 10^{-3}$ and $O(e) = 0.7$, the order of magnitude of the phase velocity of the internal waves will be, according to equation (6.1.17), 1 m/s. Consequently, a typical length of the internal tidal waves of semidiurnal period must have an order of 100 km, i.e. it is approximately two orders less than the spatial scale $(2\pi a \sin \Theta)/2$, within which the action of the tide-generating forces of the semidiurnal period manifests itself. The very fact of incommensurability of the scales of the perturbing force and the response to it casts doubt on the possibility of the direct influence of the tide-generating forces on the internal

---

† A special case when $e = \pm 1$ is not considered here.

tidal waves. There is also independent evidence of the invalidity of the hypothesis under discussion. For instance, directly measured data point to the fact that the characteristics of the internal tides appear to be incoherent with the tide-generating force, which is evidenced by obvious changes in the phase difference between the observed oscillations of water temperatures at a fixed level and the statical tide.

The most likely explanation for the mechanism of generation of the internal tidal waves seems to be that suggested by Cox and Sandstrom (1962). They considered the internal tidal waves to be the reaction of a stratified ocean to the propagation of surface tidal waves over the irregular sea bottom. Their theory accounts for the fact that the internal tides can be generated throughout. Its essence is as follows: disturbance to the flow by bottom irregularities results in velocity divergence. Due to the conditions of continuity and incompressibility of sea water, the latter leads to the appearance of vertical velocities and those, in their turn, result in the deformation of the isopicnal surfaces. In this way Cox and Sandstrom interpret the mechanism of generation of the internal waves with tidal period. According to their conclusions, confirmed later by Munk (1966), the flow of energy ε_{0n} from the surface wave ($n = 0$) to the internal mode numbered n, referred to the mass of the water column of a unit section and the height D, for great values of n is

$$\varepsilon_{0n} = \frac{1}{2}\pi g |\kappa_n|^3 n^{-1} N(0) \frac{(1+e^2)}{(1-e^2)} D^{-2} A_0^2 S(|\kappa_n|). \qquad (6.1.18)$$

If now the spectral density of the bottom irregularities $S(|\kappa_n|)$ is set to be inversely proportional to the third porter of the modulus of the wave number $|\kappa_n|$ with the coefficient $B = 1$ m (spectral analysis of the sea-bottom relief confirms this assumption at least within the range $10^{-4} < |\kappa| < 10^{-3}\,\mathrm{m}^{-1}$), then

$$\varepsilon_{0n} = \frac{1}{2}\pi \frac{(1+e^2)}{(1-e^2)} n^{-1} g N(0) D^{-2} A_0^2 B, \qquad (6.1.19)$$

where A_0 is the amplitude of the surface tide; N_0 is the value of the Brunt–Väisälä frequency at the free surface of the ocean, and other notations are the same.

In this case with $D = 4000$ m and $A_0 = 0.5$ m we have

$$\varepsilon_{0n} = 1.2 \times 10^{-6} n^{-1} \qquad (6.1.20)$$

Thus the summarized energy flow to the first 20 modes equals 4×10^{-1} erg/(g·s) which totals 5×10^{18} erg/s when evaluated for the World Ocean.

The above estimate shows the outflow of tidal energy into internal waves and its subsequent dissipation to be approximately two orders greater than the dissipation of the tidal energy in the open ocean due to the mechanism of the turbulent friction (see Section 4.6). Even this amount of energy outflow must not, however, result in any noticeable change in the surface tidal oscillations. Indeed, let us make use of Munk's estimate and calculate the energy $\langle \mathcal{E}_w \rangle$ dissipated in internal waves at the semidiurnal period. It is 2×10^{22} erg, which is two orders less than the potential energy $\langle \mathcal{E}_p \rangle$ of the semidiurnal lunar tides in the World Ocean [Hendershott (1972) estimated $\langle \mathcal{E}_p \rangle = 2.07 \times 10^{24}$ erg]. Having found $\langle \mathcal{E}_w \rangle$ and $\langle \mathcal{E}_p \rangle$ one can estimate the difference between the surface tidal oscillations in stratified and homogeneous oceans. For this purpose equate $\langle \mathcal{E}_w \rangle$ to the difference of $\langle \mathcal{E}_p \rangle$ values in homogeneous and stratified oceans. We obtain $\langle \mathcal{E}_w \rangle / \langle \mathcal{E}_p \rangle \approx 2\delta\zeta/\zeta$, and the relative changes in the surface tidal oscillations $\delta\zeta/\zeta$ in a stratified ocean will amount to 5×10^{-3}. Thus, the conclusion made in Section 1.3, the density stratification of the sea has a negligibly small influence on the tides in the ocean, is confirmed.

6.2. Qualitative Analysis of the Equations for Internal Waves

Let us consider the case of free oscillations of the tidal period. Assume that in an unperturbed state a vertically stratified but horizontally homogeneous ocean is in a state of rest, the sea bottom is flat and all the perturbations of the characteristics in question are small deviations from their mean values. It has been found in the preceding section that the horizontal scale of the motion in internal waves is much less than the Earth's radius. Therefore we shall make use of orthogonal Cartesian coordinates with the axis z directed vertically upward. Then, after linearization and omitting the terms describing the dissipation processes (friction and diffusion) the equations of motion, conservation of density and continuity will take the following forms:

$$\frac{\partial \mathbf{u}}{\partial t} + \mathbf{A}_1 \mathbf{u} = -\frac{1}{\varrho_0} \cdot \nabla p; \qquad (6.2.1)$$

$$\frac{\partial w}{\partial t} = -\frac{1}{\varrho_0} \frac{\partial p}{\partial z} - \frac{g}{\varrho_0} \varrho; \qquad (6.2.2)$$

$$\frac{\partial \varrho}{\partial t} - \frac{N^2}{g/\varrho_0} w = 0; \qquad (6.2.3)$$

$$\frac{\partial w}{\partial z} + \text{div } \mathbf{u} = 0, \qquad (6.2.4)$$

where, as earlier, $\mathbf{u} = (u, v)$ is the vector of horizontal velocity; w is the vertical component of the velocity; p and ϱ are the perturbations of the pressure and density; $\varrho_0 = \varrho_0(z)$ is the density in an undisturbed state; $\mathbf{A}_1$ is the

VERTICAL STRUCTURE OF INTERNAL TIDAL WAVES

Coriolis matrix;

$$N(z) = (g\mathfrak{E})^{1/2} \approx \left(-\frac{g}{\varrho_0}\frac{d\varrho_0}{dz}\right)^{1/2}$$

is the Brunt–Väisälä frequency.

Note that writing down the system (6.2.1)—(6.2.4) we have neglected the effects of the Coriolis terms containing $\cos\varphi$ and those of the water compressibility. The first assumption, being of a formal character, was made only for the sake of simplification, the second had the objective of filtering out the acoustic waves (see Chapter 1 for details). At the same time, in the equation of motion for the vertical component of the velocity we have preserved the term describing changes of w in time. It will be shown below that the omission of this term can result in distortion of the oscillation spectrum only in the range of large wave numbers.

For the case of a boundless ocean, the system of boundary conditions, for which the solution of equations (6.2.1)—(6.2.4) is valid, will contain only the condition of continuity of pressure and the kinematic relation at the free surface

$$p = g\varrho_0\zeta, \quad w = \frac{\partial \zeta}{\partial t} \quad \text{at} \quad z = D \tag{6.2.5}$$

and the condition of impermeability of the sea bottom

$$w = 0 \quad \text{at} \quad z = 0. \tag{6.2.6}$$

As usual, we shall set no initial conditions. Instead, we merely specify that all the sought functions are harmonic in time:

$$(\mathbf{u}, w, p, \varrho) = \text{real } (\bar{\mathbf{u}}, \bar{w}, \bar{p}, \bar{\varrho}) \exp(i\sigma t). \tag{6.2.7}$$

Here the oscillation frequency σ at a stable or neutral stratification ($N \geqslant 0$) can assume only real values, otherwise (in the case of infinitely growing and damping with time solutions) the law of energy conservation will not be fulfilled.

Now substitute equation (6.2.7) into equation (6.2.1)—(6.2.4), combine equation (6.2.2) and (6.2.3) and obtain

$$(i\delta\sigma + \mathbf{A}_1)\bar{\mathbf{u}} = -\frac{1}{\varrho_1}\cdot\nabla\bar{p}; \tag{6.2.8}$$

$$(\sigma^2 - N^2)\bar{w} - \frac{i\sigma}{\varrho_0}\frac{\partial \bar{p}}{\partial z} = 0; \tag{6.2.9}$$

$$\frac{\partial \bar{w}}{\partial z} + \text{div } \bar{\mathbf{u}} = 0. \tag{6.2.10}$$

In equations (6.2.8)–(6.2.10) we separate the independent variables. For this purpose assume that

$$\bar{p} = \mathcal{P}(z)\Pi(\mathbf{x});$$

$$\bar{w} = i\sigma \mathcal{W}(z)\Pi(\mathbf{x});$$

$$\bar{\mathbf{u}} = \frac{1}{\varrho_0}\mathcal{P}(z)\mathcal{U}(\mathbf{x}). \tag{6.2.11}$$

Substitute equations (6.2.11) into equations (6.2.8)–(6.2.10) and, introducing the separation constant $\hbar$ (which is connected with the depth of a dynamically equivalent ocean h by the relation $\hbar = 1/gh$) obtain the two subsystems

$$(i\sigma + \mathbf{A}_1)\mathcal{U} = -\nabla \Pi;$$

$$i\sigma \hbar \Pi + \operatorname{div} \mathcal{U} = 0 \tag{6.2.12}$$

$$(\sigma^2 - N^2)\mathcal{W} - \frac{1}{\varrho_0}\frac{d\mathcal{P}}{dz} = 0;$$

$$\frac{d\mathcal{W}}{dz} - \frac{\hbar}{\varrho_0}\mathcal{P} = 0, \tag{6.2.13}$$

the first describing the horizontal and the second the vertical structure of oscillations.

The boundary conditions (6.2.5), (6.2.6), together with equations (6.2.11) and (6.2.13), are transformed as follows:

$$\frac{d\mathcal{W}}{dz} - g\mathcal{W}\hbar = 0 \quad \text{at} \quad z = D, \tag{6.2.14}$$

$$\mathcal{W} = 0 \quad \text{at} \quad z = 0.$$

For a given distribution of the Brunt–Väisälä frequency and for a fixed value of the Coriolis parameter each of the subsystems (6.2.12), (6.2.13) contains two variable parameters (σ and $\hbar$), the solution of the above subsystems obey the chosen boundary conditions only at a definite combination of σ and $\hbar$. Thus, the problem is reduced to finding all possible combinations of the eigenvalues of σ and $\hbar$. The latter form on the plane $(\sigma, \hbar)$ two sets of curves (which are called eigencurves) corresponding to the problems of the vertical and horizontal structure of oscillations. Their intersection affords the required values σ and $\hbar$ at which both subsystems have non-trivial solutions.

Let us consider first the problem of the vertical structure of oscillations, setting $N = \text{const}$ for the sake of simplicity. Neglect the vertical changes of

VERTICAL STRUCTURE OF INTERNAL TIDAL WAVES

ϱ_0 in the second term of equation (6.2.13) and reduce subsystem (6.2.13) to a single equation with respect to $\mathcal{W}$:

$$\frac{d^2}{dz^2} + \hbar(N^2 - \sigma^2)\mathcal{W} = 0. \tag{6.2.15}$$

Its solution obeys boundary conditions (6.2.14) if at $\hbar > 0$, $N^2 < \sigma^2$ and a fixed σ the eigenvalues are the roots of the equation

$$\frac{(\sigma^2 - N^2)^{1/2}}{g\hbar^{1/2}} = \tanh \sqrt{\hbar(\sigma^2 - N^2)}\, D. \tag{6.2.16}$$

At $\hbar > 0$ and $N^2 > \sigma^2$ eigenvalues of $\hbar$ are determined by the equation

$$\frac{(N^2 - \sigma^2)^{1/2}}{g\hbar^{1/2}} = \tan \sqrt{\hbar(N^2 - \sigma^2)}\, D. \tag{6.2.17}$$

The analogous equations at $\hbar < 0$, $N^2 < \sigma^2$ and $\hbar < 0$, $N^2 > \sigma^2$ will have the form

$$-\frac{(\sigma^2 - N^2)^{1/2}}{g(-\hbar)^{1/2}} = \tan \sqrt{-\hbar(\sigma^2 - N^2)}\, D \tag{6.2.18}$$

and

$$-\frac{(N^2 - \sigma^2)^{1/2}}{g(-\hbar)^{1/2}} = \tanh \sqrt{-\hbar(N^2 - \sigma^2)}\, D. \tag{6.2.19}$$

The scheme of graphic solution of equations (6.2.16)–(6.2.19) is shown in Fig. 6.3 (a–d). The figure shows that equation (6.2.19) has no roots at all, equation (6.2.16) has only one root, while equation (6.2.17) has a denumerable set of roots $\hbar_0, \hbar_1, \hbar_2, \ldots$ and $\hbar_{-1}, \hbar_{-2}, \ldots$, where the positive indices correspond to equation (6.2.17) and the negative indices to equation (6.2.18).

Let us qualitatively analyse the forms of the eigencurves $\hbar_0(\sigma)$, $\hbar_1(\sigma)$, $\hbar_{-1}(\sigma)$, ... and their position on the plane $(\hbar, \sigma)$. If we set $\sigma \to \infty$ then equation (6.2.16) yields $\hbar \approx \sigma^2/g^2$. At $\sigma \to 0$ equation (6.2.16) affords $\hbar_0 = 1/gD$. According to equation (6.2.17) the error of replacing $\hbar_0$ by its approximate value $1/gD$ is N^2D/g. The values of $\hbar_1, \hbar_2, \ldots$ and $\hbar_{-1}, \hbar_{-2}, \ldots$ are found to the same accuracy if the right sides of equation (6.2.17), (6.2.18) are set equal zero. In this case

$$\hbar_n = \frac{(n\pi)^2}{D^2(N^2 - \sigma^2)}, \quad n = 1, 2, 3, \ldots; \tag{6.2.20}$$

$$\hbar_n = -\frac{(n\pi)^2}{D^2(\sigma^2 - N^2)}, \quad n = -1, -2, -3, \ldots. \tag{6.2.21}$$

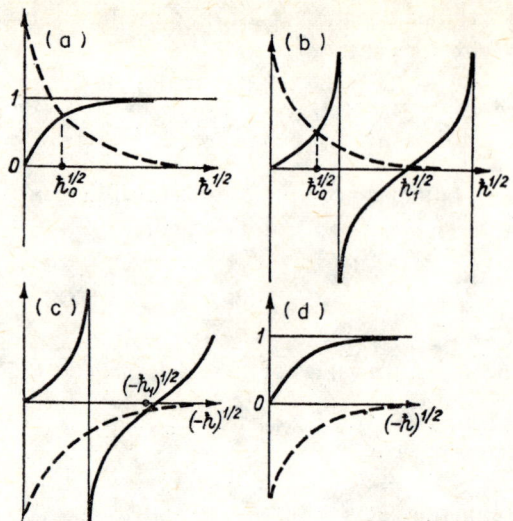

FIG. 6.3. Scheme of a graphic solution of Eqs. (6.2.16)–(6.2.19) (from Kamenkovitch, 1973). The left-hand side of the equations is depicted by a dashed curve, the right-hand side by a solid curve.

The above conclusions are illustrated by Fig. 6.4 which represents the eigencurves of the problem on the vertical structure of the oscillations.

Let us now consider subsystem (6.2.12). Applying to the first equation of this subsystem the operator of horizontal divergence and then combining it with the second equation of the subsystem affords the Laplace tidal equation on a plane

$$\Delta \Pi + (\sigma^2 - 4\omega_z^2) \hbar \Pi = 0. \qquad (6.2.22)$$

Let us look for its solution in the following form:

$$\Pi \sim \exp i\boldsymbol{\kappa} \cdot \mathbf{x}.$$

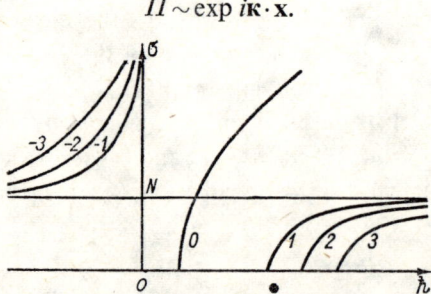

FIG. 6.4. Schematic representation of the eigencurves of the problem on the vertical structure of oscillations at $\varrho_0 = $ const and $N = $ const (from Kamenkovitch, 1973). Each curve is denoted by its number n. The picture is symmetrical with respect to the axis $\sigma = 0$.

VERTICAL STRUCTURE OF INTERNAL TIDAL WAVES 261

Substituting this expression into equation (6.2.22) yields

$$\hbar = \frac{|\kappa|^2}{\sigma^2 - 4\omega_z^2}. \qquad (6.2.23)$$

The eigencurves of equation (6.2.22) are presented in Fig. 6.5.

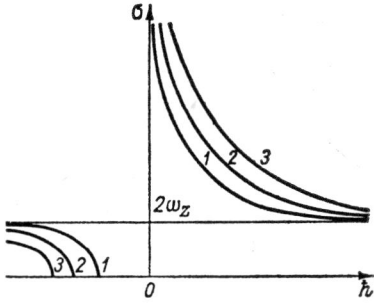

FIG. 6.5. Schematic representation of the eigencurves of Laplace's tidal equation on a plane (from Kamenkovitch, 1973). Numbers near the curves correspond to defined values of $|\kappa|$. The picture is symmetrical with respect to the axis $\sigma = 0$.

Superimpose Fig. 6.4 on Fig. 6.5. In this case the solutions of system (6.2.1)–(6.2.4), corresponding to the points of intersection of the eigencurve $\hbar_0$ with the eigencurvees of equation (6.2.23), will describe the *surface* gravitational waves with the frequencies $\sigma^2 > 4\omega_z^2$. These are called surface waves since their generation is related with the existence of perturbation of the ocean surface. Indeed, if $\zeta = 0$, then in line with the second condition of equation (6.2.5) $\mathcal{W} = 0$. In this case, however, the eigencurve $\hbar_0$ coincides with the y-axis in Fig. 6.4 and, hence, it cannot intersect the eigencurves of the Laplace tidal equation, the latter fact being tantamount to dissappearence of surface gravitational waves. These waves should not be confused with the *internal* gravitational waves to which correspond the points of intersection of the curves of equations (6.2.20), (6.2.21) and (6.2.23). They intersect only in the case when $\hbar > 0$ and $N > 2\omega_z$, showing that the internal gravitational waves fill the interval $[N, 2\omega_z]$. Omitting $\hbar$ from equations (6.2.20) and (6.2.23) affords an approximate dispersion relation for the internal gravitational waves at $N = $ const:

$$\sigma^2 = \frac{4\omega_z^2 \left(\frac{n\pi}{D}\right)^2 + N^2 |\kappa|}{\left(\frac{n\pi}{D}\right)^2 + |\kappa|^2}, \qquad (6.2.24)$$

which is graphically presented in Fig. 6.6.

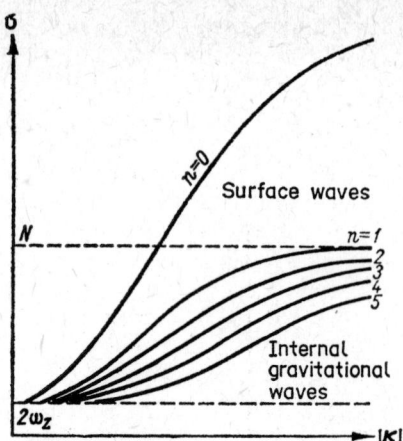

FIG. 6.6. Schematic representation of the eigencurves of problem (6.2.1)—(6.2.4) with $\varrho_0 = \text{const}$ and $N = \text{const}$. Numbers near the curves show the number of the nodes of the corresponding eigenfunctions.

Note that one of the peculiarities of internal gravitational waves is the fact that they are practically independent of the vertical displacements of the free surface of the ocean. This fact can be easily seen by comparing the orders of magnitudes of both terms of the first equality in equation (6.2.14). Thus, while investigating internal gravitational waves the free surface of the ocean can be interpreted as a flat lid subjected to no vertical deformations.

6.3. Vertical Structure of Internal Tidal Waves in a Realistically Stratified Ocean

Now we are going to dwell on the problem of the vertical structure of oscillations with an arbitrary vertical distribution of mean density in the ocean. Let us limit ourselves to the case when the internal gravitational waves are observed throughout the oceanic thickness. In the preceding section the frequency of such waves has been shown to be less than the minimal possible value of the Brunt–Väisälä frequency. This assumption is quite valid when applied to low-frequency internal tidal waves.

From now on it will be more convenient to operate with the equation for pressure perturbations. Therefore we reduce system (6.2.13) to a single equation for $\mathcal{P}$ taking into account at the same time that ϱ_0 and N are variable functions of the vertical coordinate and that the separation constant $\hbar$ is determined by equation (6.2.23). Thus, we have

$$\frac{d}{dz} \frac{\sigma^2 - 4\omega_z^2}{N^2 - \sigma^2} \frac{d\mathcal{P}}{dz} + \frac{N^2}{g} \frac{\sigma^2 - 4\omega_z^2}{N^2 - \sigma^2} \frac{d\mathcal{P}}{dz} + |\kappa|^2 \mathcal{P} = 0. \quad (6.3.1)$$

VERTICAL STRUCTURE OF INTERNAL TIDAL WAVES 263

Boundary conditions (6.2.14) will be rewritten as follows:

$$\mathcal{P} = 0 \quad \text{at} \quad z = D;$$

$$\frac{d\mathcal{P}}{dz} = 0 \quad \text{at} \quad z = 0. \tag{6.3.2}$$

Introduce the dimensionless variables choosing the Coriolis parameter $2\omega_z$ as a characteristic scale of frequency and the depth of the ocean D as a characteristic scale of length. Denote the dimensionless variables by the same symbols and reduce problem (6.3.1), (6.3.2) to the following form:

$$\frac{d}{dz}\frac{\sigma^2-1}{N^2-\sigma^2}\frac{d\mathcal{P}}{dz} + \mu N^2 \frac{\sigma^2-1}{N^2-\sigma^2}\frac{d\mathcal{P}}{dz} + |\kappa|^2\mathcal{P} = 0; \tag{6.3.3}$$

$$\mathcal{P} = 0 \quad \text{at} \quad z = 1$$

$$\frac{d\mathcal{P}}{dz} = 0 \quad \text{at} \quad z = 0, \tag{6.3.4}$$

where μ is the dimensionless parameter equal to $4\omega_z^2 D/g$.

Since equation (6.3.3) is not self-conjugate, make use of the equality

$$(\mathcal{P}^*, \mathfrak{L}\mathcal{P} + |\kappa|^2\mathcal{P}) = (\mathcal{P}, \mathfrak{L}^*\mathcal{P}^* + |\kappa|^2\mathcal{P}^*)$$

and introduce a Lagrangian conjugate problem for the function $\mathcal{P}^*$. Here $\mathfrak{L}$ is the differential operator of the principal problem (6.3.3), (6.3.4); $\mathfrak{L}^*$ is the conjugate operator with respect to $\mathcal{P}$; (,) is the scalar product which equals

$$\int_0^1 \mathcal{P}\mathcal{P}^* \, dz$$

for the conjugate functions $\mathcal{P}$ and $\mathcal{P}^*$ set within the change interval of z from 0 to 1.

Setting that the functions $\mathcal{P}$ and $\mathcal{P}^*$ are real affords the following conjugate problem:

$$\frac{d}{dz}\frac{\sigma^2-1}{N^2-\sigma^2}\frac{d\mathcal{P}^*}{dz} - \mu \frac{d}{dz} N^2 \frac{\sigma^2-1}{N^2-\sigma^2}\mathcal{P}^* + |\kappa|^2\mathcal{P} = 0; \tag{6.3.5}$$

$$\mathcal{P}^* = 0 \quad \text{at} \quad z = 1;$$

$$\frac{d\mathcal{P}^*}{dz} - \mu N^2 \mathcal{P}^* = 0 \quad \text{at} \quad z = 0. \tag{6.3.6}$$

Let us now investigate the spectrum of the eigenvalues of the principal and conjugate problems. For this purpose, multiply all the terms of equation (6.3.3) by the multiplier

$$\exp \mu \int_0^1 N^2 \, dz.$$

As a result, the equation is reduced to a self-conjugate one

$$\frac{d}{dz}\left[\frac{\sigma^2-1}{N^2-\sigma^2}\exp\left(\mu\int_0^1 N^2\,dz\right)\frac{d\mathcal{P}}{dz}\right] + |\kappa|^2\mathcal{P}\exp\mu\int_0^1 N^2\,dz = 0. \quad (6.3.7)$$

Since the coefficients of this equation are real and the coefficient at $\mathcal{P}$ does not change its sign throughout the range of z values from 0 to 1, then all the eigenvalues of the problem will be real. This conclusion, as noted earlier, is drawn from the very formulation of the problem which does not admit either energy sources or sinks and hence cannot have infinitely growing or damped solutions.

Let us demonstrate that the eigenvalues of the problem are positive. Indeed, the quadratic functional

$$\mathfrak{K}[\mathcal{P},\mathcal{P}] = -\int_0^1 \mathcal{P}\frac{d}{dz}\frac{\sigma^2-1}{N^2-\sigma^2}\exp\left(\mu\int_0^1 N^2\,dz\right)\frac{d\mathcal{P}}{dz}\,dz$$

according to conditions (6.3.4) equals

$$\int_0^1 \frac{\sigma^2-1}{N^2-\sigma^2}\exp\left(\mu\int_0^1 N^2\,dz\right)\left(\frac{d\mathcal{P}}{dz}\right)^2 dz.$$

Remember that for non-degenerate internal gravitational waves the coefficient $(\sigma^2-1)/(N^2-\sigma^2)$ throughout the z segment from 0 to 1 can be only positive and that the multiplier

$$\exp \mu \int_0^1 N^2 \, dz$$

is always greater than zero. Hence for any function which is not equal to the constant, $\mathfrak{K}[\mathcal{P},\mathcal{P}]$ is also greater than zero. Thus, all the eigenvalues of problem (6.3.3), (6.3.4) are positive.

Completeness of the system of the eigenfunctions for problem (6.3.3), (6.3.4), which is an ordinary Sturm–Liouville problem, does not require proof. To determine these eigenfunctions and the corresponding eigenvalues, it is necessary to specify the profile of the Brunt–Väisälä frequency. While solving problems (6.3.3), (6.3.4) and (6.3.5), (6.3.6), however, the use of any real dependence for $N(z)$ necessitates numerical methods. Let us consider one of them in detail.

Divide the interval $0 \leq z \leq 1$ into k segments, denoting the boundaries between the segments by the integers $0, 1, 2, \ldots, j, \ldots k$. Refer the level with the index "0" to $z = 0$, locating the level with index "k" at $z = 1$. Select also, within each of the segments, the mean level which will be denoted by a fractional index.

Introduce then the following notation:

$$I = s^2 \frac{d\mathcal{P}}{dz}, \qquad (6.3.8)$$

where $s^2 = (\sigma^2 - 1)/(N^2 - \sigma^2)$. Integrating equation (6.3.8) between the limits $(j-1, j)$ assuming that I and s^2 belong to the class of continuous functions we find that approximately

$$I_{j-1/2} = s^2_{j-1/2} \frac{\mathcal{P}_j - \mathcal{P}_{j-1}}{z_j - z_{j-1}}. \qquad (6.3.9)$$

Similarly,

$$I_{j+1/2} = s^2_{j+1/2} \frac{\mathcal{P}_{j+1} - \mathcal{P}_j}{z_{j+1} - z_j}. \qquad (6.3.10)$$

Integrate now the principal equation (6.3.3) from $z_{j-1/2}$ to $z_{j+1/2}$. Then, assuming that $\mathcal{P}$ is also a continuous function and taking into account equations (6.3.9), (6.3.10), we have

$$s^2_{j+1/2} \frac{\mathcal{P}_{j+1} - \mathcal{P}_j}{z_{j+1} - z_j} + s^2_{j-1/2} \frac{\mathcal{P}_j - \mathcal{P}_{j-1}}{z_j - z_{j-1}}$$
$$+ \mu(N^2 s^2)_j (\mathcal{P}_{j+1/2} - \mathcal{P}_{j-1/2}) + |\kappa|^2 \mathcal{P}_j (z_{j+1/2} - z_{j-1/2}) = 0.$$

The functions $\mathcal{P}$ with the fractional indices contained in the above equation can be excluded provided the following interpolation formula is used:

$$\mathcal{P}_{j+1/2} = \mathcal{P}_{j-1/2} + \frac{z_{j+1/2} - z_{j-1/2}}{z_{j+1} - z_{j-1}} (\mathcal{P}_{j+1} - \mathcal{P}_{j-1}).$$

This formula is found by expanding the function $\mathcal{P}$ in a Taylor series in the vicinity of the point with the fractional index.

Thus, we have a three-point difference equation

$$-a_j \mathcal{P}_{j+1} + b_j \mathcal{P}_j - c_j \mathcal{P}_{j-1} = |\kappa|^2 \mathcal{P}_j, \qquad (6.3.11)$$

where

$$a_j = \frac{1}{z_{j+1/2} - z_{j-1/2}} \left[\frac{s^2_{j+1/2}}{z_{j+1} - z_j} + \mu(N^2 s^2)_j \frac{z_{j+1/2} - z_{j-1/2}}{z_{j+1} - z_{j-1}} \right];$$

$$b_j = \frac{1}{z_{j+1/2} - z_{j-1/2}} \left[\frac{s^2_{j+1/2}}{z_{j+1} - z_j} + \frac{s^2_{j-1/2}}{z_j - z_{j-1}} \right];$$

$$c_j = \frac{1}{z_{j+1/2} - z_{j-1/2}} \left[\frac{s^2_{j-1/2}}{z_j - z_{j-1}} - \mu(N^2 s^2)_j \frac{z_{j+1/2} - z_{j-1/2}}{z_{j+1} - z_{j-1}} \right].$$

The system of difference equations (6.3.11) must be closed by taking account of the boundary conditions (6.3.4). The first of them can be satisfied if in the equation for the level $k-1$ we set $\mathcal{P}_k$ equal to zero. Then equation (6.3.11) for this level will be rewritten as follows:

$$b_{k-1} \mathcal{P}_{k-1} - c_{k-1} \mathcal{P}_{k-2} = |\kappa|^2 \mathcal{P}_{k-1}, \qquad (6.3.12)$$

where

$$b_{k-1} = \frac{1}{z_{k-1/2} - z_{k-3/2}} \left[\frac{s^2_{k-1/2}}{z_k - z_{k-1}} + \mu(N^2 s^2)_{k-1} \frac{z_{k-1/2} - z_{k-3/2}}{z_k - z_{k-2}} \right];$$

$$c_{k-1} = \frac{1}{z_{k-1/2} - z_{k-3/2}} \left[\frac{s^2_{k-3/2}}{z_{k-1} - z_{k-2}} - \mu(N^2 s^2)_{k-1} \frac{z_{k-1/2} - z_{k-3/2}}{z_k - z_{k-2}} \right].$$

To satisfy the second boundary condition (6.3.4), let us introduce the imaginary level $j = -1/2$ and integrate equation (6.3.3) between the limits $(z_{-1/2}, z_{1/2})$. Express then the value of the function $\mathcal{P}$ at the point with the index $j = -1/2$ through the value of this function at the point with index $j = 1/2$ and its derivative at the point with the index $j = 0$. In line with (6.3.4), the condition $d\mathcal{P}/dz|_0 = 0$ yields

$$-a_0 \mathcal{P}_1 + b_0 \mathcal{P}_0 = |\kappa|^2 \mathcal{P}_0, \qquad (6.3.13)$$

where

$$a_0 = b_0 = \frac{1}{z_{1/2} - z_{-1/2}} \frac{s^2_{1/2}}{z_1}.$$

Now the system of algebraic equations (6.3.11)–(6.3.13) is complete. Let us rewrite it in the matrix form

$$\Lambda \mathcal{P} = |\kappa|^2 \mathcal{P}, \qquad (6.3.14)$$

VERTICAL STRUCTURE OF INTERNAL TIDAL WAVES   267

where the coefficient matrix Λ takes the form of a Jacobian matrix:

$$\Lambda = \begin{bmatrix} b_0 & -a_0 & 0 & \ldots & 0 & 0 & 0 \\ -c_1 & b_1 & -a_1 & \ldots & 0 & 0 & 0 \\ \cdot & \cdot & \cdot & \cdot & \cdot & \cdot & \cdot \\ \cdot & \cdot & \cdot & \cdot & \cdot & \cdot & \cdot \\ \cdot & \cdot & \cdot & \cdot & \cdot & \cdot & \cdot \\ 0 & 0 & 0 & \ldots & -c_{k-1} & b_{k-2} & -a_{k-2} \\ 0 & 0 & 0 & \ldots & 0 & -c_{k-1} & b_{k-1} \end{bmatrix},$$

and $\mathcal{P}$ is the vector with the components $\{\mathcal{P}_j\}$.

If now we introduce the conjugate equation

$$\Lambda^* \mathcal{P}^* = |\kappa|^2 \mathcal{P}^*, \qquad (6.3.15)$$

where Λ^* is the transposed matrix, then the solution of equations (6.3.14), (6.3.15) affords the biorthogonal system of the eigenfunctions $\{\mathcal{P}_n\}$, $\{\mathcal{P}_n^*\}$ and the system of eigenvalues $|\kappa_n|^2$.

Let us make use of a typical distribution of the Brunt–Väisälä frequency in the ocean as presented in Fig. 6.7. The results of calculating the eigenvalues, as well as the eigenfunctions for the two values of the dimensionless frequency of oscillations are given in Figs. 6.8 and 6.9. The qualitative similarity of the eigencurves

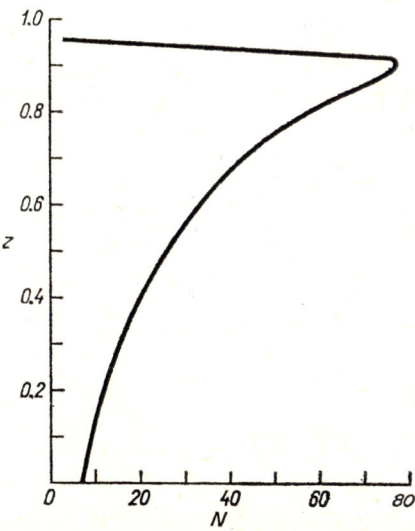

FIG. 6.7. Typical distribution of the dimensionless Brunt–Väisälä frequency in the ocean (from Eckart, 1960).

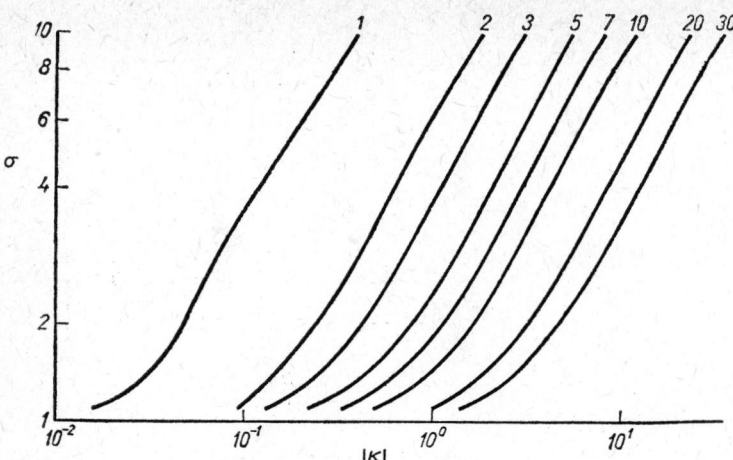

FIG. 6.8. Eigencurves of the problem on the vertical structure of internal tidal waves at a typical distribution of the Brunt-Väisälä frequency in the ocean. For notations see Fig. 6.6.

presented in Figs. 6.6 and 6.8 is evident. The first eigenfunction, describing the lowest internal mode, has the same sign throughout (see Fig. 6.9), which implies that the whole thickness of the ocean moves in the horizontal plane as a whole. The next eigenfunction changes its sign, indicating that here a change in the phase of the horizontal velocity takes place. With increasing mode-number the change of phase becomes more frequent and the vertical structure of the field of the horizontal velocity becomes more and more complex.

Having at our disposal a whole set of eigenfunctions $\{\mathcal{P}_n\}$ and positive eigenvalues $|\varkappa_n|^2$, we can find a solution of the primary system of equations (6.2.8)—(6.2.10), amplified by certain boundary conditions, in the form of the following expansion by eigenfunctions $\mathcal{P}_n$ and $d\mathcal{P}_n/dz$:

$$\bar{p} = \sum_n \Pi_n(\mathbf{x})\mathcal{P}_n(z);$$

$$\bar{w} = \sum_n W_n(\mathbf{x})\frac{d\mathcal{P}_n}{dz};$$

$$\bar{\mathbf{u}} = \sum_n \mathcal{U}_n(\mathbf{x})\mathcal{P}_n(z);$$

$$\bar{\varrho} = \sum_n P_n(\mathbf{x})\frac{d\mathcal{P}_n}{dz},$$

where the Fourier coefficients Π_n, W_n, $\mathcal{U}_n$ and P_n are determined either by the corresponding expressions obtained from equations (6.2.8)—(6.2.10) or by

VERTICAL STRUCTURE OF INTERNAL TIDAL WAVES 269

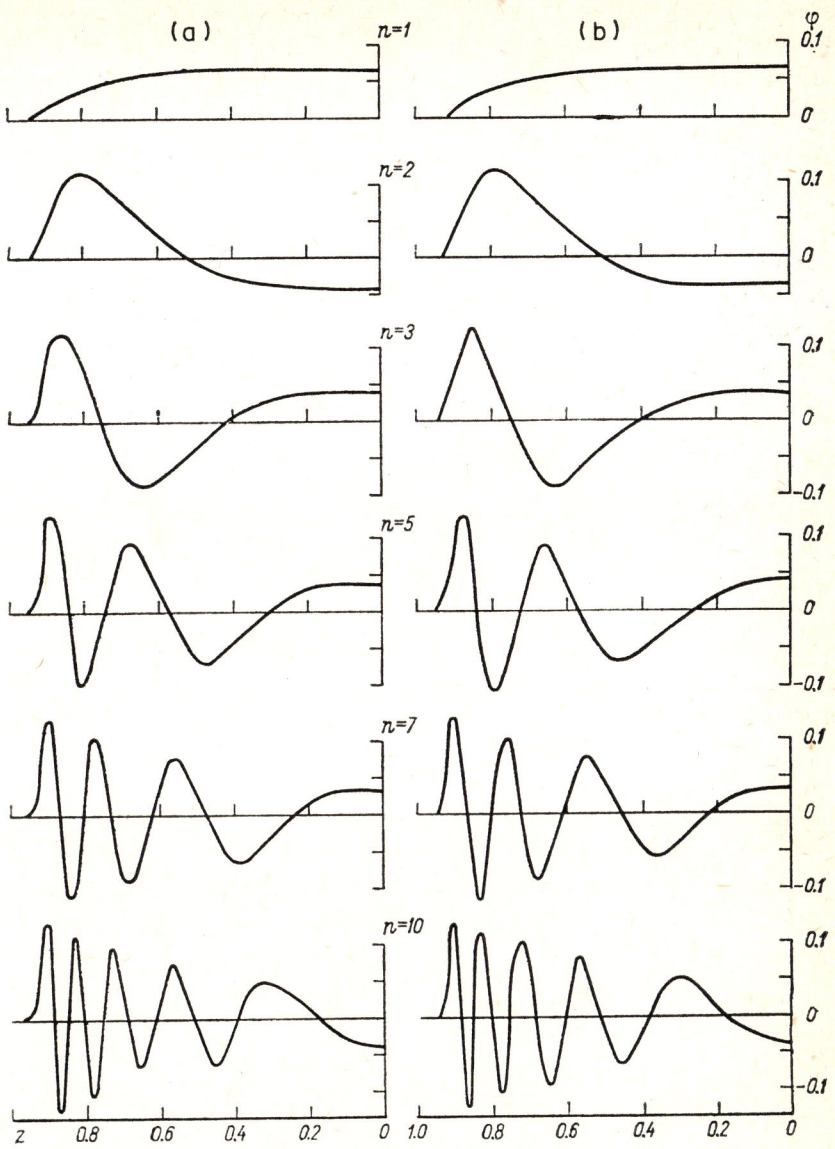

FIG. 6.9. Eigenfunctions of the pressure field. The dimensionless oscillation frequency equals 1.4 (a) and 5.6 (b).

use of observational data. In the latter case, when the measurements are carried out at several levels, the Fourier coefficients are calculated by the method of least squares.

An example of analogous calculations of harmonic constants of the tidal velocity for the M_2 wave in the North Atlantics is presented in Fig. 6.10. On the left can be seen the first four modes of oscillations (including the barotropic mode, $n = 0$), on the right the result of their superposition is shown. Figure 6.10 also presents observational data. One can see that in order to reproduce the basic regularities of the vertical structure of the velocity field, it is sufficient to use only four eigenfunctions.

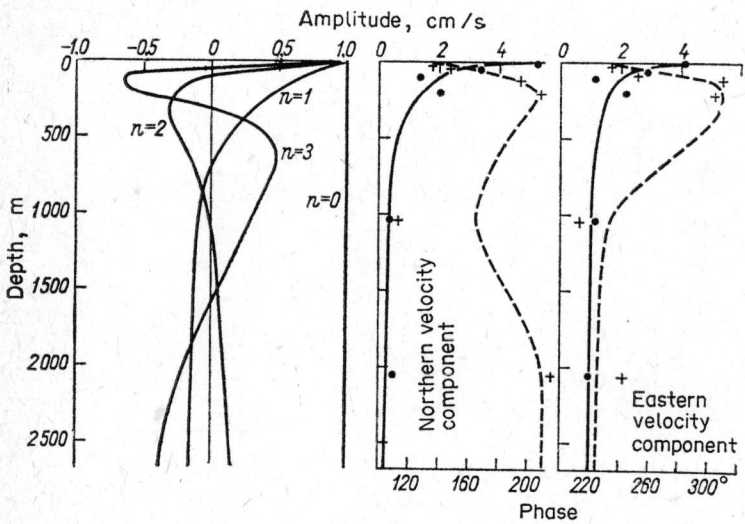

FIG. 6.10. Eigenfunctions of the tidal velocity (the M_2 wave) and the result of their superposition (by Magaard and McKee, 1973).
Solid lines (figures in the centre and to the right) represent the amplitudes, dashed lines are the phases; solid circles and crosses are the observational data of the corresponding characteristics at the station "D" (39°10′N, 70°W).

6.4. References

A detailed review of the analytical solutions of the problem on the vertical structure of internal gravitational waves for various density models is given in the monograph by Krauss (1966) (see also Ivanov, 1964). Phillips' estimation of the above solutions was borrowed from his book (Phillips, 1966).

Measurements of the internal waves carried out from research vessels are discussed in the book by Defant (1961). The analysis of observational data can be found in the above-mentioned books by Defant and Krauss and in the review by La Fond (1962).

The hypothesis of the internal waves generation by the motion of a stratified fluid over an irregular bottom was put forward by Zeilon (1912). Later he proved experimentally that the mechanism of generation of internal waves with the tidal period in the ocean can be similar (Zeilon, 1934). A simple model of this phenomenon was elaborated by Rattray (1960) and later modified by Ichije (1963). The papers by Long (1953, 1955) and Yih (1960 (a, b)) are directly related to the discussed hypothesis.

The second local hypothesis attributing the existence of internal tidal waves in the ocean to their resonance interaction with the tide-generating forces in the region of the "critical" latitudes was simultaneously suggested by Defant (1950) and Haurwitz (1950).

It was Pettersson (1930, 1933a, 1934, 1935) who suggested that the internal tidal waves in the ocean are generated by the direct effect of the vertical component of the tide-generating forces. The idea was criticized in the paper by Defant (1932). The effect of the horizontal component of the tide-generating forces on the internal tidal waves was also discussed by Rzhonsnitskij and Fux (1965).

The problem of free waves in a two-layer fluid, as has been noted in the text, is a classical one. While discussing it we have made use of the monograph by Kagan (1968).

The first reports that the water temperature oscillations induced by the internal waves have random phases can be found in the papers by Haurwitz, Stommel and Munk (1959) and La Fond (1962). Radok, Munk and Isaaks (1967) ascribed this result to the influence of the variability in time of the density structure and velocity field in the ocean. In particular, they obtained the coherence, characterizing the degree of constancy of the phase difference between the temperature oscillations at a depth of 150 m and the barotropic tide, was 0.3 for the harmonics M_2 of the tide-generating forces, the maximum value of the coherence being equal to unity, the minimum to zero.

A qualitative analysis of the equations for internal waves, written with an assumption that the Brunt–Väisälä frequency is constant with depth, is given, in particular, in the books by Eckart (1960) and Kamenkovitch (1973). In the main, we have followed the latter work.

The method of calculating the eigenvalues and eigenfunctions at an arbitrary profile of the Brunt–Väisälä frequency discussed here was suggested in the paper by Marchuk and Kagan (1970). The above method as applied to high-frequency internal waves was developed by Kuzin (1972). The properties of progressive internal waves propagating in a realistically stratified ocean were also investigated by Krupnin (1968) and Odulo (1971). The completeness of the system of eigenfunctions of the Sturm–Liouville problem was demonstrated, for instance, in the book by Kollatz (1963).

The solution to the problem of the vertical structure of internal gravitational waves at an arbitrary vertical distribution of the mean density was first obtained by Fjeldstad (1933) (see also Fjeldstad, 1963), the so-called method of the Cauchy problem having been used in this case to determine the eigenfunctions.

Bibliography

BAINES, P. G. (1979) The generation of internal tides by flat-bump topography. *Deep-Sea Res.* **20**, pp. 179–205.
BARTELS, J. (1967) Gezeitenkräfte. *Handbuch der Physik*, **48**, Springer-Verlag, Berlin, pp. 734–774.
BEREZKIN, V. A. (1947) *Sea Dynamics*. Gidrometeoizdat, Moscow–Leningrad, 683 pp.
BOGDANOV, K. T. (1962) Tides of the Pacific Ocean. *Trudy Inst. Okeanol.* **60**, pp. 142–160.
BOGDANOV, K. T. (1966) Distribution of tidal waves in the Southern Ocean. In *Antarctica*. Nauka, Moscow, pp. 65–72.
BOGDANOV, K. T. Kim, K. V. and Magarick, V. A. (1964) Numerical solution of the tidal hydrodynamical equations for the Pacific Ocean by the electronic computing machine BECM-2. *Trudy Inst. Okeanol.*, **75**, pp. 73–98.
BOGDANOV, K. T. and Magarick, V. A. (1967) Numerical solution of the problem of distribution of the semidiurnal tidal waves (M_2 and S_2) in the World Ocean. *Dokl. AN SSSR,* **172**, pp. 1315–1317.
BOWDEN, K. F. (1955) Some observations of turbulence near the sea bed in a tidal current. *Quart. J. Roy. Met. Soc.* **81**, pp. 640–641.
BOWDEN, K. F. (1962) Measurements of turbulence near the sea bed in a tidal current. *J. Geophys. Res.* **67**, pp. 3181–3186.
BOWDEN, K. F. (1965) Turbulence. In *The Sea*, I, Ed. M. N. Hill. Intersci. Publ.
BOWDEN K. F. and PROUDMAN, J. (1949) Observations of the turbulent fluctuations of a tidal current. *Proc.Roy.Soc. London*, A **199**, pp. 311–327.
BOWDEN, K. F. and FAIRBAIRN, L. A. (1952a) A determination of the frictional forces in a tidal current. *Proc. Roy. Soc. London*, A **214**, pp. 371–392.
BOWDEN, K. F. and FAIRBAIRN, L. A. (1952b) Further observations of the turbulent fluctuations in a tidal current. *Philos. Trans. Roy. Soc. London*, A **244**, pp. 335–392.
BOWDEN, K. F. and FAIRBAIRN, L. A. (1956) Measurements of the turbulent fluctuations and Reynold stresses in a tidal current. *Proc. Roy. Soc. London*, A **237**, pp. 422–438.
BOWDEN, K. F., FAIRBAIRN, L. and HUGHES, P. (1959) The distribution of shearing stresses in a tidal current. *Geophys. J. Roy. Astron. Soc.*, **2**, pp. 288–305.
BOWDEN, K. F. and HOWE, M. R. (1963) Observations of turbulence in a tidal current. *J. Fluid Mech.* **17**, pp. 271–284.
BOWLES, P., BURNS, R. H., HUDSWELL, F. and WHIPPLE, R. T. P. (1958) Sea disposal of low activity of effluent. *Proc. 2nd Int. Conf. Peaceful Uses Atom Energy*, **18**, pp. 376–389.
BRADSHAW, P. (1962) The turbulent structure of equilibrium boundary layers. *J. Fluid Mech.*, **29**, pp. 625–645.
CARLSEN, N. A. (1967) Measurements in the turbulent wave boundary layer. *Basic Res., Progress Rep.*, No. 14, Coastal Eng. Lab., Tech. Univ. of Denmark.
CARTWRIGHT, D. E. (1971) Tides and waves in the vicinity of Saint Helena. *Philos. Trans. Roy. Soc. London*, A **270**, pp. 603–646.
CHARNOCK, H. (1959) Tidal friction from currents near the sea bed. *Geophys. J. Roy. Astron. Soc.*, **2**, pp. 215–221.
COLLINS, J. I. (1963) Inception of turbulence at the bed under periodic gravity waves. *J. Geophys. Res.*, **68**, pp. 6007–6014.

Cox, G. S. and Sandstrom, H. (1962) Coupling of internal and surface waves in water of variable depth. *J. Ocean. Soc. Japan*, 20th Ann. vol., pp. 499–513.
Darwin, G. H. (1886) On the correction to the equilibrium theory of tides for the continents. *Proc. Roy. Soc. London*, XL April, pp. 303–315.
Darwin, G. H. (1898) *The Tides and Kindered Phenomena in the Solar System*. London.
Darwin, G. H. (1902) A new theory of the tides of terrestrial oceans. *Nature*, **66**, pp. 444–445.
Davenport, A. G. (1961) The spectrum of horizontal gustiness near the ground in high winds. *Quart. J. Roy. Met. Soc.* **87**, pp. 194–211.
Defant, A. (1918) Neue Methode zur Ermittlung der Eigenschwingungen (Seiches) von abgeschlossenen Wassermassen (Seen, Buchten usw). *Ann. Hydr. Mar. Met.* **46**, 78–85.
Defant, A. (1924) Die Gezeiten des Atlantischen Ozeans und des Arktischen Meers. *Ann. Hydr. (Berlin)*, **52**, 153–166, 177–184.
Defant, A. (1932) Die Gezeiten und inneren Gezeitenwellen des Atlantischen Ozeans. *Wiss Ergebn. Deut. Atl. Exp. "Meteor" 1925–1927*, **7**, 1–318.
Defant, A. (1950) On the origin of internal tide waves in the open sea. *J. Mar. Res.* **9**, pp. 111–119.
Defant, A. (1961) *Physical Oceanography*, 2. Pergamon Press, p. 598.
Dietrich, G. (1944) Die Schwingungssysteme der half- und eintägigen Tiden in den Ozeanen. *Veröff. Inst. Meeresk. Univ. Berlin*, A **41**, pp. 1–68.
Doodson, A. T. (1958) Ocean tides. *Advances in Geophys.* **5**, pp. 117–152.
Dronkers, J. J. (1964) *Tidal Computations in Rivers and Coastal Waters*.
Duvanin, A. I. (1960) *Tides in the Sea*. Gidrometeoizdat, Leningrad, 389 pp.
Dvorkin, E. N., Kagan, B. A. and Kleshcheva, G. P. (1972) Calculation of tidal motions in the Arctic seas. *Izv. AN SSSR, Fizika atmosfery i okeana*, **8**, pp. 298–306.
Dyer, K. R. (1972) Bed shear stresses and sedimentation of sandy gravels. *Marine Geol.* **13**, pp. 31–36.
Eckart, C. (1960) *Hydrodynamics of Oceans and Atmospheres*. Pergamon Press.
Farrel, W. E. (1973) Earth tides, ocean tides and tidal loading. *Phil. Trans. Roy. Soc. London*, A **274**, pp. 253–259.
Favre, A., Gaviglio, J. and Dumas, R. (1958) Further space-time correlations of velocity in a turbulent boundary layer. *J. Fluid Mech.*, **3**, pp. 344–356.
Ferrel, W. O. (1874) *Tidal Researches*. London, p. 237.
Filloux, J. H. (1969) Bourdon tube deep-sea tide gages. *Proc. Symp. on Tsunamis and Tsunami Res.*, Honolulu.
Fjeldstad, J. (1929) Contribution to the dynamics of free progressive tidal waves. *Sci. Res. Norweg. North. Polar. Exped. "Maud", 1918–1925*, **4**, No. 3.
Fjeldstad, J. (1933) Interne Wellen. *Geofys. Publik*, **10**, No. 6, pp. 1–35.
Fjeldstad, J. (1963) Internal waves of tidal origin. I. Theory and analysis of observations. *Geofys. Publik*, **25**, No. 5, pp. 1–73.
Godunov, S. K. and Rjaben'kij, V. S. (1962) *Introduction to the Theory of Difference Scheme*. Fizmatgiz, Moscow, 340 pp.
Goldsbrough, G. R. (1927) The tides in oceans on a rotating globe, I. *Proc. Roy. Soc. London*, A **117**, pp. 692–718.
Gordeev, R. G. (1924) Numerical experiments on tidal dynamics in the World Ocean. Candidate thesis, Institute of Oceanology of the USSR Academy of Sciences, Moscow, 17 pp.
Gordeev, R. G., Kagan, B. A. and Rivkind, V. Ja. (1973) Numerical solution of the equations of tidal dynamics in the World Ocean. *Dokl. AN SSSR*, **209**, pp. 340–343.
Gordeev, R. G., Kagan, B. A. and Rivkind, V. Ja. (1924) Estimation of the rate of tidal energy dissipation in the World Ocean. *Okeanologiya*, **14**, pp. 226–229.
Gordeev, R. G., Kagan, B. A. and Rivkind, V. Ja. (1975) Numerical experiments on tidal dynamics in the World Ocean. *Izv. Akad. Nauk SSSR, Fizika atmosfery i okeana*, II, pp. 162–174.

BIBLIOGRAPHY

GRACE, S. F. (1929) Internal friction in certain tidal currents. *Proc. Roy. Soc. London*, A **124**, pp. 150–162.
GRACE, S. F. (1936) Friction in the tidal currents of the Bristol Channel. *Mon. Not. Roy. Astron. Soc., Geophys. Suppl.* **3**, p. 388.
GRACE, S. F. (1937) Friction in the tidal currents of the English Channel. *Mon. Not. Roy. Astron. Soc., Geophys. Suppl.* **4**, p. 133.
GREEN, A. W. (1968) An experimental study of the interactions between non-steady Ekman layers and an annular vortex. Doct. Diss., Dept. Meteorol., Mass. Inst. Techn.
HARRIS, R. A. (1904) Manual of tides, p. IV B: *Cotidal Lines of the World*. U. S. Coast and Geod. Survey Rep. App. No. 5, pp. 315–400.
HAURWITZ, B., (1950) Internal waves of tidal character. *Trans. Amer. Geophys. Un.* **31**, pp. 47–52.
HAURWITZ, B., STOMMEL, H. and MUNK, W. (1959) On the thermal unrest in the ocean. In *The Atmosphere and the Sea in Motion*, Ed. B. Bolin. Rockefeller Inst. Press, N. Y., pp. 74–94.
HEAPS, N. S. (1969a) A two-dimensional numerical sea model. *Phil. Trans. Roy. Soc. London*, A **265**, pp. 93–137.
HEAPS, N. S. (1969b) Some notes on tidal theory and its possible relevance to a program of deep-sea tidal measurements. *Dt. Hydrorg. Zs.* **22**, 11–25.
HELLAND-HANSEN, B. and NANSEN, F. (1909) The Norwegian Sea. *Rep. Norweg. Fish and Mar. Invest.* **2**, No. I, 2.
HENDERSHOTT, M. C. (1972) The effects of solid Earth deformation on global ocean tides. *Geophys. J. Roy. Astron. Soc.* **29**, pp. 389–402.
HENDERSHOTT, M. C. (1973) Ocean tides. *Trans. Amer. Geophys. Un.* **54**, pp. 76–86.
HENDERSHOTT, M. C. and MUNK, W. (1970) Tides. *Ann. Rev. of Fluid Mech.* **2**, pp. 205–224.
HANSEN, W. (1952) Gezeiten und Gezeitenströme der halb-tägigen Hauptmondtide M_2 in der Nordsee. *Dt. Hydrogr. Zs., Ergänzungsheft*, pp. 1–46.
HANSEN, W. (1956) Theorie zur Errechnung des Wasserstandes und der Strömungen in Randmeeren nebst Anwendungen. *Tellus*, **8**, pp. 287–300.
HANSEN, W. (1962) Hydrodynamical methods applied to oceanographic problems. *Proc. Symp. Mathem. Hydrodyn. Methods of Phys. Oceanogr., Inst. Meereskunde Univ. Hamburg*, pp. 25–34.
HANSEN, W. (1966) The reproduction of the motion in the sea by means of Hydrodynamical Numerical Methods. *Mitt. Inst. Meerskunde Univ. Hamburg*, No. 5, pp. 1–57.
HOUGH, S. S. (1892) On the application of harmonic analysis to the dynamical theory of the tides. *Trans. Roy. Soc. London*, A **189**, pp. 201–257.
HOUGH, S. S. (1899) *Ibid.*, A **191**, pp. 139–185.
ICHIJE, T. (1955) On the friction in the tidal current. *The Oceanogr. Mag.* **7**, pp. 55–77.
Internal waves over a continental shelf. Florida State Univ. Oceanogr. Inst. Tech. Rep., No. 3 (1963).
IRISH, J., MUNK, W. and SNODGRASS, F. (1971) M_2-amphidrome in the North-east Pacific. *Geophys. Fluid Dyn.* **2**, pp. 355–360.
IVANOV, A. A. (Edit.) (1964) *Internal Waves*. Sb. perevodov, Mir, Moscow, 300 pp.
JEFFREYS, H. (1970) *The Earth, its Origin, History and Physical Condition*, 5th ed., Cambridge Univ. Press, p. 525.
JONSSON, I. G. (1963) Measurements in the turbulent wave boundary layer. *10th Congress, I. A. H. R., London*, I, pp. 85–92.
KAGAN, B. A. (1964) On the profile of the longitudinal component of the tidal current in a deep canal. *Okeanologiya*, **4**, pp. 778–787.
KAGAN, B. A. (1968) On the structure of the tidal flow in the sea. *Izv. AN SSSR, Fizika atmosfery i okeana*, **2**, pp. 956–969 (1966).
KAGAN, B. A. (1968) *Hydrodynamic Models of the Tidal Motions in the Sea*. Gidrometeoizdat, Leningrad, 219 pp.

KAGAN, B. A. (1970) On the properties of some difference schemes employed at numerical integration of equations of tidal dynamics. *Izv. AN SSSR, Fizika atmosfery i okeana*, **6**, pp. 704–717.
KAGAN, B. A. (1971) On bottom friction in a one-dimensional tidal flow. *Izv. AN SSSR, Fizika atmosfery i okeana*, **7**, pp. 1190–1200.
KAGAN, B. A. (1972) On the resistance law in a tidal flow. *Izv. AN SSSR, Fizika atmosfery okeana*, **8**, pp. 533–542.
KAJIURA, K. (1968) A model of the bottom boundary layer in water waves. *Bull. Earthquake Res. Inst.* **46**, pp. 75–123.
KAMENKOVITCH, V. M. (1967) On the problem of the coefficients of turbulent diffusion and viscosity at large-scale motions of the ocean and atmosphere. *Izv. AN SSSR, Fizika atmosfery i okeana*, **3**, pp. 1326–1333.
KAMENKOVITCH, V. M. (1973) *Fundamentals of Ocean Dynamics*. Gidrometeoizdat, Leningrad, 240 pp.
KAZANSKY, A. B. and MONIN, A. S. (1961) On the dynamic interaction between the atmosphere and the earth's surface. *Izv. An SSSR, ser. geofiz.*, No. 5, pp. 786–788.
KEMPFF, K. N. (1968) A difference method for large-scale dynamical problems in oceanography. *Proc. Symp. Mathem.—Hydrodyn. Investigations of Phys. Processes in the Sea, Mitt. Inst. Meereskunde Univ. Hamburg*, No. 10, pp. 50–54.
KIRWAN, A. D. (1969) Formulation of constitutive equations for largescale turbulent mixing. *J. Geophys. Res.* **74**, pp. 6953–6959.
KITAJGORODSKY, S. A. (1970) *Physics of Interaction between the Atmosphere and Ocean*. Gidrometeoizdat, Leningrad, 281 pp.
KOLESNIKOVA, V. N. and MONIN, A. S. (1965) On the oscillation spectra of meteorological fields. *Izv. AN SSSR, Fizika atmosfery i okeana*, **1**, pp. 653–669.
KOLLATZ, L. (1969) *Eigenwertaufgaben mit technischen Anwendungen*. Akad. Verlag.
KRAUSS, W. (1980), *Interne Wellen*. Gebrüder Borntraeger, Berlin.
KRUPNIN, V. D. (1969) On a property of internal waves. *Akustich. Zurnal*, **15**, issue 1, pp. 83–91.
KRÜMMEL, O. (1911) *Handbuch der Ozeanographie*, 2, Stuttgart: J. Engelhorn, p. 766.
KUZIN, V. I. (1972) On the problem of gravitational internal waves in a realistically stratified ocean, In: *Numerical Models of Oceanic Circulation*, Computing Center of the Siberian Branch of the USSR Academy of Sciences, Novosibirsk, pp. 123–146.
LADYZHENSKAJA, O. A. (1970) *Mathematical Problems of Dynamics of the Viscous Incompressible Liquid*. Nauka, Moscow, 288 pp.
LAMB, H. (1925) *Hydrodynamics*, Dover Publications, N. Y.
LASKA, M. (1971) Numerical sea model investigations for barrage studies. *Archiwum Hydrotechniki*, **18**, No. 1, pp. 3–47.
LA FOND, E. C. (1941) The use of bathythermograms to determine ocean currents. *Trans. Amer. Geophys. Un.* **30**, pp. 231–237.
LA FOND, E. C. (1962) Internal waves. in: *The Sea*, I, Ed. M. N. Hill. Intersci. Publ.
LESSER, R. M. (1951) Some observations of the velocity profile near the sea floor. *Trans. Amer. Geophys. Un.* **32**, pp. 207–211.
LI, H. (1954) Stability of oscillatory laminar flow along a wall. Tech. Mem., Beach Erosion Board, No. 47.
LINEJKIN, P. S. (1937) On the theory of tides in basins and canals. *Zurnal geofiziki*, **7**, issue 1, pp. 5–47.
LONG, R. R. (1953) Some aspects of the flow of stratified fluids. I. A theoretical investigation. *Tellus*, **5**, pp. 42–57.
LONG, R. R. (1955) Some aspects of the flow of stratified fluids. III. Continuous density gradients. *Tellus*, **7**, pp. 342–357.
LONGMAN J. M. (1963) A Green's function for determining the deformation of the Earth

under surface mass loads. 2. Computations and numerical results. *J. Geophys. Res.* **68**, pp. 485–496.

LOVE, A. E. H. (1927) *A Treatise on the Mathematical Theory of Elasticity.* Cambridge Univ. Press.

MAGAARD, L. and MCKEE W. D. (1973) Semi-diurnal tidal currents at "site D". *Deep-Sea Res.* **20**, pp. 997–1009.

MARCHUK, G. I. (1961) *Methods of Nuclear Reactor Calculations.* Gosatomizdat, Moscow, 667 pp.

MARCHUK, G. I. (1969) On numerical solution of the Poincaré problem for oceanic circulations. *Dokl. AN SSSR*, **185**, pp. 1041–1044.

MARCHUK, G. I. (1973) *Methods of Digital Mathematics.* Nauka, Novosibirsk, 352 pp.

MARCHUK, G. I. (1974) Conjugate equations of tidal dynamics and the perturbation theory. Preprint, Computing Center of the Siberian Branch of the USSR Academy of Sciences, Novosibirsk, 18 pp.

MARCHUK, G. I., GORDEEV, R. G., KAGAN, B. A. and RIVKIND V. J. (1972) A numerical method of the solution of tidal dynamics equations and the results of its experimental verification. Preprint, Computing Center of the Siberian Branch of the USSR Academy of Sciences, Novosibirsk, 78 pp.

MARCHUK, G. I., GORDEEV, R. G., KAGAN, B. A. and RIVKIND, V. J. (1973) A numerical method for the solution of tidal dynamics equations and the results of its application. *J. Comput. Physics*, **13**, pp. 15–34.

MARCHUK, G. I. and KAGAN, B. A. (1970) Internal gravitational waves in a really stratified ocean. *Izv. Akad. Nauk SSSR, Fizika atmosfery i okeana*, **6**, pp. 412–422.

MARCHUK, G. I., KAGAN, B. A. and TAMSALU, R. E. (1969) Numerical method of tidal motion calculations in the marginal seas. *Izv. AN SSSR, Fizika atmosfery i okeana*, **5**, pp. 694–703.

MARMER, H. A. (1932) Tides and tidal currents. *Bull. Nat. Res. Council*, No. 85, *Phys. of the Earth*, **5**, pp. 229–309.

MOSBY, H. (1947) Experiments on turbulence and friction near the bottom of the sea. *Bergens Mus. Årbok, Naturvitens. rekke*, No. 3, pp. 1–6.

MOSBY, H. (1949) Experiments on bottom friction. *Bergens Mus. Årbok, Naturvitens. rekke*, No. 10, pp. 1–12.

MUNK, W. H. (1966) Abyssal recipes. *Deep-Sea Res.* **13**, pp. 707–730.

MUNK, W. H. and MCDONALD, G. T. F. (1960) *The Rotation of the Earth. A Geophysical Discussion.* Cambridge Univ. Press.

MUNK, W., Snodgrass, F. and WIMBUSH, M. (1970) Tides off-shore: transition from California coastal to deep-sea waters. *Geophys. Fluid Dyn.* **1**, pp. 161–235.

NAN'NITI, T. (1956) On the structure of ocean currents. II. The turbulent fluctuations in a tidal current. *Papers in Met. and Geophys.* **7**, No. 2, pp. 161–170.

NANSEN, F. (1902) *Oceanography of the North Polar Basin. The Norwegian North Polar Expedition, 1893–96*, 3.

NEKRASOV, A. V. (1973) On the reflection of tidal waves from the shelf zone. *Okeanologiya*, **13**, pp. 210–215.

NIKURADZE, J. (1953) Strömungsgesetze in rauhen Röhren. *Verhandl. Deut. Ing. Forsch.*, No. 361, pp. 1–22.

ODULO, A. B. (1971) Vertical structure of planetary waves in a stratified ocean, in: *Morskije geofizicheskije issledovanija*, No. 6, MGI AN USSR, Sevastopol, pp. 82–94.

PEKERIS, C. L. and Accad, Y. (1962) Solution of Laplace's equations for the M_2-tide in the World Ocean (abstract). *Abstracts of Papers Int. Ass. Phys. Oceanogr., XIV Gen. Assem., IUGG,* **5**, p. 85

PEKERIS, C. L. and Accad, Y. (1969) Solution of Laplace's equations for the M_2-tide in the World Ocean. *Philos. Trans. Roy. Soc. London*, A **265**, pp. 413–436.

PETTERSSON, O. (1907) *Strumstudien vid Osterjons portar.* Svenska hydrografisk, Biologiska—Komm., 3.

PETTERSSON, O. (1930) The tidal force. *Geogr. Ann.*, Book 4, Stockholm, pp. 261–262.
PETTERSSON, O. (1933a) Interne Gezeitenwellen. *Rapp. Cons. Expl. Mer., Copenhagen*, **82**, pp. 1–26.
PETTERSSON, O. (1933b) Tiddvattnets Problem, I—IV. *Arkiv. Math., Astr., Fys., Stockholm*, Bd. 23 A, No. 23, pp. 1–9.
PETTERSSON, O. (1934) *Ibid.*, Bd. 24 A, No. 16, pp. 1–30, No 17, pp. 1–16.
PETTERSSON, O. (1935) *Ibid.*, Bd. 25 A, No. 1, pp. I–II.
PHILLIPS, O. M. (1966) *The Dynamics of the Upper Ocean.* Cambridge Univ. Press.
PLATZMAN, G. W. (1971) Ocean tides and related waves. *Lectures in Applied Mathematics*, 14. *Mathematical Problems in the Geophysical Sciences*, pt. 2. Ed. W. H. Reid, pp. 239–291.
PLATZMAN, G. W. (1972) North Atlantic Ocean: Preliminary description of normal modes. *Science*, **178**, pp. 156–157.
PNUELI, A. and PEKERIS, C. L. (1968) Free tidal oscillations in rotating flat basins of the form of rectangles and sectors of circles. *Phil. Trans. Roy. Soc. London*, A **263**, pp. 149–171.
POINCARÉ, H. (1895) Sur l'equilibre et les mouvements des mers. *J. Math. Pures Appl.* 1, No. 2, pp. 57–102.
PROUDMAN, J. (1923) Summary of the meeting for the discussion of geophysical subjects, 1923, Nov. 5. *Observatory* (a monthly review of astronomy), **46**, pp. 368–372.
PROUDMAN, J. (1941) The effect of coastal friction on the tides. *Mon. Not. Roy. Astron. Soc., Geophys. Suppl.* **5**, pp. 23–26.
PROUDMAN, J. (1944) The tides of the Atlantic Ocean. *Mon. Not. Roy. Astron. Soc.* **104**, pp. 244–256.
PROUDMAN, J. (1953) *Dynamical Oceanography.* Methuen, London, Wiley, N.Y.
PRÜFER, G. (1939) Die Gezeiten des Indischen Ozean. *Veröffentl. Inst. Meeresk. Univ. Berlin*, A 37, 1–56.
RADOK, R., MUNK, W. and ISAACS, Y. (1967) A note on mid-ocean internal tides. *Deep-Sea Res.* **14**, pp. 121–124.
RATTRAY, M. (1960) On the coastal generation of internal tides. *Tellus*, **12**, pp. 54–62.
RICHARDSON, L. F. (1922) *Weather Prediction by Numerical Process.* Cambridge Univ. Press, p. 236.
RZHONSITCHKIJ, V. B. and FUX, V. R. (1965) On the origin of internal tidal waves. In *Studies of Tidal Phenomena in a Non-homogenous Sea.* Gidrometeoizdat, Leningrad, pp. 148–157.
SEE, T. J. J. (1927) New dynamical wave-theory of the tides. Hydrorg. Office U. S. Navy, No. 207, Washington
SRETENSKIJ, L. N. (1936) *A Theory of Wave Motions of Liquids.* ONTI, Moscow–Leningrad, 303 pp.
STERNBERG, R. W. (1966) Boundary layer observation in a tidal current. *J. Geophys. Res.* **71**, pp. 2175–2178.
STERNBERG, R. W. (1968) Friction factors in tidal channels with differing bed roughness. *Marine Geol.* **6**, pp. 243–260.
STERNBERG, R. W. (1970) Field measurements of the hydrodynamic roughness of the deep-sea boundary. *Deep-Sea Res.* **17**, pp. 413–420.
STERNECK, R. (1920) Die Gezeiten der Ozeans. *Sitzber. Akad. Wiss. Wien*, **129**, 131–150.
STOKER, J. J. (1957) *Water Waves. The Mathematical Theory with Applications.* N.Y.–London, Intersci. Publ.
SUMMERFIELD, W. (1972) Circular islands as resonators of long-wave energy. *Phil. Trans. Roy. Soc. London*, A **272**, pp. 361–402.
SVERDRUP, H. U. (1926) Dynamics of tides on the North Siberian shelf. *Geophys. Publik.* **4**, No. 5, pp. 1–75.
SVERDRUP, H. U., JOHNSON, M. W. and FLEMING, R. H. (1942) *The Oceans, their Physics, Chemistry and General Biology.* N.Y., p. 1087.

TATRO, P. R. and MOLLO-CHRISTIANSEN, E. L. (1967) Experiments on Ekman layer instability. *J. Fluid Mech.* **28**, pp. 531–543.

TAYLOR, G. I. (1918) Tidal friction in the Irish Sea. *Phil. Trans. Roy. Soc. London*, A **220**, pp. 1–33.

TAYLOR, R. J. (1958) Thermal structures in the lowest layers of the atmosphere. *Australian J. Phys.* II, pp. 168–176.

THOMSON, W. and TAIT, P. G. (1883) *Treatise of Natural Philosophy*, I, pt. 2, p. 527.

THORADE, H. (1931) Probleme der Wasserwellen. *Probl. Kosm. Phys.* **13–14**, pp. 219.

TIRON, K. D., SERGEEV, YU. N. and MICHURIN, A. N. (1967) The tidal chart of the Pacific, Atlantic and Indian Oceans. *Vestnik LGU*, No. 24, pp. 123–135.

TITOV, V. B. and SHESTERIKOV, N. P. (1964) The distribution and character of the tidal wave in the South Ocean. *Bull. Sov. Antarctich. ekspeditchii*, No. 47, pp. 35–39.

UENO, T. (1964) Theoretical studies on tidal waves travelling over the rotating globe. *Oceanogr. Mag.* **15**, pp. 99–111, **16**, pp. 47–51.

VAGER, B. G. and KAGAN, B. A. (1969) Dynamics of the turbulent boundary layer in a tidal current. *Izv. Akad. Nauk SSSR, Fizika atmosfery i okeana*, **5**, pp. 168–179.

VAGER, B. G. and KAGAN, B. A. (1971) Vertical structure and turbulence regime in a stratified boundary layer of a tidal current. *Izv. Akad. Nauk SSSR, Fizika atmosfery i okeana*, **7**, pp. 766–777.

VAN DER HOVEN, J. (1957) Power spectrum of horizontal wind in the frequency range from 0.0007 to 900 cycles per hour. *J. Meteorol.* **14**, pp. 160–164.

VAPNYAR, D. U. (1962) The method and some results of calculating friction characteristics from high and low tide observations. *Izv. Akad. Nauk SSSR, ser. geofiz.*, No. 12, pp. 1804–1814.

VILLAIN, G. (1952) Cartes des lignes cotidales dans les oceans. *Ann. Hydrorg. (Paris)*, **4**, pp. 269–388.

VINCENT, G. E. (1957) Contribution to the study of sediment transport on a horizontal bed due to wave action. *Proc. 6th Conf. Coastal Eng., Florida*, pp. 326–335.

VOIT, S. S. (1970) The theory of tidal waves. In *Mekhanika v SSSR za 50 let*. Nauka, Moscow, **2**, pp. 79–84.

VOLTSINGER, N. E. and PJASKOVSKY, R. V. (1968) *Basic Oceanological Problems of the Theory of Shallow Water*. Gidrometeoizdat, Leningrad, 300 pp.

WARBURG, H. D., *Tides and Tidal Streams*. Cambridge Univ. Press, p. 12.

WEATHERLEY, G. L. (1972) A study of bottom boundary layer of the Florida current. *J. Phys. Oceanogr.* **2**, pp. 54–72.

WHEWELL, W. (1833) Essay towards a first approximation to a map of cotidal lines. *Trans. Roy. Soc. London*, pp. 147–236.

WILLIAMS, G. P. (1972) Friction term formulation and convective instability in a shallow atmosphere. *J. Atmosph. Sci.* **29**, pp. 870–876.

WIMBUSH, M. and MUNK, W. (1971) The benthic boundary layer. In *The Sea* **4**, pt. 1, Ed. A. E. Maxwell, pub. John Wiley & Sons, pp. 731–758.

YIH, C. S. (1960a) Gravity waves in a stratified fluid. *J. Fluid Mech.* **8**, pp. 481–508.

YIH, C. S. (1960b) Exact solutions for steady two-dimensional flow of a stratified fluid. *J. Fluid Mech.* **9**, pp. 161–174.

ZAHEL, W. (1970) Die Reproduktion Gezeitenbedingter Bewegungsvorgänge im Weltozean Mittels des Hydrodynamisch-Numerischen Verfahrens. *Mitteil. Inst. Meereskunde Univ. Hamburg*, No. 17, pp. 1–50.

ZALESNY, V. B. and POSPELOV, B. V. (1972) Calculations of the harmonic components of the K_1 and M_2 tide in the La Manche Strait, in *Numerical Models of Oceanic Circulations*. Computing Center of the Siberian Branch of the USSR Academy of Sciences, Novosibirsk, pp. 147–166.

ZALESNY, V. B. and TAMSALU, R. E. (1972) Numerical method of tide calculations in the marginal seas. In *Internal Waves in the Oceans*. Computing Center of the Siberian Branch of the USSR Academy of Sciences, Novosibirsk, pp. 168–180.

ZALESNY, V. B., TAMSALU, R. E., PRIVALOVA, I. V. and SGIBNEVA, L. A. (1972) On the calculation of tides in the marginal seas. *Izv. AN SSSR, Fizika atmosfery i okeana*, **8**, pp. 108–112.

ZEILON, N. (1912) On tidal boundary waves and related hydrodynamical problems. *Kungl. Svenska Vetens. Akad. Handling*, **47**, No. 4, pp. 1–46 (1912).

ZEILON, N. (1934) Experiments on boundary tides. *Medd. Göteborgs Högskolas Oceanogr. Inst.* III, No. 10, pp. 1–8.

ZILITINKEVITCH, S. S. (1970) *Dynamics of the Atmosphere Boundary Layer*. Gidrometeoizdat, Leningrad, 290 pp.

ZILITINKEVITCH, S. S., LAIHTMAN, D. L. and MONIN, A. S. (1967) Dynamics of the atmosphere boundary layer. *Izv. AN SSSR, Fizika atmosfery i okeana*, **3**, pp. 297–333.

ZUBOV, N. N. (1933) *Elementary Studies of Tides in the Sea*. Izd. GMK SSSR, Moscow, 125 pp.

Additional References

ACCAD, Y. and PEKERIS, C. L. (1978) Solution of the tidal equations for M_2 and S_2 tides in the World Ocean from a knowledge of the tidal potential alone. *Phil. Trans. Roy. Soc. London*, **A290**, 235-266.
ANWAR, H. O. (1981) A study of the turbulence structure in a tidal flow. *Estuar., Coast. Shelf Sci.*, **13**, 373-387.
BAINES, P. G. (1982) On internal tide generation models. *Deep-Sea Res.*, **29**, 307-338.
BELL, T. H. (1975) Topographically generated internal waves in the open ocean. *J. Geophys. Res.*, **80**, 320-327.
BOWDEN, K. F. (1978) Physical problems of the benthic boundary layer. *Geophys. Surveys*, **3**, 255-296.
BOWDEN, K. F. and FERGUSON, S. P. (1980) Variations with height of the turbulence in a tidally-induced bottom boundary layer. In: *Marine turbulence*. Ed. J. C. J. Nihoul, Elsevier Sci. Publ. Co., Amsterdam, pp. 259-286.
CARTWRIGHT, D. E. (1977) Ocean tides. *Rep. Progr. Phys.*, **40**, 665-708.
CARTWRIGHT, D. E., EDDEN, A. C., SPENCER, R. and VASSIE, J. M. (1980) The tides of the northeast Atlantic ocean. *Phil. Trans. Roy. Soc., London*, **A298**, 87-139.
CARTWRIGHT, D. E., ZETLER, B. D. and HAMON, N. D. V. (1979) Pelagic tidal constants. *IAPSO Publ. Sci.*, No. 30, 65 pp.
CLARKE, A. J. and BATTISTI, D. S. (1981) The effect of continental shelves on tides. *Deep-Sea Res.*, **28**, 665-682.
ESTES, R. H. (1977) A computer software system for the generation of global ocean tides including self-gravitation and crustal loading effects. Final Report Business and Technical Systems, Inc., Seabrook, 60 pp.
ESTES, R. H. (1980) A simulation of global ocean tide recovery using altimeter data with systematic orbit error. *Mar. Geol.*, **3**, 75-139.
GARRETT, C. J. R. (1974) Normal modes of the Bay of Fundy and Gulf of Maine. *Can. J. Earth Sci.*, **11**, 549-556.
GARRETT, C. J. H. (1975) Tides in gulfs. *Deep-Sea Res.*, **22**, 23-35.
GOLITSIN, G. S. and DIKIY, L. A. (1966) Eigenoscillations of planetary atmospheres depending on the velocity of planet's rotation. *Izv. Akad. Nauk SSSR, Phys. Atmosph. Ocean*, **2**, 225-235.
GORDEEV, R. G., KAGAN, B. A. and POLYAKOV, E. V. (1977) The effects of loading and self-attraction on global ocean tides: The model and the results of a numerical experiment. *J. Phys. Oceanogr.*, **7**, 161-170.
GOTLIB, V. YU. and KAGAN, B. A. (1979) On parameterization of the shelf effects in simulation of ocean tides. *Izv. Akad. Nauk SSSR, Fizika atmosfery i okeana* **15**, 425-434.
GOTLIB, V. YU. and KAGAN, B. A. (1980) Resonance periods of the World Ocean. *Dokl. AN SSSR*, **251**, 710-713.
GOTLIB, V. YU. and KAGAN, B. A. (1981) Numerical simulation of tides in the World Ocean: 1. Parameterization of the shelf effects. *Dt. Hydrogr. Z.*, **34**, 273-283.
GOTLIB, V. YU. and KAGAN, B. A. (1982) Numerical simulation of tides in the World Ocean: 2. Experiments of the sensitivity of the solution to choice of the shelf effect parameterization and to variations in shelf parameters. *Dt. Hydrogr. Z.*, **35**, 1-14.
GOTLIB, V. YU. and KAGAN, B. A. (1982) Numerical simulation of tides in the World Ocean: 3. A solution to the spectral problem. *Dt. Hydrogr. Z.*, **35**, 47-58.
HEATHERSHAW, A. D. (1974) Bursting phenomena in the sea. *Nature*, **273**, 394-395.

ADDITIONAL REFERENCES

HEATHERSHAW, A. D. (1976) Measurements of turbulence in the Irish Sea benthic boundary layer. In: *Benthic boundary layer.* Ed. I. M. McCave, Plenum Publ. Co., pp. 11–31.

HEATHERSHAW, A. D. (1979) The turbulence structure of the bottom boundary layer in a tidal current. *Geophys. J. Roy. Astron. Soc.,* **58,** 395–430.

HEATHERSHAW, A. D. and SIMPSON J. H. (1978) The sampling variability of the Reynolds stress and drag coefficient measurements. *Estuar. Coast. Mar. Sci.,* **6,** 263–274.

HINO, M., SAWAMOTO, M. and TAKASU S. (1976) Experiments on transition to turbulence in an oscillatory pipe flow. *J. Fluid Mech.,* **75,** 193–207.

HOUGH, S. S. (1898) On the application of harmonic analysis to the dynamical theory of the tides. II. On the general integration of Laplace's tidal equations. *Phil. Trans. Roy. Soc., London,* **A191,** 139–185.

JOHNS, B. (1967) Tidal flow and mass transport in a slowly converging estuary. *Geophys. J. Roy. Astron. Soc.,* **13,** 377–386.

JOHNS, B. (1968) Some effects of topography on the tidal flow in a river estuary. *Geophys. J. Roy. Astron. Soc.,* **15,** 501–507.

JOHNS, B. (1969) On the representation of the Reynolds stress in a tidal estuary. *Geophys. J. Roy. Astron. Soc.,* **17,** 39–44.

JOHNS, B. (1975) The form of the velocity profile in a turbulent shear wave boundary layer. *J. Geophys. Res.,* **80,** 5109–5112.

JOHNS, B. (1978) The modelling of tidal flow in a channel using a turbulence energy closure scheme. *J. Phys. Oceanogr.,* **8,** 1042–1049.

JONSSON, I. G. (1980) A new approach to oscillatory rough turbulent boundary layers. *Ocean Engng.,* **7,** 109–152.

KAGAN, B. A. (1977) *Global Interaction between Ocean and Earth Tides.* Leningrad, Gidrometeoizdat Publ. House, 45 pp.

KAGAN, B. A. (1981) The dynamics of the oceanic benthic boundary layer. In: *Itogi nauki i tekhniki. Okeanologia,* **6,** 81–189.

KERCZEK, C. and DAVIS, S. H. (1972) The stability of oscillatory Stokes layers. *Studies in Appl. Math.,* **51,** 239–252.

KERCZEK, C. and DAVIS, S. H. (1974) Linear stability theory of oscillatory Stokes layers. *J. Fluid Mech.,* **62,** 753–773.

LAMBECK, K. (1980) *The Earth's variable rotation.* Cambridge Univ. Press, 449 pp.

LAPLACE, P. S. Recherches sur plusieurs points du systéme du monde. *Mem. Acad. Roy. Sci., Paris,* **88,** 75–182.

LIU, S. K. and LEENDERTSE, J. J. (1979) A three-dimensional model for estuaries and coastal seas: Bristol bay simulations. Rand, R-2405-NOAA, 121 pp.

LONGUET-HIGGINS, M. S. (1968) The eigenfunctions of Laplace's tidal equations over a sphere. *Phil. Trans. Roy. Soc., London,* **A262,** 511–607.

LONGUET-HIGGINS, M. S. and POND, G. S. (1970) The free oscillations of fluid on a hemisphere bounded by meridians of longitude. *Phil. Trans. Roy. Soc., London,* **A226,** 193–223.

LUTHER, B. S. and WUNSCH, C. (1975) Tidal charts of the central Pacific Ocean. *J. Phys. Oceanogr.,* **5,** 222–230.

MARGULES, M. (1893) Luftbewegungen in einer rotierenden Scharoidschale. *Sitzungsber. Acad. Wiss., Wien,* **102,** 11–56.

MCCAMMON, C. and WUNSCH, C. (1977) Tidal charts of the Indian Ocean north of 15°S. *J. Geophys. Res.,* **82,** 5993–5998.

MERKLI, P. and THOMANN, H. (1975) Transition to turbulence in oscillatory pipe flow. *J. Fluid Mech.,* **68,** 567–575.

MILES, J. W. (1971) Resonant response of harbors: an equivalent circuit analysis. *J. Fluid Mech.,* **46,** 241–266.

MUNK, W. H. (1968) Once again—tidal friction. *Quart. J. Roy. Astron. Soc.,* **9,** 352–375.

NISHIDA, M. (1980) Improved tidal charts for the western part of the North Pacific Ocean. *Rep. Hydrogr. Res.,* No. 15, 55–70.

ADDITIONAL REFERENCES

Papa, L. (1977) The free oscillations of the Ligurian Sea computed by the HN-method. *Dt. Hydrogr. Z.*, **30**, 81–90.

Parke, M. E. and Hendershott, M. C. (1980) M_2, S_2, K_1 models of the global ocean tides on an elastic Earth. *Mar. Geod.*, **3**, 379–408.

Piño Griver, F. (1970) On the cotidal configuration along the Pacific coast of Mexico. *Geofis. Internat.*, **10**, 17–35.

Platzman, G. W. (1972) Two-dimensional free oscillations in natural basins. *J. Phys. Oceanogr.*, **2**, 117–130.

Platzman, G. W. (1975) Normal modes of the Atlantic and Indian oceans. *J. Phys. Oceanogr.*, **5**, 201–221.

Platzman, G. W. (1978) Normal modes of the World Ocean. Part 1. Design of a finite-element barotropic model. *J. Phys. Oceanogr.*, **8**, 323–343.

Platzman, G. W., Curtis, G. A., Hansen, K. S. and Slater R. D. (1981) Normal modes of the World Ocean. Part II: Description of modes in the period range 8 to 80 hours. *J. Phys. Oceanogr.*, **11**, 579–603.

Protasov, A. V. (1979) A numerical method for the solution of the problem of free oscillations of the World Ocean to barotropic approximation. *Meteorologia i gidrologia*, No. 6, 57–66.

Proudman, J. (1916) On the dynamical equations of the tides. *Proc. London Math. Soc.*, 2nd Ser., **18**, 1–68.

Proudman, J. (1941) The effect of coastal friction on the tides. *Month. Not. Roy. Astron. Soc. Geophys. Suppl.*, **5**, 23–26.

Rao, D. B. (1966) Free gravitational oscillations in rotating rectangular basins. *J. Fluid Mech.*, **25**, 523–555.

Rao, D. B., Mortimer, C. H. and Schwab, D. J. (1976) Surface normal modes of Lake Michigan: calculations compared with spectra of observed water level fluctuations. *J. Phys. Oceanogr.*, **6**, 575–586.

Richards, K. J. (1982) Modeling the benthic boundary layer. *J. Phys. Oceanogr.*, **12**, 428–439.

Rodriguez, J. H. G. (1981) Über die Bestimmung von reibungslosen barotropen Eigenschwingungen des Weltozeans mittels der Lanczos Method. Diss. Univ. Hamburg, 25 pp.

Schwab, D. J. (1977) Internal free oscillations in Lake Ontario. *Limnol. and Oceanogr.*, **22**, 700–708.

Schwiderski, E. W. (1980) Ocean tides, part II: A hydrodynamical interpolation model. *Mar. Geod.*, **3**, 219–255.

Soulsby, R. L. (1977) Similarity scaling of turbulence spectra in marine and atmospheric boundary layers. *J. Phys. Oceanogr.*, **7**, 934–937.

Street, R. O. (1933) The tides in a hemispherical ocean bounded by a continental shelf along meridian. *Month. Not. Roy. Astron. Soc., Geophys. Suppl.*, **3**, 163–167.

Tee, K. T. (1979) The structure of three-dimensional tide-generating currents. Part 1: oscillating currents. *J. Phys. Oceanogr.*, **9**, 930–944.

Vassie, J. (1982) Tides and low frequency variations in the equatorial Atlantic. *Oceanol. Acta*, **5**, 3–6.

Weatherly, G. L. (1975) A numerical study of time-dependent turbulent Ekman layers over horizontal and sloping bottoms. *J. Phys. Oceanogr.*, **5**, 288–299.

Weatherly, G. L. and Martin, P. J. (1928) On the structure and dynamics of the oceanic bottom boundary layer. *J. Phys. Oceanogr.*, **8**, 557–570.

Webb, D. J. (1980) Tides and tidal friction in a hemispherical ocean centred at the equator. *Geophys. J. R. astr. Soc.*, **61**, 573–600.

Won, I. J., Kuo, J. T. and Jachens, R. C. (1978) Mapping ocean tides with satellites: a computer simulation. *J. Geophys. Res.*, **83**, 5947–5960.

Wübber, C. and Krauss, W. (1979) The two-dimensional seiches of the Baltic Sea. *Oceanol. Acta*, **2**, 435–446.

ADDITIONAL REFERENCES

WUNCH, C. (1975) Internal tides in the ocean. *Rev. Geophys. Space Phys.*, **1**, 167–182.

ZAHEL, W. (1977) A global hydrodynamic numerical 1°-model of the ocean tides; the oscillation system of the M_2-tide and its distribution of energy dissipation. *Ann. Geophys., Paris*, **33**, 31–40.

ZAHEL, W. (1980) Mathematical modelling of global interaction between ocean tides and Earth tides. *Phys. Earth Planet. Inter.*, **21**, 202–217.

Appendix
Notes added to the proofs

Chapter 1

Derivation of the expression for the tidal potential and a compact presentation of the existing techniques for the prediction of tides can be found in a review paper by Cartwright (1977).

The effects of loading and self-attraction of ocean tides and their role in the global interaction between ocean and earth tides have been investigated by Gordeev *et al.* (1977), Estes (1977), Accad and Pekeris (1978), Parke and Hendershott (1980) and Zahel (1980). It has been shown that allowance for the effects of loading and self-attraction of ocean tides does not lead to a radical reconstruction of the spatial structure of ocean tides. Incidentally, this is hardly surprising because otherwise it would be difficult to explain why tidal maps of the World Ocean obtained in the framework of the traditional formulation of the problem (i.e. in the absence of the above effects) are consistent qualitatively with observational data.

Shelf effects can be taken into account in two ways in modelling global tides: by direct simulation of the shelf phenomena on a grid with high resolution or by using additional relations connecting the shelf phenomena with explicitly calculated characteristics of tides in the open ocean. In the sense of uniformity of the description of tides in the ocean-shelf system, the first way is certainly preferable. However, the experience of its realization for the World Ocean (see Zahel, 1977) indicates that even one degree resolution of the grid does not provide the required accuracy. There remains the second possibility—to parameterize processes occurring on the shelf and then include them in a global tidal model.

At present, two approaches to the problem of parameterizing the shelf effect have come into being. They are reduced to derivation of an impedance boundary condition, i.e. a relation between level oscillations and velocity of tidal flow on the shelf edge. In this case, boundary conditions can be classified as local or integral, depending on the kind of shelf parameters used—local values of the depth, width and coefficient of bottom friction, or their values within the entire length of the shelf zone. The former originate in a paper by Proudman

(1941) and the latter, in that of Miles (1971). Different techniques for parameterizing shelf effects of the local and integral types have been suggested by Garrett (1975), Gotlib and Kagan (1979, 1981) and Clarke and Battisti (1981). Investigation into the influence of shelf effects on tides in the open ocean was initiated by Street (1933) and Proudman (1941) who obtained a solution to the problem of tides in a semi-spherical ocean of constant depth with a narrow shelf zone at the equator. The role of shelf effects in the formation of semidiurnal tides in the World Ocean of real configuration has been studied by Gotlib and Kagan (1982).

Chapter 2

Proof of uniqueness of the solution to a periodic boundary-value problem of tidal dynamics in the presence of the effects of loading and self-attraction of ocean tides is given in Gordeev *et al.* (1977).

Eigenoscillations of a spherical ocean of constant depth were first investigated by Laplace (1775). This investigation was further developed in papers by Margules (1893), Hough (1898), Golitsin and Dikiy (1966) and Longuet-Higgins (1968).

The frequency spectrum of eigenoscillations of a semi-spherical ocean has been studied by Longuet-Higgins and Pond (1970) and by Webb (1980) and that of an ocean having the form of a rectangle and of a circular sector, by Pnueli and Pekeris (1968).

All papers devoted to the solution for the spectral problem for natural basins can be divided into two groups. In papers of the first group, this problem reduces to two simpler spectral problems for the potential and stream function for the transport field, whose solution is, in turn, found using an approximate procedure. This method was suggested by Proudman (1916) and realized for the Great Lakes by Rao (1966), Rao *et al.* (1976) and Schwab (1977). It was used also with application to the Atlantic–Indian ocean system by Platzman (1978) and to the entire World Ocean by Platzman *et al.* (1981).

The second group combines papers where the spectral problem is solved through direct inversion of the difference analogue of Laplace's tidal operator. These papers are distinguished by the choice of different procedures for solving the set of algebraic equations. Thus, Platzman (1975) used the Lanczos process in defining the frequencies and form of eigenoscillations of the Atlantic–Indian ocean system. The same method was used by Rodriguez (1981) in solving the spectral problem for the entire World Ocean. A similar problem, but for a limited range of frequencies was solved using the method of inverse iterations by Protasov (1979).

APPENDIX 287

The second group includes also papers in which the spectral problem is solved as a periodic boundary-value problem for forced oscillations with the form and frequency adjustment of the exciting force to the form and frequency of the separated free oscillation. This method suggested by Platzman and referred to as the method of resonance iterations was tested for the North Atlantic, Great Lakes, the Gulf of Mexico and the Gulf of Maine-Bay of Fundy system by Platzman (1972) and Garrett (1974).

Another method is based on numerical solution of the initial boundary-value problem for free oscillations excited either by the initial disturbance or by the mass force acting during a finite time interval. This method has found application to the study of free oscillations of the Ligurian and Baltic seas (see Papa, 1977; Wübber and Krauss, 1979). It was used also by Gotlib and Kagan (1980, 1982) in the investigation of the frequency spectrum and the eigenoscillation structure for the World Ocean in realistic configuration.

Chapter 4

New data on the distribution of amplitudes and phases of the M_2-tidal wave are presented in Luther and Wunsch (1975), McCammon and Wunsch (1977), Piño (1970), Cartwright et al. (1980), Nishida (1980) and Vassie (1982). The values of tidal harmonic constants in 108 points of deep-sea measurements in different regions of the World Ocean have been summarized in Cartwright et al. (1979).

These data confirm the existence of an amphidromy in the region of the Society Islands in the equatorial Pacific, in the north-west Indian Ocean, Caribbean Sea, and in the central North Atlantic.

Use of satellite altimetry in the study of the spatial structure of ocean tides have been shown to be quite promising in Won et al. (1978) and Estes (1980). New global tidal maps of the M_2-wave obtained in the framework of a semi-empirical approach are given by Schwiderski (1980), Parke and Hendershott (1980) and in the framework of a theoretical approach, by Estes (1977), Accad and Pekeris (1978) and Zahel (1977).

Results of the solution to the spectral problem for the World Ocean in realistic configuration given in Gotlib and Kagan (1980, 1982), Platzman et al. (1981) and Rodriguez (1981) indicate that in the semidiurnal and diurnal spectral bands there are modes whose periods are close to those in the main harmonics of the tidal potential. In particular, these modes have periods 11.96; 12.50; 12.69; 12.82 h in the semidiurnal and 22.49; 23.88; 25.90 h in the diurnal bands of the spectrum. Furthermore, the spatial structure of the elevations corresponding to these modes resembles very closely the spatial structure of forced tidal waves of appropriate periodicity. This confirms the possibility of resonance excitation of semidiurnal and diurnal tides in the World Ocean.

The problem of ocean tide energetics and the estimates of tidal energy dissipation in the ocean are discussed by Munk (1968), Kagan (1977), Cartwright (1977), and Lambeck (1980).

Chapter 5

Systematic presentation of experimental and theoretical investigations into the benthic boundary layer in a tidal flow is also offered in review papers by Bowden (1978) and Kagan (1981). In the latter is given a summary of experimental estimates of the critical Reynolds number. These estimates appear to be significantly larger than the theoretical estimates found by Kerczek and Davis (1972, 1974) for two-dimensional and three-dimensional perturbations in an oscillating flow. A possible reason for the above discrepancy is associated with the fact that theoretical estimates of the minimum value of the critical Reynolds number define the limit of stability of the motion considered rather than the transition from a laminar to a turbulent regime.

Extensive studies into the structure of turbulence in the benthic boundary layer of a tidal flow have been conducted by Heathershaw (1976, 1979); see also Bowden and Ferguson (1980).

The conclusion that relative intensity of the longitudinal velocity fluctuations is independent of the mean velocity of tidal flow is not valid during the entire tidal period. This is associated with intermittency of turbulence in the oscillating boundary layer. According to data from laboratory measurements (see Merkli and Thomann, 1975), turbulence in the Stokes boundary layer occurs when the velocity becomes maximum and disappears as the direction of flow changes. Hino et al. (1976) have separated two stages of laminar regime and three stages of turbulent regime within an oscillatory cycle of velocity and have shown that transition from one stage to the other takes place at definite values of the Reynolds number.

The one-dimensional spectra of velocity fluctuations in the benthic boundary layer of tidal flow follow the "$-5/3$" law at lower wave numbers than is predicted by the theory of locally isotropic turbulence. A similar peculiarity is inherent also in the spectra of temperature fluctuations (see Bowden, 1978; Soulsby, 1977; Heathershaw, 1979; Wimbush and Munk, 1971). The reason for this is not clear.

Intermittency of turbulence in the benthic boundary layer of tidal flow leads to a strong variability of the Reynolds stress with time. This variability is doubled if separate samples are not independent statistically (see Heathershaw and Simpson, 1978). Information about the frequency distribution of the bursts of turbulence, their contribution to the Reynolds stress and length can be found in Heathershaw (1974) and Anwar (1981). Changes in the properties of turbulence with height within the benthic boundary layer of tidal flow are discussed also by Bowden and Ferguson (1980).

Johns (1967, 1968, 1969) and Tee (1979) came out with a series of papers on the vertical structure of the benthic boundary layer with an *a priori* assignment of the vertical eddy viscosity which, conceptually, are close to the investigations described in this chapter. Models of the benthic boundary layer based on closure of the dynamical equations using semi-empirical hypotheses have been suggested also by Johns (1975, 1978), Weatherly (1975), Weatherly and Martin (1978), Liu and Leendertse (1979). Comparative analysis of the models of this type indicates that the more refined is the model, the more assumptions and adjustment parameters it contains. If the solution appears to be sensitive to their choice, these models are less advantageous, for example, than models based on an *a priori* assignment of the vertical eddy viscosity. This situation can be avoided by constructing of models which include evolution equations for the components of the Reynolds stress tensor. Richard's (1982) model can serve as an example.

Derivation of the resistance law for one-dimensional oscillating boundary layer is given in Jonson (1980).

Chapter 6

The main features of the generation and distribution of internal tidal waves in the World Ocean have been described by Wunsch (1975), Bell (1975) and Baines (1982) who have also estimated the energy of internal tidal waves and its rate of dissipation.

Index

Bottom boundary layer 177–247, 289
  under neutral stratification 214–223, 247
  under stratified flow 223–233, 247
  over rough and smooth surfaces 180–184
  roughness parameters for 182–183
  theoretical models of 204–233
  friction coefficient – see drag law
Boundary conditions 31–34
Boundary values (method) 76–84, 109
Brunt–Väisälä frequency 224–225, 258, 265, 267, 268, 271

Clairaut constant 17
Convergence of numerical methods 104–106, 108
Cotidal charts (World Ocean) – empirical 110–121
  by Dietrich/Villain 118–121, 123, 151
  by Harris 112–113, 121, 123, 170
  by Sterneck 113–117, 123
  by Whewell 110–112, 168–169
  Atlantic Ocean 124–126, 169–171
  Indian Ocean 123–124, 166, 170
  Pacific Ocean 121–123, 169–170
  Southern Ocean 126–128, 168–169
Cotidal charts – semi empirical 143–147, 287
  by Bogdanov & Magarik 144, 150–154
  by Gordeev, Kagan, Rivkind 137, 151–154, 175, 286
  by Hendershott 143–147, 150–154
  by Schwiderski, Parke & Hendershott 287
  by Tiron, Sergeev, Michurin 145, 151–154
Cotidal charts – theoretical 147–150
  by Gotlib & Kagan 286–287
  by Pekeris & Accad 136, 148, 287
  by Zahel 149, 287
  from numerical experiments 155–171
  minimal velocity from 178–179
  comparison with island data 139–143
  comparison with empirical charts 150–154

Darwin constant 16–17
  correction 17
  nomenclature 21–23
Doodson constant 16–17
Drag law (bottom friction coefficient) 37–38, 77, 83, 233–247
  geostrophic friction coefficient 238–242
Dynamic equations 36–40
  a priori estimates 45–52, 87–89, 90–91
  conjugate equations 63–67, 75
  continuity equation 37, 76, 84
  convergence 91–93, 104–106, 108
  existence theorem 52–59
  existence of periodic solutions 59–63
  perturbation theory 67–71
  spaces – Hilbert 59–60, 64
    – Sobolev 40
  spectral problem (normal modes) 71–75, 286–287

Earth (terrestrial) tides 6, 26–30, 35, 285
Eigenoscillations – see normal modes
Ellipses of tidal flow 208–211
Energy dissipation 11, 171–176, 255–256, 288
Equilibrium (tidal) shape 17

Factors of potential – astronomic 16, 18
  – geodetic (Doodson) 16–17
Forces –
  attractive 1–4
  centrifugal 5
  Coriolis 5, 23–24, 204–206, 212, 222
  frictional 6–11
  generating (tide-) 3
  gravitational 5
  hydrostatic 4–5
  potentials of 11–23, 26–29
Flow profiles 184–189
  ellipses of 208, 211
  spectrum of 189, 199–204, 288

INDEX

Friction coefficient — see drag law
HN method — see numerical methods

Internal tides (waves)   248–272, 289
  basic equations   256–261
  eigenvalues   263–272
  generation of   248–256
  vertical structure   262–270

Kelvin waves   32, 125, 169

Loading and self-attraction   29, 35, 285
Love numbers   27–29, 35

Nodal modulation   21–22
Nodal lines (Sterneck)   115–116
Normal modes (spectral problem)   71–75, 286–287
Numerical methods   76–109
  boundary values   76–84, 109
  convergence   91–93, 104–106, 108
  fractional steps   95–103, 109
  Galerkin   63, 130
  HN method   84–95, 109
  modified HN   89–93
  modelling world ocean   128–143
  solvability   77–79
  stability   99–104, 132–135

Periods, fundamental   19–21
Poincaré waves   32, 125, 169
Potential (tidal)   11–23
  of deformation   26–29
  harmonic analysis of   18–23, 34
  nodal modulations of   21–22

Red-Wharf Bay, experiments at   185, 191–194
Reynolds number   178–182, 246, 288
  critical   178–179, 246
Reynolds stresses   7–10, 202, 288–289
  deviatoric   9
Richardson number   178
Rossby number   204, 238–245
Rough and smooth surfaces   180–184
Roughness parameter   182–183

Satellite altimetry   287
Shelf effects   31–35, 36, 285–286
Southern ocean as energy source   168–171, 175–176

Turbulence   189–194, 288–289
  in bottom boundary layer   178
  coefficient   205–229
  correlation function   194–199, 247
  spectra   199–204, 288
Turbulent viscosity coefficients   8–9, 34